Advanced Modeling of Thermodynamic Cycles of Nuclear Power Plants

Advanced Modeling of Thermodynamic Cycles of Nuclear Power Plants offers a comprehensive resource on Thermodynamic Cycles of Nuclear Power Plants (TCNPPs) modeling and provides detailed studies of TCNPPs for current and emerging reactors, including Small Modular Reactors.

After covering reactor physics and operating principles, this book examines water-steam cycles using advanced tools like the pinch method and exergy balances to optimize performance. Subsequent chapters cover gas turbine and supercritical CO_2 cycles, combined cycles, and nuclear applications, including desalination, hydrogen production, and cogeneration installations achieving superior efficiencies. Using Thermoptim software's systemic modeling approach, readers can accurately represent nuclear energy system peculiarities.

This book will benefit mechanical, nuclear, and energy engineering researchers, professionals and graduate students in the fields of applied thermodynamics, nuclear power plants, and energy systems.

Renaud Gicquel served as a Full Professor and later a Distinguished Professor at PSL University, France, directing its Centre for Energy Studies (1987–2004). A graduate of École des Mines de Paris, France, with a doctorate from Paris VI University, France, Dr. Gicquel's passion for integrating technology with thermodynamics education led him to develop Thermoptim, an innovative software widely used for energy system simulation and teaching, and he is the author of over 80 papers and three books. He has also created three MOOCs, democratizing access to advanced energy concepts globally. Dr. Gicquel's career spans numerous notable international roles: Special Assistant to the UN Secretary General for the Conference on New and Renewable Energy Sources (1980), senior positions at France's Ministry for Research and CNRS, and co-founding the Mediterranean Energy Observatory (1986).

Advanced Modeling of Thermodynamic Cycles of Nuclear Power Plants

Volume 2

Renaud Gicquel

CRC Press
Taylor & Francis Group
Boca Raton London New York

CRC Press is an imprint of the
Taylor & Francis Group, an **informa** business

Designed cover image: Pixabay; Renaud Gicquel

First edition published 2026
by CRC Press
2385 NW Executive Center Drive, Suite 320, Boca Raton FL 33431

and by CRC Press
4 Park Square, Milton Park, Abingdon, Oxon, OX14 4RN

CRC Press is an imprint of Taylor & Francis Group, LLC

ISBN: 978-1-032-99787-2 (hbk)
ISBN: 978-1-032-99789-6 (pbk)
ISBN: 978-1-003-60602-4 (ebk)

DOI: 10.1201/9781003606024

Typeset in Times
by codeMantra

Access the Support Material: www.routledge.com/9781032997872

To my wife and lifelong companion

To our three daughters

To our grand children

Contents

Disclaimer

This book is an educational text designed to teach advanced thermodynamic modeling principles and methods. While it emphasizes technological realism and practical applications, it is not intended to serve as an industrial design handbook.

The thermodynamic data, constants, and correlations used throughout this book are selected for pedagogical clarity and may differ from those found in certified industrial databases. They should not be used as reference values for professional engineering calculations.

The Thermoptim software used is a pedagogical tool that facilitates the learning of complex concepts. The examples and exercises in this book were developed and validated using Thermoptim versions 2.52, 2.62, 2.72, and 2.82. Results obtained with other versions may differ. In all cases, Thermoptim's results are illustrative and do not match the precision or validation required by professional engineering design tools used in industrial practice.

Readers engaged in actual system design should therefore consult appropriate industrial software and professional engineering resources.

The author and publisher accept no liability for any consequence arising from the use of the methods, data, or software results presented in this book in a professional or industrial context.

ADDITIONAL RESOURCES

Numerous digital resources for teaching energy systems have been progressively compiled into a portal called Thermoptim-UNIT (www.thermoptim.org), whose content is freely accessible with few exceptions, such as solutions to selected exercises and problems.

This book's website (https://server.s4e2.com/crc/tcnpp/) provides access to various supplementary resources, including:

- Links to digital versions of selected figures
- Updated footnote links in case of changes
- Application examples of developments presented in specific chapters, available as guided explorations or examples of Thermoptim models with their associated files
- Each chapter opens with a landing page featuring an overview and abstract, along with a dedicated Table of Contents and chapter-specific resources.
- Theoretical and methodological supplements, either on the portal or dedicated pages

Since the complete list of resources was not finalized at the time of publication, please refer to the website during your reading to discover all available materials. Some resources will be reserved for registered book purchasers.

Preface

ENGINEERING THERMODYNAMICS: ADVANCED MODELING OF ENERGY SYSTEMS AND NUCLEAR CYCLES

In nearly all courses of thermodynamics applied to energy systems worldwide, the chosen approach does not take into account the technological characteristics of components such as compressors or heat exchangers. As a result, it does not allow the study of the behavior of these systems when their operating conditions vary, for example, when the ambient temperature changes or when the load they are subjected to evolves. This is known as the study of the operation of energy systems under off-design conditions.

Moreover, the thermodynamic cycles that are studied are generally only caricatures of those implemented in reality, as learners do not have the tools that would allow them to model them correctly. This is particularly true for Thermodynamic Cycles of Nuclear Power Plants (TCNPPs), some of which are relatively complex due to the peculiarities of this energy source.

The originality of this set of books lies in addressing these two subjects, which are not covered in the thermodynamics textbooks available today. This gap has several causes:

First, curricula have gradually been simplified over the past decades, limiting the time available for in-depth studies, notably the study of the operation of energy systems under off-design conditions.

Second, this reduction in curricula has mainly affected technological issues, with most teachers favoring theoretical approaches with which learners are more familiar.

Finally, the degree of difficulty of these subjects is significantly higher than merely learning the basics of thermodynamics, which are already relatively complex.

As a result, these modeling difficulties are evaded in most courses.

In our book *Energy Systems: A New Approach to Engineering Thermodynamics*, published in 2021, we presented a method for learning thermodynamics as applied to energy systems that is simplified in terms of equations. This approach is designed to meet the needs of learners with limited backgrounds in mathematics and physics, and who do not wish to engage heavily in modeling.[1]

The cycle studies in these three books are conducted using simulation software called Thermoptim, which allows for the modeling of a wide range of systems while faithfully representing technological reality.

Modeling is performed systemically by graphically linking icons that represent components. Since the equations are encapsulated within the modules corresponding to each component, learners don't need to constantly rewrite them. This allows them more time to focus on the architecture of the cycles, the interactions between different components, and so on. Consequently, this approach fosters a way of reasoning for learners that is distinctly different from the traditional methods used in most thermodynamics textbooks.

The first volume elaborates on the analysis of the behavior of the basic components and explains how customized components can be modeled and coupled to Thermoptim through a mechanism named external classes. Several case studies show how these developments can be put into practice in simple cycles.

The second volume shows how they can be used to model complex thermodynamic cycles coupled to nuclear reactors.

Building on a more advanced scientific level than the 2021 edition, this new set of books complements the previous work on two key fronts, addressing important aspects that were intentionally omitted in 2021 to maintain a simplified approach.

While preserving the same systemic approach, this new series expands on the earlier work by exploring more in-depth modeling, which was previously limited to the basics, and by thoroughly examining TCNPPs.

Structured in two volumes, it equips readers with the tools needed to master these timely and, as we will demonstrate, highly accessible subjects with the right resources.

ADVANCED MODELING OF THERMODYNAMIC ENERGY COMPONENTS AND SYSTEMS

The objective of the first volume is to present a modeling approach for the main components used in energy systems (pumps, compressors, expansion devices, heat exchangers, combustion chambers, etc.) that is detailed enough to account for their key technological characteristics (a dynamic compressor does not behave like a positive displacement compressor).

At the same time, this approach is suitable for integration into a global system that can involve dozens of such components. These models offer intermediate complexity, positioned between general models that do not consider technological specificities and highly detailed finite element models, which completely describe geometry but are not suitable for systemic analyses.

Chapters 1 and 2 provide a theoretical background to introduce the notations that will be used later.

Chapters 3 and 4 are dedicated to the modeling of different components (pumps, compressors, expansion devices, heat exchangers, combustion chambers), taking into account their technological specificities.

Chapters 6–10 address the behavior of systems under off-design conditions.

ADVANCED MODELING OF THERMODYNAMIC CYCLES OF NUCLEAR POWER PLANTS

Recent events have sparked renewed interest in nuclear energy, particularly the so-called Small Modular Reactors, which are much smaller (between 30 and 340 MWe) than the large-scale pressurized water reactors that make up the majority of the world's nuclear fleet (from 900 to 1,650 MWe).

The second volume of this set of books is devoted to the study of most TCNPPs.

The first chapter explores the evolution of applied thermodynamics pedagogy, highlighting two main teaching paradigms that effectively equip students to be fully operational by the end of the course. The developments presented in this book demonstrate how to bridge the gap between these complementary approaches.

Chapter 2 explores the physical phenomena involved in nuclear reactors and present the main types of current and future nuclear reactors. In this chapter, we provide only a brief presentation of the associated thermodynamic cycles, indicating only their type and capacity, the pressure and temperature levels reached, and the efficiencies achieved. Further details are provided in the rest of the book.

Chapters 3–5 mainly deal with the water-steam cycles that are the most used or considered for nuclear power plants. They show learners how they can use powerful tools such as the pinch method or exergy balances built thanks to exergy structures to improve the performance of these cycles.

Chapter 6 is dedicated to cycles involving gas turbines and those with supercritical CO_2.

Chapter 7 begins by presenting combined cycles, which usually combine a gas turbine cycle and a steam cycle. Then, it introduces technologies that could constitute a significant application for nuclear reactors, namely seawater desalination and large-scale hydrogen production, and it concludes with cogeneration installations that have the advantage of leading to the best overall efficiencies.

Chapter 8 presents 11 case studies relating to TCNPPs.

AI Acknowledgments

Claude was used for linguistic revision purposes. ChatGPT, Grok, Perplexity, Gemini, Mistral and Claude were used to determine translations of technical terms from French to English.

About the Author

Prof. Renaud Gicquel (ORCID ID: 0000-0001-7960-0874) has published several books:

- In 1992, *Introduction aux Problèmes Énergétiques Globaux* (Economica). A revised version was published in 2013 (in collaboration with May Gicquel) by Presses des Mines and in English by CRC Press Balkema under the title *Introduction to Global Energy Issues.*
- In 2001, *Systèmes Énergétiques*, Volumes 1 and 2 (Presses des Mines). A revised edition was published in 2009 with a third volume on advanced cycles and innovative low-environmental-impact systems. A new version was published in 2023 by Presses des Mines. The 2009 revised version was translated into English and published in 2012 by CRC Press Taylor & Francis as *Energy Systems: A New Approach to Engineering Thermodynamics.* A second edition was published in 2021.

Renaud's special interest and passion is combining thermodynamics and energy system education with modern information technology tools. To this end, he has developed various software packages to facilitate the teaching and learning of applied thermodynamics and the simulation of energy systems, including Thermoptim software, distributed by S4E2 (https://server.s4e2.com/jml/index.php/en/).

CAREER

Renaud Gicquel is now retired. The main part of Prof. Gicquel's career took place at École des Mines de Paris (PSL University, France), where he led the Centre for Energy Studies from September 1987 to December 2004. Appointed as a Full Professor in July 1986 and later as a Distinguished Professor, he taught applied thermodynamics, energy system modeling, and global energy issues. He has developed three MOOCs on these topics.

He was trained as an engineer at École des Mines de Paris, France, and earned his doctorate in engineering from Paris VI University. He started his career in 1975 as a researcher on solar energy at École des Mines, before being appointed Special Assistant to the Secretary General of the United Nations Conference on New and Renewable Sources of Energy in New York in 1980. He then became Head of Multilateral Issues in the International Affairs Service of the Ministry for Research and Technology in Paris in 1982. From 1983 to 1985, he was Adviser for International Issues at the National Center for Scientific Research (CNRS). In 1986, he co-founded the Mediterranean Energy Observatory with the late Michel Grenon. In 1990, Dr. Gicquel created the ARTEMIS group in Nantes, a research body for thermal energy research, together with the University of Nantes and ISITEM (now Polytech). He also served as Deputy Director at École des Mines de Nantes for three years beginning in 1991.

List of Symbols

A	area m^2
A_c	free flow area m^2
B_o	Boiling number
C	concentration
C	speed $m\ s^{-1}$
C	fluid velocity $m\ s^{-1}$
C_s	speed of sound $m\ s^{-1}$
C_p	molar heat capacity at constant pressure $J\ mol^{-1}\ K^{-1}$
c_p	specific heat capacity at constant pressure $J\ kg^{-1}\ K^{-1}$
C_v	molar heat capacity at constant volume $J\ mol^{-1}\ K^{-1}$
c_v	specific heat at constant volume $J\ kg^{-1}\ K^{-1}$
D	diameter m
d_h	hydraulic diameter m
e	thickness m
f	friction coefficient
f	surface factor
F	molar flow rate $mol\ s^{-1}$
G	molar Gibbs energy $J\ mol^{-1}$
g	mass Gibbs energy $J\ kg^{-1}$
g	acceleration due to gravity $m\ s^{-2}$
H	molar enthalpy $J\ mol^{-1}$
h	mass enthalpy $J\ kg^{-1}$
h	heat transfer coefficient $J\ K^{-1}m^{-2}$
h	height difference m
h_f	enthalpy of formation $J\ mol^{-1}$
i	van't Hoff factor of the solute
I	current density A
k	polytropic coefficient
k	dissociation rate
K	kinetic energy $J\ kg^{-1}$
K_{eb}	ebullioscopy constant $K\ kg\ mol^{-1}$
K_p	equilibrium constant
K_0	cone or Stodola constant
L	enthalpy of evaporation $J\ kg^{-1}$
L	length m
M	molar mass $kg\ mol^{-1}$
Ma	Mach number
m	mass kg
n	number of moles
N	rotation speed $tr\ mn^{-1}$
N_u	Nusselt number
NTU	number of transfer units

P	pressure Pa or bar
Pr	Prandtl number
p	wetted perimeter
P	power W
Q	heat J kg^{-1}
R	universal constant 8,314 J mol^{-1} K^{-1}
R	ratio of heat capacity rates of a heat exchanger
r	ideal gas constant J K^{-1}
r	compression ratio
RR	relative roughness
R_e	Reynolds number
S	molar entropy J mol^{-1} K^{-1}
s	mass entropy J kg^{-1} K^{-1}
S	surface m^2
T	thermodynamic temperature K
t	temperature °C or K
u	specific internal energy J kg^{-1}
U	peripheral speed m s^{-1}
U	heat transfer coefficient W $K^{-1}m^{-2}$
v	volume m^3
v	specific volume m^3 kg^{-1}
V	volume m^3
V	molar volume m^3 mol^{-1}
V	voltage V
W	specific mechanical energy J kg^{-1}
W	power density W kg^{-1}
x	length m
x	solute mass concentration
x	vapor quality
x	mole fraction
x_h	specific exergy J kg^{-1}
x_q	specific heat-exergy J kg_{-1}
y	mass fraction
z	altitude m
d	exact differential
δ	differential form
Δ	variation

INDICES

a	related to admission
c	related to the critical point
c	corrected
c	related to compressor
c	related to cogeneration
c	cold

cc	related to combined cycle
d	related to expansion
g	related to the gas turbine
g	related to the gas phase
h	hot
hum	humid
i	related to inlet
l	related to liquid phase
lm	logarithm mean
p	related to polytropic
r	reduced (relative to the critical value)
r	related to discharge
s	related to saturation
s	related to isentropic
s	related to the output
sat	related to saturation
t	total
t	related to the turbine
u	useful
v	related to the steam cycle
vap	related to the vapor

GREEK SYMBOLS

α	Baumann coefficient
β	losses
β	compression or expansion ratio exponent
Δ	difference
ε	degree of reaction
ε	effectiveness
ε	bleed fraction
ε	initial expansion ratio
ε	roughness m
φ	heat flux J kg^{-1}
γ	ratio of thermal capacities
η	efficiency, effectiveness
φ	flow factor
λ	thermal conductivity W $K^{-1}m^{-1}$
μ	viscosity kg $s^{-1}m^{-1}$
ω	rotation speed rd s^{-1}
π	heat created by irreversibilities J kg^{-1}
π	osmotic pressure Pa ou bar
θ	Carnot factor
θ	ratio of the turbine inlet temperature at the compressor inlet temperature
ρ	density kg m^{-3}
ρ	volumetric compression ratio

τ	specific shaft work J kg^{-1}
τ	hydrogen utilization rate
τ	recirculation rate

Acronyms

ABWR	Advanced Boiling Water Reactor
AGR	Advanced Gas-cooled Reactor
ASHRAE	American Society of Heating, Refrigerating, and Air-Conditioning Engineers
ASN	Autorité de Sûreté Nucléaires
ATR	Autothermal Reforming
BWR	Boiling Water Reactor
CANDU	Canada Deuterium Uranium
CANFLEX	CANDU FLEXible
CC	Composite Curves
CC	Combined Cycle
CCGT	Closed Cycle Gas Turbine
CCHP	Combined Cooling, Heating, and Power Generation
CFRP	Components, Functions, Reference Processes
CHP	Combined Heat and Power
CTP	CTP Lib library
DCS	Distillation and Condensation System
DOE	United States Department of Energy
EDF	Electricité de France
EES	Engineering Equation Solver
EPR	European Pressurized Reactor
ESBWR	Economic Simplified Boiling Water Reactor
FMSR	Fast Molten Salt Reactor
FNR	Fast Neutron Reactor
GCR	Gas-Cooled Reactor
GDR	German Democratic Republic
GFR	Gas-Cooled Fast Reactor
GIF	Generation IV International Forum
GOR	Gain Output Ratio
GT-MHR	Gas Turbine Modular Helium-Cooled Reactor
HP	High Pressure
HRSG	Heat Recovery Steam Generator
HT	High-Temperature Reactor
HTE	High-Temperature Electrolysis
HTGR	High-Temperature Gas-Cooled Reactor
HTR	High-Temperature Reactor
HTR-PM	High-Temperature Reactor Pebble Bed Module
IAEA	International Atomic Energy Agency
IHX	Intermediate Heat Exchanger
IP	Intermediate-Pressure
LFR	Lead-Cooled Fast Reactor
LHV	Lower Heating Value

LMTD	Logarithmic Mean Temperature Difference
LP	Low Pressure
LPS	Low-Pressure Section
LT	Low Temperature
LTE	Low-Temperature Electrolysis
MED	Multi-Effect Distillation
MIT	Massachusetts Institute of Technology
MOX	Mixed Oxides of Uranium and Plutonium
MPS	Medium-Pressure Section
MSF	Multistage Flash
MSR	Moisture Separator Reheater
MSR	Molten Salt Reactor
MVC	Mechanical Vapor Compression
NASA	National Aeronautics and Space Administration
NPP	Nuclear Power Plant
NRC	U.S. Nuclear Regulatory Commission
NTU	Number of Transfer Units
NUSCLE	Nuclear Secondary Circuit Lite Emulator
ORC	Organic Rankine Cycle
PBMR	Pebble Bed Modular Reactors
PCHE	Printed Circuit Heat Exchanger
PDU	Productive and Dissipative Unit
PEM	Polymer Electrolyte Membrane
PFBR	Prototype Fast Breeder Reactor
PFD	Process Flow Diagram
PHWR	Pressurized Heavy Water Reactor
POX	Partial Oxidation
PR	Performance Ratio
PTA	Problem Table Algorithm
PWR	Pressurized Water Reactor
RBMK	Reaktor Bolshoy Moshchnosti Kanalny
RNR	Rapid Neutron Reactor
RO	Reverse Osmosis
sCO_2	Supercritical CO_2
SCWR	Supercritical Water-Cooled Reactor
SDR	Steam-to-Distillate Ratio
SEC	Specific Energy Consumption
SFR	Sodium-Cooled Fast Reactor
SG	Steam Generator
SMR	Small Modular Reactor
SMR	Steam Methane Reforming
SOEC	Solid Oxide Electrolysis Cell
SOFC	Solid Oxide Fuel Cell
STEP	Supercritical Transformational Electric Power
SVG	Scalable Vector Graphics
TCNPP	Thermodynamic Cycles of Nuclear Power Plants

TPP	Thermal Power Plant
TRISO	Tri-Structural Isotropic
TRL	Technology Readiness Level
TSG	Turbine Stage Group
TVC	Thermal Vapor Compression
UA	Product of the exchanger surface A by the overall heat transfer coefficient U
UNGG	Uranium Naturel Gaz et Graphite
UO_2	Uranium Dioxide
URL	Uniform Resource Locator
USA	United States of America
USSR	Union of Soviet Socialist Republics
VHTR	Very High-Temperature Reactor
VVER	Vodo-Vodyanoi Energetichesky Reactor
WCR	Water-Cooled Reactor

Conversion Factors

To Obtain	Multiply	By	Reverse Factor
Meter (m)	inch (in)	2.540000E−02	3.937008E+01
Centimeter (cm)	inch (in)	2.540000E+00	3.937008E−01
Meter (m)	foot (ft)	3.048000E−01	3.280840E+00
Square meter (m^2)	square inch (in^2)	6.451600E−04	1.550003E+03
Square centimeter (cm^2)	square inch (in^2)	6.451600E+00	1.550003E−01
Cubic meter (m^3)	cubic foot (ft^3)	2.831685E−02	3.531466E+01
Cubic meter (m^3)	cubic inch (in^3)	1.638706E−05	6.102376E+04
Meter per second (m/s)	foot per second (ft/s)	3.048000E−01	3.280840E+00
Meter per second (m/s)	inch per second (in/s)	2.540000E−02	3.937008E+01
Meter per second (m/s)	kilometer per hour (km/h)	2.777778E−01	3.600000E+00
Meter per second (m/s)	mile per hour (mi/h)	4.470400E−01	2.236936E+00
Kilometer per hour (km/h)	mile per hour (mi/h)	1.609344E+00	6.213712E−01
Radian per second (rad/s)	rpm (revolution per minute) (r/min)	1.047198E−01	9.549292E+00
Gram (g)	ounce (oz)	2.834952E+01	3.527397E−02
Kilogram (kg)	pound (lb)	4.535924E−01	2.204622E+00
Kilogram per cubic meter (kg/m^3)	pound per cubic foot (lb/ft^3)	1.601846E+01	6.242797E−02
Kilogram per cubic meter (kg/m^3)	pound per cubic inch (lb/in^3)	2.767990E+04	3.612730E−05
Pascal (Pa)	pound-force per square foot (lbf/ft^2)	4.788026E+01	2.088543E−02
Pascal (Pa)	pound-force per square inch (psi) (lbf/in^2)	6.894757E+03	1.450377E−04
Pascal (Pa)	bar (bar)	1.000000E+05	1.000000E−05
Pascal (Pa)	atmosphere, standard (atm)	1.013250E+05	9.869233E−06
Watt (W)	British thermal unit per hour (Btu/h)	2.930711E−01	3.412141E+00

(Continued)

(*Continued*)

To Obtain	Multiply	By	Reverse Factor
Watt (W)	British thermal unit per second (Btu/s)	1.055056E+03	9.478170E–04
Watt (W)	ton of refrigeration (12 000 Btu/h)	3.516853E+03	2.843451E–04
Joule per kilogram kelvin [J/(kg · K)]	British thermal unit per pound degree Fahrenheit [Btu/(lb · °F)]	4.186800E+03	2.388459E–04
Watt per square meter (W/m^2)	British thermal unit per square foot hour [$Btu/(ft^2 \cdot h)$]	3.154591E+00	3.169983E–01
Watt per square meter (W/m^2)	British thermal unit per square foot second [$Btu/(ft^2 \cdot s)$]	1.135653E+04	8.805507E–05
Joule per square meter (J/m^2)	British thermal unit per square foot (Btu/ft^2)	1.135653E+04	8.805507E–05
Joule per cubic meter (J/m^3)	British thermal unit per cubic foot (Btu/ft^3)	3.725895E+04	2.683919E–05
Joule per kilogram (J/kg)	British thermal unit per pound (Btu/lb)	2.326000E+03	4.299226E–04
Joule per kelvin (J/K)	British thermal unit per degree Fahrenheit (Btu/°F)	1.899101E+03	5.265649E–04
Watt per square meter kelvin [$W/(m^2 \cdot K)$]	British thermal unit per hour square foot degree Fahrenheit [$Btu/(h \cdot ft^2 \cdot °F)$]	5.678263E+00	1.761102E–01
Watt per square meter kelvin [$W/(m^2 \cdot K)$]	British thermal unit per second square foot degree Fahrenheit [$Btu/(s \cdot ft^2 \cdot °F)$]	2.044175E+04	4.891949E–05
Joule (J)	watt hour (W · h)	3.600000E+03	2.777778E–04
Joule (J)	British thermal unit (Btu)	1.055056E+03	9.478170E-04
Joule (J)	foot pound-force (ft · lbf)	1.355818E+00	7.375621E–01
Kelvin (K)	degree Celsius (°C)	T/K = t/°C + 273.15	T/°C = t/K – 273.15
Degree Celsius (°C)	degree Fahrenheit (°F)	t/°C = (t/°F – 32)/1.8	t/°F = 1.8 t/°F + 32
Kelvin (K)	degree Fahrenheit (°F)	T/K = (t/°F + 459.67)/1.8	T/°F = 1.8 t/K – 459.67
Kelvin (K)	degree Rankine (°R)	T/K = (T/°R)/1.8	T/°R = 1.8 T/K
Degree Celsius (°C)	kelvin (K)	t/°C = T/K – 273.15	T/K = t/°C + 273.15

These values have been obtained from NIST Guide to the SI.[2]

NOTES

1 https://direns.minesparis.psl.eu/Sites/Thopt/en/co/secondEditionEnerSyst.html.
2 http://physics.nist.gov/Pubs/SP811/appenB9.html.

Introduction

INTRODUCTION

In nearly all thermodynamics courses applied to energy systems worldwide, the thermodynamic cycles studied are generally only simplified representations of those implemented in practice, as learners lack the tools needed to model them correctly. This is particularly true for Thermodynamic Cycles of Nuclear Power Plants (TCNPPs), some of which are relatively complex due to the specific characteristics of this energy source.

The goal of this book is to teach learners how to model such thermodynamic cycles themselves. This book is therefore a textbook and not a handbook. Its primary purpose being educational, the presentation of concepts and methods for their implementation proceeds progressively, especially in the first four chapters, starting from fundamentals and advancing to relatively complex detailed models.

I.1 A UNIQUE WORK

This book fills a significant gap in the current literature on applied thermodynamics. To date, there is no comprehensive reference work that specifically addresses the modeling of thermodynamic cycles associated with nuclear reactors, despite their crucial role—particularly in the context of low-carbon electricity generation.

To our knowledge, the only publication dedicated to such cycles is the book *Les cycles thermodynamiques des centrales nucléaires*, edited by Thibaud Normand, Jessica Andreani, and Vincent Tejedor, and published by Presses des Mines, which already presented several model examples as early as September 2010. Other existing studies on this topic rely on simplifications that considerably limit their scope.

Even textbooks commonly used in the training of nuclear engineers—such as *Nuclear Systems* by Todreas and Kazimi—cover secondary cycles only superficially, giving much greater emphasis to the primary circuit.

As for general thermodynamics textbooks (Moran et al., 2018; Çengel & Boles, 2006; Annamalai et al., 2011, etc.), even when applied to power production cycles, they also only present simplified cycles, even for thermal power plants. In his book, *Thermal Power Plant: Design and Operation*, Sarkar addresses the case of the feedwater train of a triple extraction regenerative cycle, but this leads him to elaborate several pages of equations for a single example.

A thorough bibliographic review reveals that the most significant contemporary references are those by Professor Igor Pioro of the Faculty of Engineering and Applied Science at Ontario Tech University and his co-authors. However, as they themselves acknowledge, the cycles they present are simplified representations. This is understandable, given that their contributions are not specifically centered on

DOI: 10.1201/9781003606024-1

thermodynamic cycle modeling but rather encompass multiple dimensions, including reactor technologies, fluid thermophysical properties, and overall system behavior.

Professor Pioro is notably the editor of the *Handbook of Generation IV Nuclear Reactors* published by Elsevier, which is a substantial repository of information and references on the reactors discussed in this book. We will refer to it several times as the G4NR handbook.

Although thermodynamics plays a central role in the education of energy engineers, the cycles studied in most curricula are generally presented in highly simplified—and sometimes even oversimplified—forms, far removed from industrial realities. This tendency toward simplification is even more pronounced in the case of nuclear cycles, whose complexity is often incompatible with conventional pedagogical approaches.

The originality of this work lies in its ability to make advanced modeling accessible through a rigorous dual systems and analytical approach, supported by powerful methodological tools. It notably relies on the Thermoptim software, designed to represent complex energy systems both faithfully and pedagogically. This book thus targets a broad audience of scientists, educators, engineers, and students eager to understand and master the thermodynamics of nuclear reactor secondary circuits.

Please note that, despite their emphasis on technological realism, these books are manuals intended for learning the discipline and not industrial handbooks. The same applies to the Thermoptim software, whose results do not match the precision of professional tools used by industrial engineering offices. The process flow diagrams presented are only examples of possible operating regimes intended to illustrate the cycles.

To emphasize the importance of the skills discussed in this volume, it is worth noting that, beyond the innovation possibilities enabled by the ability to model thermodynamic cycles precisely, the existence of this capability within a company provides strategic advantages that are often underappreciated, such as:

- the ability to critically evaluate announcements, to distinguish truth from falsehood by estimating the gap between theoretical promises and operational realities.
- the ability to engage with industrialists and manufacturers in the field, to debate their claims, and to earn their respect through competence and counter-expertise, particularly during the drafting of specifications.
- when analyzing responses to calls for tenders, the ability to conduct thorough critical analysis to identify the best offers in a reasoned manner.
- since accurate measurements are often very difficult to obtain upon receiving equipment, the ability to interpret the limited available data by comparing it with simulation results.

I.2 AN INDUSTRIAL AND STRATEGIC CHALLENGE

In the current context marked by the climate crisis, the growing demand for energy, particularly electricity, and geopolitical uncertainties, nuclear reactors represent a reliable, low-carbon, competitive, and scalable solution. They serve as essential complements to intermittent renewable energies and play a structural role in energy transition and sovereignty policies.

This book addresses several major, yet often overlooked, challenges:

- Optimizing secondary circuits is essential for improving the energy efficiency, safety, and profitability of nuclear power plants (NPPs), particularly in the context of developing Small Modular Reactors.
- Unlike conventional thermal power plants, NPP cycles present specific technological constraints. For example, in steam cycles, expansion in the wet region significantly impacts cycle architecture. For gas cycles, their closed nature requires specific thermodynamic optimization strategies.
- Nuclear secondary circuits are poorly documented in open literature, not only due to their complexity but also because of their strategic nature. This book provides rare insight into an opaque technological field, grounded in deep technical expertise and realistic case studies.
- The gradual retirement of engineers who designed the first generations of NPPs risks leading to a loss of skills, knowledge, and expertise. This work also aims to preserve and transmit this technical knowledge by proposing a renewed pedagogical approach.
- While numerous nuclear cycle architectures have been proposed over the decades, very few have been implemented. Apart from a few prototypes, only steam thermodynamic cycles are currently in operation. Why such a gap? This book offers insights into this critical question for the future of nuclear energy.
- It is impossible to answer this question without addressing the issue of the technological maturity of the various competing technologies. To do this, we will use the concept of Technology Readiness Level (TRL), initially developed by NASA for its own needs and since then widely used worldwide.

I.3 PEDAGOGICAL AND METHODOLOGICAL CONTEXT

Several observations explain why nuclear reactor thermodynamic cycles are rarely studied in depth in engineering curricula:

- Training programs have progressively reduced the emphasis on complex case studies, often favoring more generic or cross-disciplinary knowledge.
- Educators often prioritize abstract analytical modeling, distant from industrial realities they may not fully master, limiting learners' ability to understand concrete systems.
- NPP cycles require dual expertise in advanced thermodynamics and system modeling, which is rarely combined in traditional training.
- It is also critical to consider technical and industrial constraints: only certain operations are technologically feasible, limiting available options, while fluid properties impose additional constraints. A purely theoretical approach is insufficient, especially since some phenomena involved in the processes are not fully understood or modeled.
- Finally, economic considerations must also be taken into account. Some improvements are technically feasible but not implemented due to excessive costs.

In our previous work, *Energy Systems: A New Approach to Engineering Thermodynamics*, published in 2021, we proposed a learning method based on reducing mathematical formalism without sacrificing scientific rigor, designed to meet the needs of learners with limited backgrounds in mathematics and physics who do not wish to engage in advanced modeling.

In this two-volume set, of which this is the second volume, we complement this simplified approach by enhancing the presentation of component models, particularly in terms of their equation formulation.

After a review of thermodynamic fundamentals, the first volume delves into the analysis of the **behavior of the most commonly used components** and explains how customized components can be modeled and coupled with Thermoptim through a mechanism called external classes. It then addresses a particularly challenging issue: the sizing of components and the study of systems under **off-design** conditions. Several case studies illustrate how these developments can be applied to simple cycles. As its scope is much broader than NPP cycles alone, the examples presented in that volume are deliberately more diverse.

This volume demonstrates, among other things, how the developments from the first volume can be used to model in detail complex thermodynamic cycles coupled with nuclear reactors. It assumes familiarity with these fundamentals and does not revisit them. Those who do not have access to the first volume can find some of the developments presented there on the Thermoptim-UNIT portal.

It should also be noted that we were forced to limit the number of illustrations of the various technologies for several reasons. Indeed, for editorial reasons, the total number was limited and we preferred to prioritize screenshots related to the models we present, so that readers can visualize them without difficulty. Furthermore, due to the confidential nature of many illustrations, it would have been difficult to obtain reproduction permissions for the most interesting ones. Lastly, other works, such as the G4NR handbook, contain numerous illustrations, and interested readers can easily acquire them if they wish.

I.4 SYSTEMIC AND ANALYTICAL MODELING SUPPORTED BY POWERFUL METHODOLOGICAL TOOLS

One of the major strengths of this book is its full integration of the Thermoptim software, which enables the construction of realistic models of complex energy systems through simple graphical assembly of components.

Each component encapsulates the equations necessary for its modeling, reducing the risk of formulation errors and allowing users to focus on the conceptual aspects of the system. A key software feature enables the automatic generation of equations associated with a model defined primarily graphically, paving the way for more advanced analyses in external computational environments.

This **dual approach** (graphical and analytical) facilitates progressive learning of complex systems, fostering an engineering mindset grounded in both physical intuition and analytical rigor.

I.5 TECHNOLOGY READINESS LEVEL

TRLs are a standardized method to assess the maturity of a given technology. Originally developed by NASA in the 1970s–1980s, this nine-level scale is now widely used across research, industry, and innovation project management.

The scale begins at TRL 1, representing the observation and reporting of basic principles of a highly experimental technology. As development progresses, it moves through laboratory validation phases (TRL 2-4), prototype demonstration in relevant environments (TRL 5-6), and demonstration in actual operational conditions (TRL 7-8). TRL 9 corresponds to a fully mature technology proven and deployed successfully in production or operational service.

TRLs provide a common language for engineers, researchers, and investors to estimate risks, plan development, reduce commercialization times, and optimize investments. We will use them several times to compare the diffusion prospects of various technologies.

I.6 CONTENT OF THIS BOOK

This book is organized around eight main chapters, complemented by eleven case studies enabling in-depth analysis of concrete implementations:

1. Methodology for Studying the Systems and Adopted Pedagogical Paradigm: understanding complex thermodynamic cycles, such as those associated with nuclear reactors, relies on both rigorous analysis of energy systems and an adapted pedagogical approach.
2. Nuclear Reactors and Technological Contexts: presentation of the fundamentals of nuclear fission, introduction to the various reactor types (AGR, PWR, BWR, SCWR, FNR, SFR, LFR, HTR, VHTR, MSR, FMSR, Small Modular Reactors, etc.), and the main characteristics of associated cycles (temperatures, pressures, power outputs, efficiencies).
3. Steam and Gas Power Cycles: Understanding, Analysis, Methodologies: introduction to fundamental thermodynamic concepts and tools for analyzing power cycles, emphasizing methods for improving efficiency.
4. High Temperature Steam Cycles: detailed study of water-steam cycles coupled with nuclear reactors, from simple Rankine cycle diagrams to the most complex configurations, incorporating tools like exergy analysis.
5. Medium Temperature Steam Cycles: analysis of main types of Water Cooled Nuclear Reactor Cycles.
6. Closed Gas and Supercritical CO_2 Cycles: exploration of innovative gas cycles adapted to certain advanced reactors, considering technical challenges and application prospects.
7. Specific Applications and Perspectives: analysis of possibilities offered by combined cycles, desalination, hydrogen production, or cogeneration.
8. Case Studies: eleven case studies mainly dedicated to actual thermodynamic cycles of NPPs.
9. The last chapter attempts a synthesis and perspective.

For the bibliography, we have made a specific choice: we only include references that we have actually used in our writing work, as well as some general books. We have deliberately omitted exhaustive bibliographic lists to avoid overwhelming the reader with a mass of references that are now easily accessible through modern databases.

Many digital resources for teaching energy systems have been gradually collected into a portal called Thermoptim-UNIT (www.thermoptim.org), whose content is freely accessible, with few exceptions, including solutions to some exercises and problems.

This book's website[1] provides access to various supplementary resources, including:

- Links to digital versions of selected figures
- Updated footnote links in case of changes
- Application examples of developments presented in specific chapters, available as guided explorations or small Thermoptim projects with their associated files
- Theoretical and methodological supplements, either on the portal or dedicated pages

Since the complete list of resources was not finalized at the time of publication, please refer to the website during your reading to discover all available materials. Some resources will be reserved for registered book purchasers.

1.7 CONCLUSION

By combining scientific rigor, active pedagogy, and realistic modeling, this book constitutes an unprecedented reference for educators, engineers, researchers, and students seeking to deepen their understanding of TCNPPs.

It offers a concrete response to the challenges of training, knowledge transmission, and technological innovation, equipping readers with a robust methodological approach and powerful digital tools to design, analyze, and improve the energy cycles of 21st-century nuclear technology.

This entire book is illustrated with numerous examples of cycles modeled using Thermoptim, giving readers access to models whose structure or parameters can be customized to perform various simulations.

The list of these examples, of varying difficulty, is provided in a portal called Thermoptim-UNIT (www.thermoptim.org) along with comments and suggestions on how they can be used in teaching, depending on context and educational objectives.

More generally, many digital resources for teaching energy systems have been gradually collected into this portal, whose content is freely accessible, with few exceptions, including solutions to some exercises and problems.

Since this is a manual and not a handbook, we have deliberately limited the references, citing mainly only those that were directly useful for the development of the chapter content.

NOTE

1 https://server.s4e2.com/crc/tcnpp/.

BIBLIOGRAPHY

Annamalai, K., Puri, I.K. & Jog, M.A. (2011). *Advanced Thermodynamics Engineering*, 2nd ed. CRC Press, Boca Raton, FL. ISBN 978-1-4398-0572-5.

Balmer, R.T. (2011). *Modern Engineering Thermodynamics*, 1st ed. Academic Press (Elsevier), New York. ISBN 978-0-12-374996-3.

Çengel, Y.A. & Boles, M.A. (2006). *Thermodynamics: An Engineering Approach (SI Units)*, 5th ed. McGraw-Hill, New York. ISBN 978-0070606593.

Gicquel, R. (2021). *Energy Systems: A New Approach to Engineering Thermodynamics*, 2nd ed. CRC Press, Boca Raton, FL. ISBN 978-0367545129. DOI: 10.1201/9781003175629.

Moran, M.J., Shapiro, H.N., Boettner, D.D. & Bailey, M.B. (2018). *Fundamentals of Engineering Thermodynamics*, 9th ed. John Wiley & Sons, New York. ISBN 978-1-119-39138-8.

Normand, T., Andreani, J. & Tejedor, V. (eds.) (2010). *Les Cycles Thermodynamiques Des Centrales Nucléaires.* Presses des Mines, New York. ISBN 978-2911256332.

Pioro, I.L. (ed.) (2023). *Handbook of Generation IV Nuclear Reactors*, 2nd ed. Woodhead Publishing (Elsevier), New York. ISBN 978-0-12-820588-4.

Sarkar, D.K. (2015). *Thermal Power Plant: Design and Operation*, 1st ed. Elsevier, New York. ISBN 978-0-12-801575-9.

Singh, O. (2009). *Applied Thermodynamics*, 3rd ed. New Age International Publishers, New York. ISBN 978-8122426029.

Todreas, N.E. & Kazimi, M.S. (2021a). *Nuclear System, Vol. 1: Thermal Hydraulic Fundamentals*, 3rd ed. CRC Press, Boca Raton, FL. ISBN 9781351030502.

Todreas, N.E. & Kazimi, M.S. (2021b). *Nuclear System, Vol. 2: Elements of Thermal Hydraulic Design*, 2nd ed. CRC Press, Boca Raton, FL. ISBN 9780429157608.

1 Systems Approach and Innovative Teaching of Thermodynamic Cycles of Nuclear Power Plants with Thermoptim

1.1 ANALYSIS OF ENERGY SYSTEMS

1.1.1 Introduction to Energy Systems

An energy system is an organized set of components designed to convert energy from one form to another, for example, to generate useful work, produce electricity, provide cooling, or pump heat. These systems implement elementary thermodynamic processes, the most common of which are:

- **Compression**, increasing the pressure of a fluid.
- **Expansion**, decreasing the pressure of a fluid, used to generate mechanical work.
- **Heat exchange**, transferring thermal energy between fluids or with the environment.
- **Throttling**, decreasing the pressure of a fluid without producing work.

To analyze these systems, a classic thermodynamic diagram is generally used, sometimes called a **process flow diagram**, representing the different components and their interconnections. This diagram is often close to the actual physical diagram, but for in-depth thermodynamic analysis, it is preferable to reason in terms of functional structure rather than limiting oneself to the physical arrangement, as the same set of mechanical parts can perform several successive functions. For example, in a diesel engine, the cylinder, cylinder head, and piston assembly first compresses the incoming air, then performs combustion, and finally expands the combustion gases to produce work.

The functional structure can be described using a methodological approach with two complementary parts: graphical systems modeling that links components through connections such as piping or thermal couplings, and analytical modeling of each component to describe their thermodynamic behavior. The Thermoptim software, which will be used throughout this book, is a modeling environment that enables the implementation of this dual approach.[1]

DOI: 10.1201/9781003606024-2

The functional structure can be graphically represented by a functional diagram, which highlights the components, their roles, and the processes undergone by the working fluids, as well as the thermodynamic state of these fluids (pressure, temperature, enthalpy, etc.).

However, this approach does not allow for evaluating irreversibilities. To do so, an exergy structure is defined, which analyzes the distribution and evolution of exergy flows in the system.

The exergy structure describes how exergy is supplied, transformed, recycled, and lost. It enables obtaining two particularly useful tools to guide system optimization:

- The **exergy diagram**, which graphically illustrates the inputs, conversions, recyclings, and losses of exergy.
- The **exergy balance**, which is a table showing the exergy resources and products within each functional component, allowing for calculating their exergy efficiency and the magnitude of their irreversibilities, in absolute value and as a percentage.

The functional and exergy structures thus offer two complementary conceptual perspectives, while their respective diagrams constitute the graphical representations. These two structures are dual: the functional structure describes the physical organization and processes, while the exergy structure evaluates the efficiency of exergy use and irreversibilities, offering a complementary view focused on energy quality.

1.1.2 Illustration: The Steam Power Plant

1.1.2.1 General Presentation

A steam power plant is a machine that converts heat provided by a boiler into electricity (Figure 1.1.1). In many countries, a significant proportion of the power plant fleet consists of such installations, often called thermal power plants. In our case, the boiler receives its heat from a nuclear reactor, for example, an advanced gas-cooled reactor.

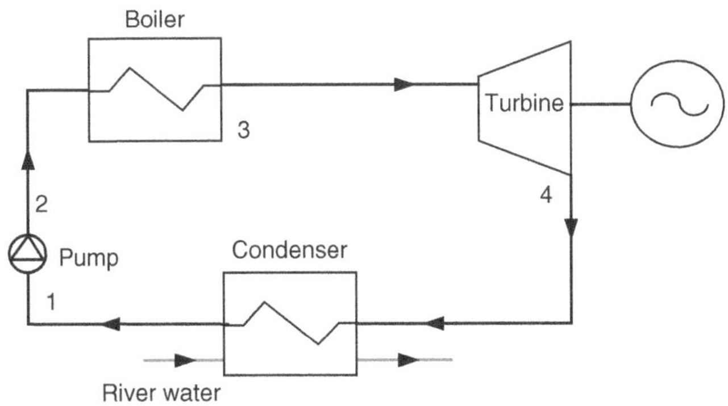

FIGURE 1.1.1 Schematic diagram of a steam power plant.

The operation of the power plant can be broken down into four successive stages:

1. The first stage occurs in the boiler, where heat is transferred to pressurized water, which transforms into steam.
2. The second stage concerns electricity production: the steam is expanded in a turbine that drives an alternator, producing electrical current transported to consumption sites via high-voltage lines.
3. The third stage is the condensation of the steam exiting the turbine into liquid water.
4. The fourth stage consists of repressurizing this water before redirecting it to the boiler.

The Hirn cycle (or Rankine cycle with superheat) uses a condensable fluid, cooled to a temperature and pressure sufficient for it to be fully liquefied before compression. Under these conditions, the compression work is almost negligible compared to the expansion work (whereas it represents about 60% in a gas turbine, as we will see). The compressed liquid is vaporized and superheated in the boiler by heat exchange with the hot source, then expanded and condensed. The two-phase state of the fluid during condensation and vaporization phases is highly favorable for heat transfer.

1.1.2.2 Operating Characteristics

The operation of a steam power plant presents several characteristics:

- The components are interconnected by the working fluids (water or steam) and by the mechanical or electrical link between the turbine and the feedwater pump, the latter being driven by part of the work produced by the turbine.
- It involves a working fluid. Here, it is water or steam, but in other systems, such as a gas turbine, if we abstract the fuel which can be liquid, the fluids are gaseous mixtures (air and combustion gases). The thermodynamic properties (pressure, temperature, enthalpy) of the working fluid(s) vary depending on the processes undergone. Unlike steam cycles, where the fluid changes state (liquid-vapor), the fluids in a gas turbine remain in the gaseous state.
- The compression and expansion processes are essential, as they allow the conversion between fluid energy and mechanical work.
- The irreversibilities that occur in heat exchangers severely penalize the cycle's efficiency and must therefore be reduced as much as possible.

1.1.3 Functional Structure and Diagram of the Steam Power Plant

The functional structure conceptually describes the sequence of components and the thermodynamic processes of the fluids.

The corresponding functional diagram in Figures 1.1.2 and 1.1.3 shows, starting from the lower left corner and moving clockwise: the pump, the nuclear boiler broken down into three elements (economizer, evaporator, superheater), the turbine, and the condenser.

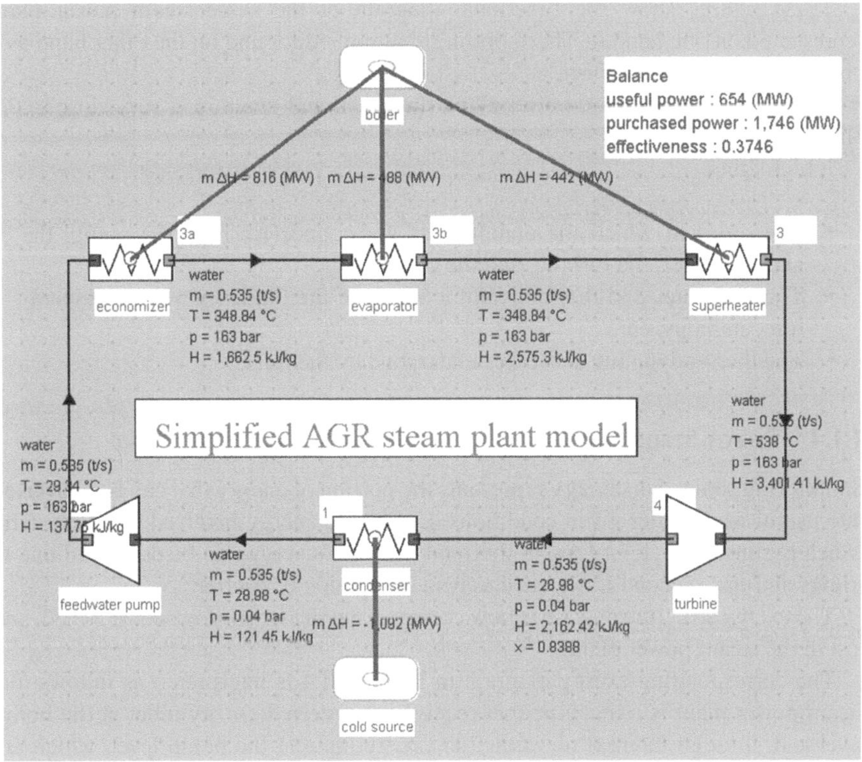

FIGURE 1.1.2 Thermoptim functional diagram of a steam power plant.

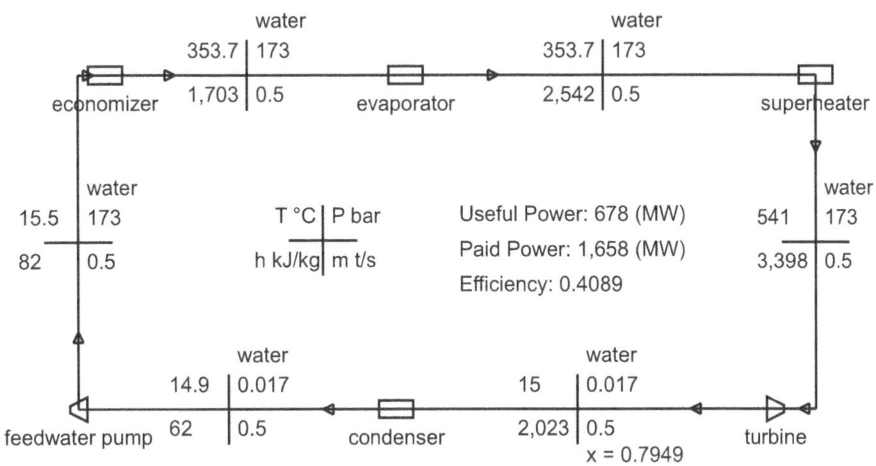

FIGURE 1.1.3 Functional diagram of a steam power plant as a flowsheet.

These figures show the functional diagram of the same steam power plant represented on one hand in Thermoptim's diagram editor and on the other hand as a process diagram or flowsheet.

In this case of the steam power plant, the functional diagram is very close to the physical diagram.

It highlights:

- The order in which the components are connected: here the pump, the nuclear boiler, the turbine, and the condenser.
- The flow rates and thermodynamic states of the fluids (pressure, temperature, enthalpy, etc.).
- The thermodynamic processes undergone by the fluids.

1.1.4 Exergy Structure and Diagram of the Steam Power Plant

Simplifying somewhat, exergy represents the portion of energy that can be converted into useful work under given conditions. Unlike the energy involved in this system, which is conserved (First Law of thermodynamics), exergy can be destroyed due to irreversibilities (Second Law), such as heat dissipation or friction.

The exergy structure describes how exergy is supplied, transformed, recycled, and lost in the steam power plant.

The corresponding exergy diagram in Figure 1.1.4 is interpreted as follows: the steam power plant is a machine that receives an external exergy input at the boiler level and, through internal recycling, an exergy input at the pump level, which are the productive units on the left of the screen. This exergy is partly converted into mechanical form in the turbine and partly dissipated in the condenser.

The net power corresponds to the fraction of mechanical power not recycled to the pump, which consumes 9 MW. This matches the 684 MW displayed on the functional diagram.

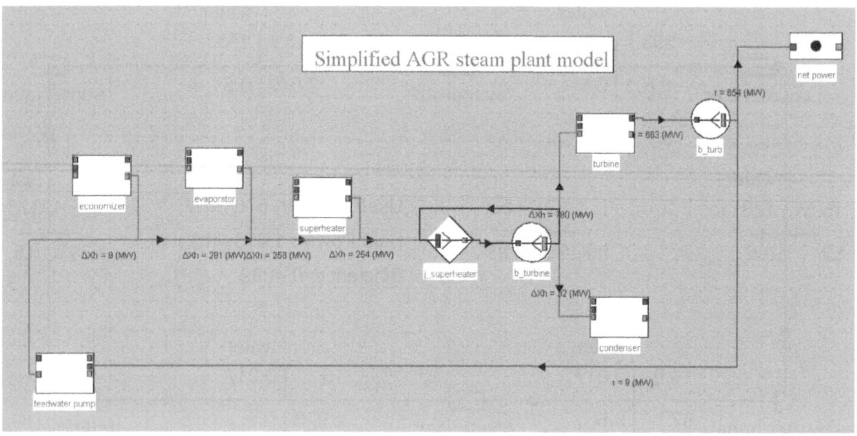

FIGURE 1.1.4 Exergy diagram of the steam power plant.

component	Resource	Product	Exergy efficiency	Irreversibilities	% total	settings
feedwater pump	8.72	8.722	100.0000%	0	0	
turbine	780.182	662.86	84.9622%	117.322	20.7304%	
condenser	32.472	0	0	32.472	5.7377%	Tk = 20.00 °C
economizer	570.008	291.426	51.1267%	278.582	49.2244%	Tk = 700.00 °C
superheater	308.832	254.356	82.3606%	54.476	9.6257%	Tk = 700.00 °C
evaporator	341.24	258.15	75.6504%	83.091	14.6819%	Tk = 700.00 °C
global	1,220.08	654.137	53.6143%	565.943	1	

FIGURE 1.1.5 Exergy balance of the steam power plant.

The exergy balance is given in Figure 1.1.5.

By analyzing the distribution of irreversibilities indicated in this balance, we note that nearly half (49.22%) occurs at the economizer level, with other losses distributed between the turbine (20.7%), the evaporator (15.5%), the superheater (9.6%), and the condenser (5.9%). The main irreversibilities in this cycle are thus related to heat exchanges.

The exergy structure thus allows for the analysis of exergy flows in the steam power plant:

- The nuclear boiler supplies thermal exergy to the system.
- The turbine converts the exergy of superheated and pressurized steam into mechanical work, usable to drive a generator or another device.
- A fraction of the mechanical work produced by the turbine is reinjected to power the pump.
- The exergy not converted into useful work is dissipated, mainly in the form of heat rejected by the condenser to the environment.

We will see that the exergy approach proves to be a powerful tool for optimizing the energy efficiency of the systems studied.

In this book, the exergy balance will often be preferred over the exergy diagram due to its richness of information, especially since editorial constraints limit the number of published figures.

1.1.5 Complementarity and Duality of Structures

As we have just seen, the functional and exergy structures offer two complementary and dual perspectives of the same energy system:

- The functional structure describes the physical organization of the system, detailing the components, their interconnections, and the thermodynamic processes undergone by the fluids.
- The exergy structure describes the quality of energy, quantifying exergy flows, conversions, and losses, and generating the exergy balance.

Their joint use allows both understanding and optimizing system operation.

1.2 A NEW PEDAGOGICAL PARADIGM FOR THERMODYNAMIC CYCLES OF NUCLEAR POWER PLANTS

The concepts presented in the first section (functional and exergy structures, exergy balances, diagrams) constitute powerful tools for analyzing and optimizing any energy system. However, acquiring mastery of these concepts, especially in the context of Thermodynamic Cycles of Nuclear Power Plants (TCNPPs), requires a pedagogical approach rethought compared to traditional methods. In this section, we outline a teaching paradigm that builds on these foundations but adapts the method and tools to the needs of today's learners.

To illustrate this point, Figure 1.2.1 presents the process diagram generated by Thermoptim of the Flamanville 3 Evolutionary Power Reactor (EPR), a third-generation French pressurized water reactor commissioned at the end of 2024. For better readability in a standard page format, this diagram has been split into two superimposed parts instead of being placed side by side. The continuation of the links ending on the right of the upper part of the figure is found on the left of its lower part.

The objective of this book is to enable learners to understand and model the operation of TCNPPs to optimize their performance based on the characteristics of different conceivable reactor types. It must allow them to fully understand a model of this level of complexity.

This model involves 66 points, 68 components representing the processes undergone by the fluid, 25 nodes, and 10 heat exchangers, whereas a simplified Rankine cycle of this power plant would use only six points and six components, as we have just seen. It is clear that traditional methods for analyzing simplified cycles are not suitable for studying a real cycle.

In our previous work *Energy Systems: A New Approach to Engineering Thermodynamics* (Gicquel, 2021), we proposed a learning method based on reducing mathematical formalism without sacrificing scientific rigor. This book extends this approach by deepening the introduced concepts, while applying them to real industrial situations, with a particular emphasis on nuclear cycles.

1.2.1 Context and Pedagogical Challenges

Several factors contribute to understanding why the teaching of applied thermodynamics to energy systems, and particularly to TCNPPs, faces numerous challenges. The complexity of the discipline, the decrease in time devoted to technical teachings in training, the growing diversity of learner profiles, and the availability of new digital tools both necessitate and enable a change in pedagogical paradigm.

Engineers, whether in initial or continuing education, must master technical skills specific to these systems while being competent in non-technical aspects such as management, leadership, economics, and environmental impact. This diversification of skills reduces their time and motivation for technical subjects.

In initial training programs, the time allocated to technical disciplines is decreasing, with a predominance of theory over practical work. Learners in continuing education, often having forgotten basic scientific concepts, demand a practical pedagogical approach directly applicable to their professional needs, particularly in the nuclear industry where safety, efficiency, and sustainability are crucial.

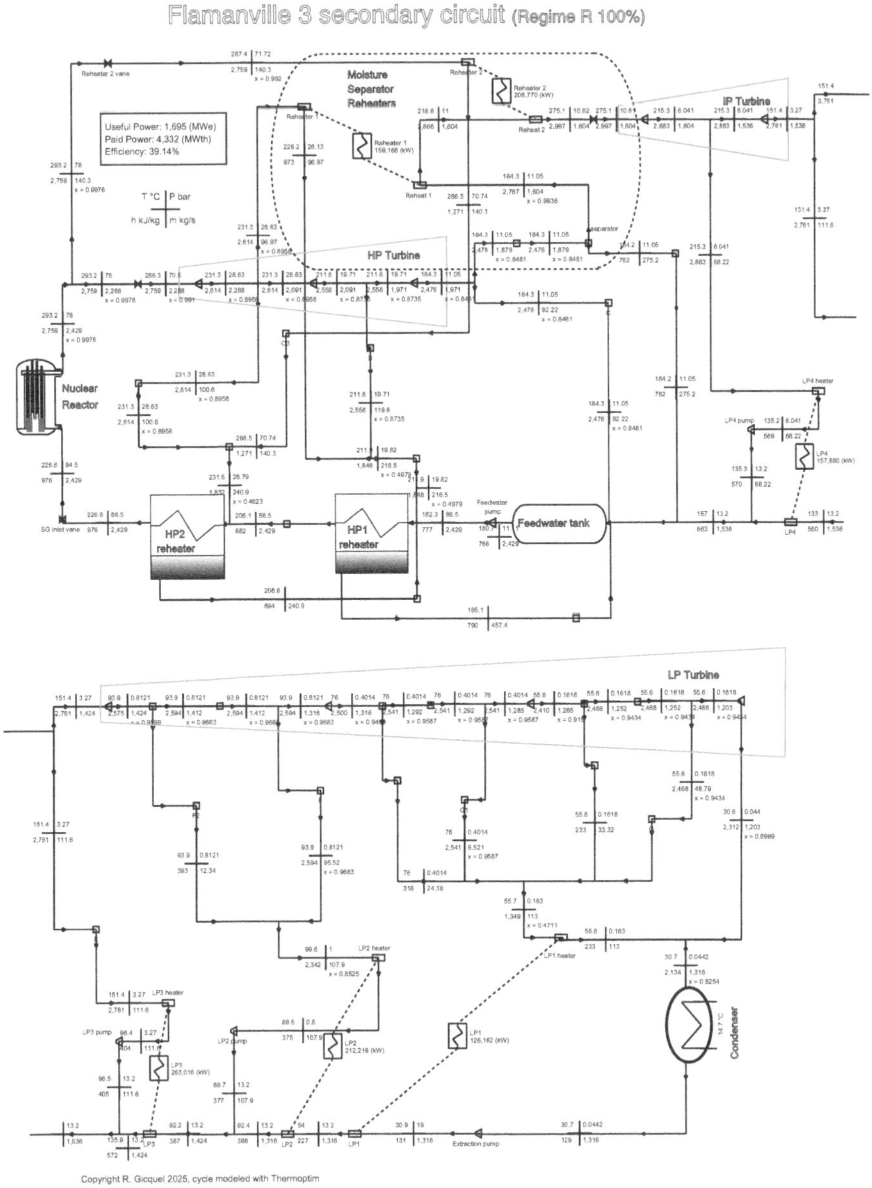

FIGURE 1.2.1 Process diagram of the Flamanville 3 power plant.

1.2.2 Limits of Classical Cycle Teaching

The study of nuclear reactor secondary cycles, such as those based on existing steam cycles, is complex due to the need to model heat transfers, fluid properties (water, steam, gases), and component performance such as turbines, condensers, pumps, or compressors. Traditional pedagogical approaches, which rely on a deductive and mathematical presentation, struggle to meet the needs of modern learners.

In these approaches, calculations of steam properties (enthalpy, entropy, pressure) are performed using steam tables or diagrams, often imprecise or laborious, requiring simplifying assumptions. This leads to models far from the reality of nuclear reactors, which demotivates learners, who do not perceive the practical relevance of the taught models, particularly in terms of energy efficiency.

Moreover, manual cycle calculations, based on interpolations in tables or diagrams, are time-consuming and prone to errors. The time devoted to equation formulation often represents the bulk of the courses, leaving little room for the study of advanced TCNPPs, such as those integrating condensate recovery or feedwater station optimization to increase overall efficiency. Learners, whether students or professionals, thus find themselves ill-prepared to design or optimize modern nuclear reactor secondary cycles, where performance and sustainability requirements are high.

1.2.3 An Innovative Pedagogical Approach

To address these challenges, we propose a pedagogical paradigm that relies on a significant reduction in equation writing and delegation of complex calculations to a simulation tool, while emphasizing qualitative understanding of component operation (turbines, condensers, pumps, compressors, steam generators) and cycle architecture analysis. The key principles are:

- A reduction in learners' cognitive load, allowing them to focus on understanding cycles and their optimization, with fluid property and component performance calculations automated.
- A dual systems approach and functional approach combining graphical system modeling and the ability to examine component behavior details.
- Practical relevance, with simulated models reflecting real operating conditions of nuclear reactors, motivating learners by having them work on concrete applications.
- An inductive approach where theory is introduced only contextually and limited to the strict minimum, facilitating learner motivation.

1.2.4 A Pedagogy Adapted to Different Audiences

The expertise reversal effect has shown that the effectiveness of learning methods varies according to the learner's level of knowledge or experience. Some pedagogical techniques, highly effective for beginners, are less so, or even counterproductive, for more advanced learners or experts.

This is why our educational paradigm is implemented in three complementary modes, adapted to learner profiles:

- The light mode is designed for beginners or professionals in continuing education. It uses the Components, Functions, Reference Processes (CFRP) approach to present basic secondary cycles, such as the simple Rankine cycle, without resorting to complex concepts like entropy. Practical activities consist of guided explorations of pre-built models. This mode suits learners with limited scientific backgrounds or seeking quick training.

- The progressive mode is intended for bachelor's students or professionals with intermediate scientific levels. It starts with the light approach, then introduces more complex concepts, such as entropy diagrams (T, s) or Mollier diagrams (h, s) and exergy balances, to analyze TCNPP performance. Learners explore pre-built models and perform practical exercises, without having to carry out complex mathematical calculations.
- The advanced mode is reserved for master's or engineering school students. It addresses learners familiar with theoretical approaches and includes a rigorous presentation of the discipline, with early introduction of entropy or Mollier diagrams and exergy balances for optimizing TCNPPs. Learners can explore or build their own models, for example, to study variants of innovative cycles, thus developing considerable autonomy.

1.2.5 CFRP Approach: An Inductive Progression

The CFRP approach begins by describing cycle architectures, such as the Rankine cycle and associated components (steam generator, turbine, condenser, pump). It shows that these cycles rely on four main functions (heat exchanges, expansions, compressions, throttling), corresponding to reference processes of the fluid. These processes are special cases of applying the First Law of thermodynamics, introduced contextually to avoid an abstract presentation.

Learners visualize the cycles in enthalpy-pressure diagrams, which illustrate the processes undergone by water or steam. This inductive method, starting from components and technologies to move toward theory, allows overcoming learners' reluctance to the complexity of NPPCs.

1.2.6 Which Equations to Teach?

In a context where the learning time for technical subjects is reduced and learners' mathematical backgrounds are often limited, the teaching of equations must be carefully considered, particularly depending on the pedagogical objective and learners' scientific level. For secondary cycles, mass and energy balance equations (First Law) are essential, but their manual resolution is generally laborious.

While we believe it is preferable to limit the use of equations as much as possible in the initial stages of training, nothing prevents their introduction in a second phase, especially if there is demand from learners.

In particular, if we adhere to the choice we made to start with the minimalist CFRP presentation without mentioning entropy, it is clear that all equations involving this state function can only be introduced in a second phase.

Some simulation tools, as explained in the following section, allow for automatically generating the equations of the studied models, facilitating complementary theoretical analyses for advanced learners wishing to optimize TCNPP performance with specific methods.

1.2.7 Comparison with the Anglo-Saxon Approach

To address the difficulties mentioned above related to the use of tables or diagrams, the Anglo-Saxon approach emphasizes learning equations, with only the calculation of fluid thermodynamic properties delegated to the computer (Moran et al., 2018; Çengel & Boles, 2006; Annamalai et al., 2011; Singh, 2009; Balmer, 2011). This paradigm aims for learners to be able, at the end of the course, to write all the equations of the thermal machines they will work on. Learners trained in this way are perfectly capable of doing so, but this can only be practical for small, simple cycles often far from the complex systems encountered in TCNPPs.

We will return in the conclusion to the limits of this educational approach.

In contrast, Thermoptim software allows for modeling realistic secondary cycles, such as those with superheat or regenerative heat recovery. Learners, even with modest scientific levels, trained according to this approach are equipped to work on real-world and large-scale systems, but the software may appear as a somewhat mysterious black box.

A synthesis can be achieved by having Thermoptim formally derive the equations associated with its models, thus allowing learners to study them at a theoretical level while preserving a systems perspective. Such an activity can find its place in the framework of the second and third modes we presented above.

The learner can start by studying, component by component, the equations implemented in their model, linking them to the theoretical and technological resources available.

In a second phase, they can work on the complete set of equations, add some if necessary, and solve it if they have an appropriate solver, which, incidentally, can prove difficult for systems involving many components.

Proceeding in this way allows in particular:

- Maintaining a systems approach in describing the model's architecture, while providing access to its equations to exploit them in other modeling environments and perform complementary treatments to those of the software, for example, optimization.
- Ensuring the consistency of equation sets for large models, which can be difficult in practice if one does not have an appropriate tool.

1.2.8 Example: Modeling a Steam Turbine

Let's illustrate this with a very simple example, that of modeling a steam turbine, detailed in the guided exploration EQUA_01. Its functional diagram is provided in Figure 1.2.2.

Consider a steam turbine supplied with a flow rate of 1,500 kg/s of saturated steam at a pressure of 70 bar, corresponding approximately to the outlet conditions of a steam generator in a Pressurized Water Reactor (PWR)-type nuclear power plant. The pressure at the turbine outlet is 35 bar, with an isentropic efficiency of 0.85.

Modeling such a turbine is simple, regardless of the modeling environment used. The model involves three components: a steam inlet, the turbine, and the exhaust of the expanded steam.

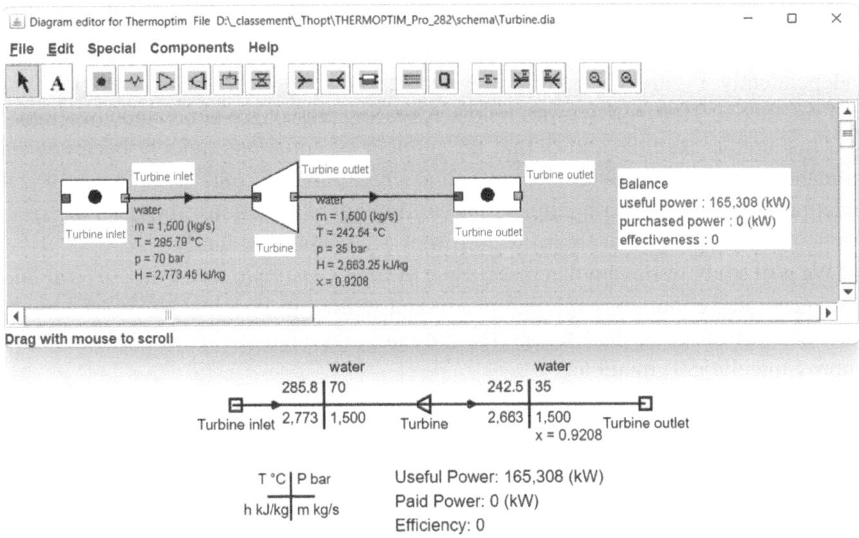

FIGURE 1.2.2 Functional diagram of the steam turbine.

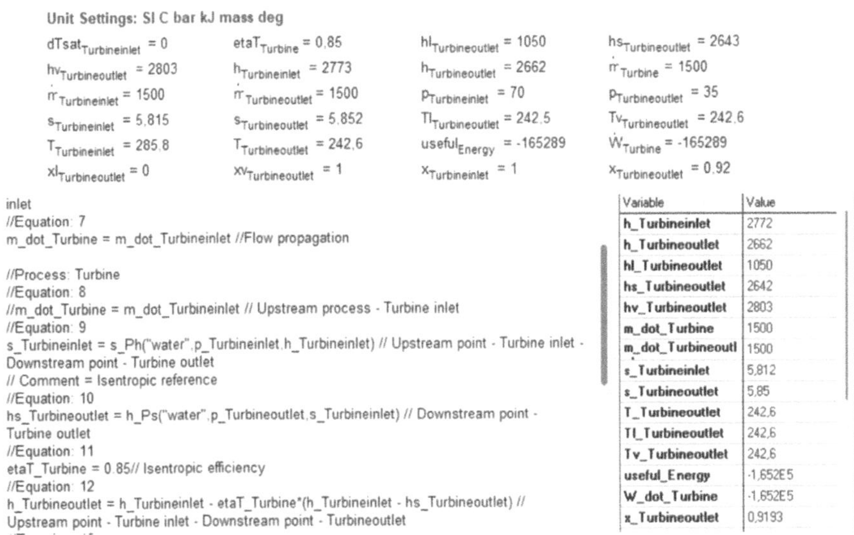

FIGURE 1.2.3 EES and Interactive Thermodynamics models of the steam turbine.

Figure 1.2.2 shows a possible model. The upper part presents the model in Thermoptim's diagram editor. The lower part represents this model as a flowsheet, displaying the thermodynamic state of the two points and the useful power in kW.

It is possible to automatically generate the set of equations corresponding to this model. Without going into details, the result is a set of 24 equations, of which only four are coupled, the others being simple value assignments.

These equations can be formatted according to the syntax of different solvers, such as Interactive Thermodynamics[2] or Engineering Equation Solver (EES),[3] and solved independently. Figure 1.2.3 shows the results obtained with EES in the upper part and Interactive Thermodynamics below, revealing part of the generated equations.

This simple example shows that the number of equations quickly becomes significant, even if value assignments (such as equations 7, 8, and 11) do not pose calculation problems. When it comes to several hundred equations, the risks of errors, particularly in naming, are significant if they are written manually.

We will study in this book a much more complex example, involving several hundred generated equations exploited in a model written in Java, the NUSCLE simulator. As an illustration, the Flamanville 3 model presented in Figure 1.2.1 comprises approximately 550 equations.

1.3 THERMOPTIM, A TOOL FOR MODELING AND TEACHING TCNPPS

The practical implementation of this pedagogical paradigm requires a simulation environment capable of integrating the notions discussed previously and making them accessible to different levels of expertise. It is in this perspective that the use of Thermoptim fits, which is both a highly accessible learning support and a high-performance engineering tool.

1.3.1 Presentation of Thermoptim

Thermoptim is simulation software allowing for modeling a wide variety of energy systems while faithfully reflecting technological reality. It provides a modeling environment comprising four interconnected elements: a diagram editor/synoptic screen, a simulator, interactive diagrams, and an optimization tool, allowing for easily varying the system's characteristics.

Once a cycle model is built in Thermoptim, it is possible to visualize the state of the various points in many thermodynamic diagrams: (T, s) entropy, (h, p) enthalpy, (h, s) Mollier, or exergy.

We presented the functional diagram of the steam power plant in the software's diagram editor (Figure 1.1.2), as well as its exergy diagram (Figure 1.1.4) in a second editor. The simulator screen is given in Figure 1.3.1, and the water enthalpy diagram in Figure 1.3.2. Other elements will be presented later.

Thermoptim allows for building cycle models by assembling graphical components corresponding to key functions (heat exchanges in the steam generator, including liquid heating, vaporization, and superheat, expansion in the turbine, condensation, liquid or gas compression, throttling). Thus, to model a Rankine or Brayton cycle, the learner places icons representing the components on a workspace, and then parameterizes their characteristics (pressure, temperature, flow rate, efficiency). Unlike the classic approach, which requires manually solving complex equations to obtain steam properties, these calculations are delegated to Thermoptim, offering increased precision and eliminating errors.

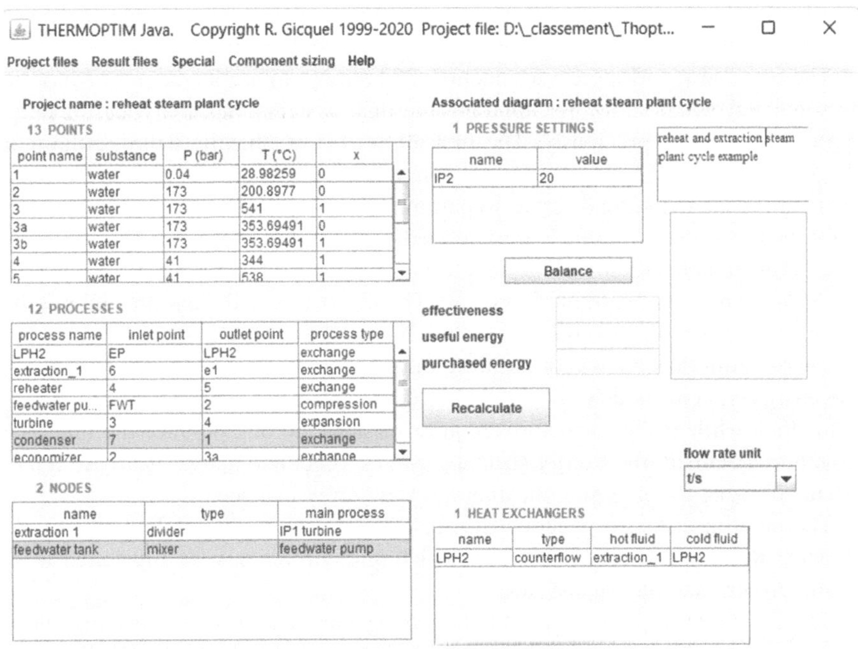

FIGURE 1.3.1 Thermoptim simulator screen.

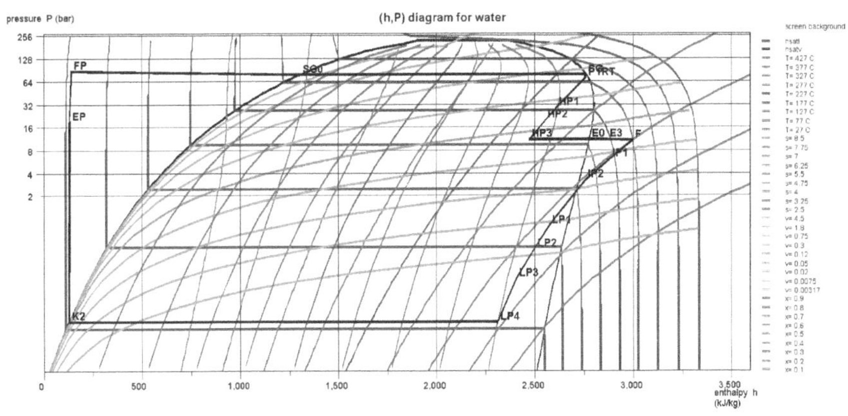

FIGURE 1.3.2 Plot of the Flamanville cycle in the (h, ln(P)) diagram.

In addition to its diagram editor, which allows for visualizing cycle architectures, Thermoptim generates process flowsheets close to those used in industry, such as the one represented in Figure 1.2.1 for the Flamanville 3 EPR. The software also allows for automatically generating the equations corresponding to the created models, offering a bridge to complementary theoretical analyses if necessary. Used in the nuclear industry by companies like EDF or Framatome, Thermoptim motivates learners by showing them that they acquire skills directly applicable to cycle design and optimization, strengthening their professional credibility.

The two main working environments of Thermoptim are its diagram editor and its simulator, which allow for applying the dual systems and analytical modeling approach discussed previously. The first allows for graphically defining the model's architecture by placing functional components on the workspace and connecting them, while the second serves to parameterize their thermodynamic properties. Once the functional structure is thus defined, a screenshot of the diagram editor provides a functional diagram in Thermoptim format, but it is also possible to generate a more classic process diagram (flowsheet). Thanks to a second diagram editor, it then becomes possible to define the exergy structure of the studied system.

Thermoptim thus allows for building functional and exergy structures of numerous energy system models. Functional structures can be created with all versions of the tool, while exergy structures require versions 2.6 or 2.8. Once the functional structure is defined, the exergy structure can be generated automatically to a large extent, allowing for obtaining the diagram and exergy balance.

Thermoptim can also generate the set of equations corresponding to a model, allowing them to be reused in other modeling environments, as illustrated in the steam turbine example we just saw.

The Thermoptim core was built to help represent the most common components: compressors, turbines, combustion chambers, expansion valves, and heat exchangers. Their thermodynamic models being almost universally accepted, it was possible to program them once and for all.

The software includes a second family of components, which take the form of external Java classes. These components (e.g., solar collectors, cooling coils, cooling towers, fuel cells, chemical reactors, etc.) are easily customizable by users, who thus have the opportunity to define themselves the phenomenological and behavior models they wish to retain.

Among these classes, we should mention the possibility of creating controller modules that allow for coordinating the sequence of calculations performed by Thermoptim. These controllers are particularly used to solve coupled equation sets, especially using an optimization library called minPack. We will see one example of this in Case Study 8.11, which presents a model of a PWR steam generator operating in off-design mode.

We would like to point out that the external classes we provide have not been fully tested, and users should carefully verify that the implemented models meet their specific needs. They are distributed in the spirit of open source, *as is*, and without any warranty of proper functioning or suitability for particular purposes.

1.3.2 Pedagogical Advantages of Thermoptim

The use of Thermoptim leads to several pedagogical improvements for teaching nuclear reactor secondary cycles:

- Focus on concepts. By automating calculations, learners concentrate on understanding physical phenomena (vaporization, expansion) and component operation (turbines, moisture separator reheater, heat exchangers).

- The graphical interface presents key concepts in intuitive screens, helping learners structure their knowledge.
- The precision of models and the engaging aspect of the interface facilitate learning, avoiding the frustration of manual calculations and resulting in increased motivation.
- Cycles can be easily visualized in the interactive thermodynamic diagrams.
- Realistic models allow for working on TCNPPs close to those of nuclear reactors, preparing learners for industrial applications.
- Thermoptim is a flexible working environment adapted to different levels of thermodynamic expertise, from exploring pre-built models for beginners to designing particular cycles for advanced learners.

As with any software, the only notable drawback is the initial learning time, which must remain reasonable. A set of guided explorations of pedagogical models allows for quick familiarization.

For a deeper understanding of these topics, two notes expand on the pedagogical use of Thermoptim: the first elaborates on the new educational paradigm it enables,[4] while the second addresses its theoretical foundations.[5]

1.3.3 Getting Started with Thermoptim

To familiarize yourself with Thermoptim, numerous resources are available on the Thermoptim-Unit portal or the website dedicated to this book:

- Guided explorations allow for progressively discovering how to model with this tool from pre-built models (https://direns.minesparis.psl.eu/Sites/Thopt/en/co/explor-scenar.html)
- Getting-started guides and Diapason sessions explain how to build your own models (https://direns.minesparis.psl.eu/Sites/Thopt/en/co/module_Logiciel_8.html)
- Four reference manuals detail all screens and possible parameterizations (https://direns.minesparis.psl.eu/Sites/Thopt/en/co/documentation_1.html)

Depending on your preferences, you can choose among these resources those that best suit you.

On this page of this book's website, you will find some guidance on how to work with Thermoptim by opening existing work files, in particular using the Thermoptim Console, a dashboard for users of the software which provides organized access to series of models distributed with the guided explorations or the books.[6]

1.4 CONCLUSION

This chapter proposes an integrated approach for the analysis and teaching of nuclear reactor thermodynamic cycles. Starting with a general presentation of energy systems, we laid the conceptual foundations necessary for understanding secondary cycles, before focusing on the pedagogical challenges specific to their teaching.

The use of Thermoptim as a central tool revolutionizes this teaching by reducing cognitive load, prioritizing an inductive progression, and offering flexibility adapted to different audiences. The combination of system analysis, innovative pedagogy, and a high-performance simulation tool constitutes an effective triptych for training in the design and optimization of nuclear secondary cycles. By strengthening the understanding of phenomena, facilitating the acquisition of practical skills, and opening the way to exploring realistic models, this approach offers learners and professionals a complete framework to meet the energy and technological challenges of the nuclear sector.

NOTES

1 A demo version of Thermoptim is freely available (https://server.s4e2.com/jml/index.php/en/thermoptim-en/download), allowing users to view existing projects and diagrams and to create small ones in order to become familiar with the tools. Naturally, it does not provide access to all software features. In particular, it does not allow saving projects, diagrams, or output result files, although it does enable the loading of projects of any size. The 2.82 demo version is sufficient to work with all the examples presented in this book.
2 https://interactive-thermodynamics.software.informer.com/.
3 https://fchartsoftware.com/ees/.
4 https://www.s4e2.com/download/peda/PedagogicalParadigmEquations.pdf.
5 https://www.s4e2.com/download/peda/PedagogieSystemesEnergetiquesThermoptim-en.pdf.
6 https://server.s4e2.com/crc/tcnpp/resources/first-steps-thopt/.

BIBLIOGRAPHY

Annamalai, K., Puri, I.K. & Jog, M.A. (2011). *Advanced Thermodynamics Engineering*, 2nd ed. CRC Press, Boca Raton, FL. ISBN 978-1-4398-0572-5.
Balmer, R.T. (2011). *Modern Engineering Thermodynamics*, 1st ed. Academic Press (Elsevier), New York. ISBN 978-0-12-374996-3.
Çengel, Y.A. & Boles, M.A. (2006). *Thermodynamics: An engineering Approach (SI Units)*, 5th ed. McGraw-Hill, New York. ISBN 978-0070606593.
Gicquel, R. (2021). *Energy Systems: A New Approach to Engineering Thermodynamics*, 2nd ed. CRC Press, Boca Raton, FL. ISBN 978-0367545129. DOI: 10.1201/9781003175629.
Moran, M.J., Shapiro, H.N., Boettner, D.D. & Bailey, M.B. (2018). *Fundamentals of Engineering Thermodynamics*, 9th ed. John Wiley & Sons, New York. ISBN 978-1-119-39138-8.
Sarkar, D.K. (2015). *Thermal Power Plant: Design and Operation*, 1st ed. Elsevier, New York. ISBN 978-0-12-801575-9.
Singh, O. (2009). *Applied Thermodynamics*, 3rd ed. New Age International Publishers, New York. ISBN 978-8122426029.

2 Physical Phenomena Involved in Nuclear Reactors, Panorama of Nuclear Reactors

2.1 PHYSICAL PHENOMENA INVOLVED IN NUCLEAR REACTORS

Nuclear reactors involve complex processes of element transformation at the subatomic level. Before explaining their operation, it is useful to revisit some basic concepts about the structure of the atom and recall the definitions of the particles that constitute it, namely protons, neutrons, electrons, isotopes, and fission.

2.1.1 Structure of an Atom

2.1.1.1 Protons, Neutrons, and Electrons

An atom is the smallest fundamental unit of matter that makes up everything around us. It consists of three main particles: protons, neutrons, and electrons, as illustrated in Figure 2.1.1.

Protons are positively charged particles located in the nucleus of the atom (labeled as *nucleus* in the figure). Each proton has a mass of approximately 1 atomic mass unit (amu).

Neutrons are also located in the nucleus of the atom, alongside protons. Neutrons have no electrical charge; they are neutral. Like protons, they have a mass of approximately 1 amu.

Electrons are negatively charged particles that orbit the nucleus of the atom. They are much lighter than protons and neutrons, with a negligible mass compared to them.

The structure of the atom is organized such that the number of protons equals the number of electrons in a neutral atom. The number of protons determines the chemical identity of the atom, represented by its atomic number. For example, hydrogen has one proton, carbon has six, oxygen has eight, and so on.

2.1.1.2 Isotopes

Isotopes of an element have the same number of protons but different numbers of neutrons. The different isotopes of an element may have similar chemical properties but different atomic masses due to the varying number of neutrons.

As an example, protium, deuterium, and tritium are three isotopes of hydrogen, the most abundant chemical element in the universe (Figure 2.1.2).

DOI: 10.1201/9781003606024-3

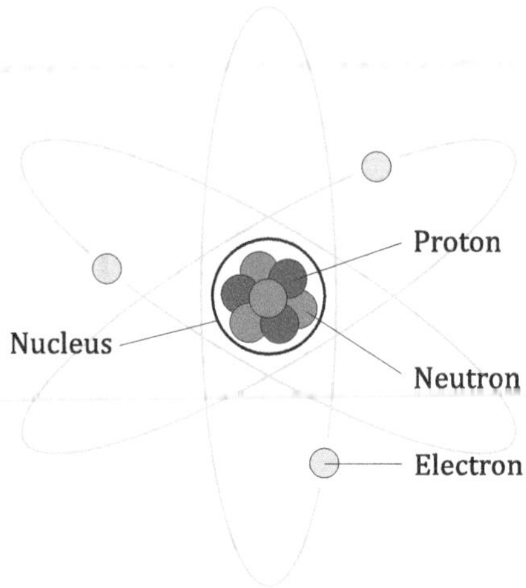

FIGURE 2.1.1 Atomic structure showing the nucleus (containing protons and neutrons) and orbiting electrons, AG Caesar, CC BY-SA 4.0 <https://creativecommons.org/licenses/by-sa/4.0>, via wikimedia commons.

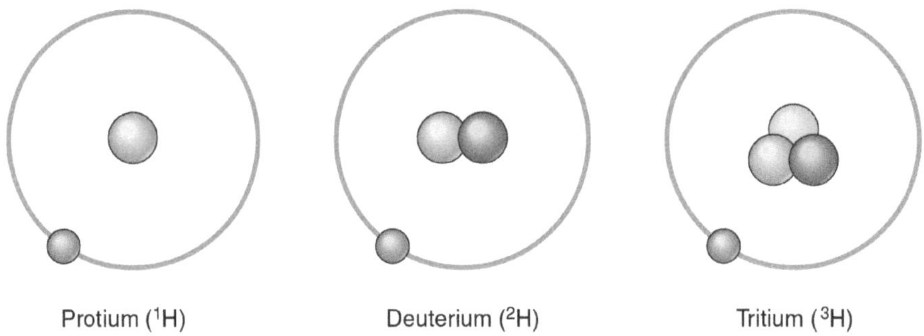

FIGURE 2.1.2 The three isotopes of hydrogen: protium, deuterium, and tritium, OpenStax College, CC BY 3.0 <https://creativecommons.org/licenses/by/3.0>, via wikimedia commons.

Protium is the most common isotope of hydrogen, with a nucleus consisting of a single proton. It contains no neutrons. Protium is the lightest isotope of all elements and constitutes approximately 99.98% of natural hydrogen.

Deuterium, also known as heavy hydrogen, is an isotope of hydrogen that contains one proton and one neutron in its nucleus. Approximately twice as heavy as protium, it is present in small quantities (0.015%) in natural hydrogen. As we will see, it is used in some nuclear reactors.

Tritium is a radioactive isotope of hydrogen that contains one proton and two neutrons in its nucleus, making it approximately three times heavier than protium. Tritium is produced in nuclear reactors and is also used in nuclear fusion reactions.

In summary, protium, deuterium, and tritium are three isotopes of hydrogen, with nuclei containing zero, one, and two neutrons, respectively. Protium is the most common and lightest isotope.

By convention, protium is often referred to as hydrogen, although, strictly speaking, hydrogen is the generic term designating all isotopes of the hydrogen element.

2.1.1.3 Neutrons: Fast and Thermal

Neutrons can be classified based on their kinetic energy into fast neutrons and slow or thermal neutrons.

Fast neutrons are neutrons with high kinetic energy, typically greater than 1 MeV (mega-electronvolt). They travel at very high speeds, on the order of several thousand kilometers per second. They are produced by nuclear fission and can induce further fission reactions. However, they have a lower probability of being absorbed by fissile nuclei compared to thermal neutrons.

Thermal neutrons, on the other hand, are neutrons with low kinetic energy, typically less than 0.1 eV (electronvolt). They travel at much lower speeds than fast neutrons, on the order of a few kilometers per second. Thermal neutrons have a higher probability of being absorbed by fissile nuclei than fast neutrons, making them more effective for maintaining a controlled chain reaction in a nuclear reactor.

The main difference between fast and thermal neutrons lies in their kinetic energy and their probability of being absorbed by fissile nuclei. Fast neutrons are more energetic and move faster but have a lower probability of being absorbed, while thermal neutrons are less energetic and move more slowly but have a higher probability of being absorbed.

Epithermal neutrons are neutrons with an energy range between approximately 0.1 eV (electronvolt) and 10 keV (kilo-electronvolts). They lie between thermal neutrons (energy below 0.1 eV) and fast neutrons (energy above 10 keV). Epithermal neutrons are neutrons that have been partially slowed down but have not yet reached thermal equilibrium with the surrounding medium.

2.1.2 Fission of Uranium

2.1.2.1 Composition of Uranium

Uranium is a chemical element with the symbol U and atomic number 92. It consists of three natural isotopes: uranium-234, uranium-235, and uranium-238.

Uranium-238 is the most abundant isotope, accounting for approximately 99.28% of natural uranium. Its nucleus consists of 92 protons and 146 neutrons.

Uranium-235 accounts for approximately 0.71% of natural uranium. Its nucleus consists of 92 protons and 143 neutrons.

Uranium-234 accounts for approximately 0.0055% of natural uranium. Its nucleus consists of 92 protons and 142 neutrons.

2.1.2.2 Fissile and Fertile Nuclei

Focusing only on the first two isotopes due to the low abundance of the third, a major difference between them is that uranium-235 has a nucleus classified as fissile, while that of uranium-238 is fertile.

A fissile nucleus is an atomic nucleus that can undergo nuclear fission when bombarded by a low-energy neutron. The most common fissile nuclei are uranium-235, uranium-233, and plutonium-239.

A fertile nucleus, on the other hand, is an atomic nucleus that can be transformed into a fissile nucleus after absorbing a neutron. The most common fertile nuclei are uranium-238 and thorium-232. When a fertile nucleus absorbs a neutron, it transforms into a radioactive isotope that can then undergo radioactive decay to form a fissile nucleus.

Uranium-235 is, in practice, the only natural uranium isotope that can undergo nuclear fission when bombarded by thermal neutrons. Uranium-238 can also undergo nuclear fission, but this requires much more energetic neutrons.

The fission of uranium is a nuclear reaction that occurs when a uranium-235 nucleus is struck by a neutron. This reaction releases a large amount of energy in the form of heat, ionizing radiation, and new neutrons.

2.1.2.3 Progression of the Fission Reaction

The phenomenon of uranium fission occurs in several stages:

1. **Initiation**: First, a neutron collides with a uranium-235 nucleus.
2. **Absorption and Instability**: The uranium-235 nucleus absorbs the neutron and becomes unstable, causing it to split into two smaller nuclei, called fission products.
3. **Energy Release**: The fission products then release a large amount of energy in the form of heat and ionizing radiation.
4. **Emission of New Neutrons**: The fission products also emit several neutrons, which can, in turn, induce further fission reactions in neighboring uranium nuclei.

This results in a chain reaction that releases a significant amount of energy, which can be used to produce electricity in a nuclear reactor.

The fission of uranium is an exothermic nuclear reaction, meaning it releases more energy than it consumes. This reaction is the basis for the operation of uranium fission nuclear reactors, which are used to produce electricity in many countries. However, uranium fission can also be used to produce nuclear weapons, highlighting the importance of safety and regulation in the use of this technology.

2.1.2.4 Breeding

In some nuclear reactors, fertile nuclei can be used to produce additional nuclear fuel by converting part of uranium-238 into plutonium-239, which is fissile. This process is called "breeding."

Breeding is thus a process in which a nuclear reactor produces more fissile material than it consumes by converting fertile isotopes (such as uranium-238 or thorium-232) into fissile material (such as plutonium-239 or uranium-233).

Under such conditions, the fuel becomes essentially inexhaustible, which explains the interest in these types of reactors.

Here are some examples of breeding reactors that will be studied later:

1. Sodium-cooled fast reactors (SFRs)
2. Lead-cooled fast reactors (LFRs)
3. Gas-cooled fast reactors (GFR)
4. Molten salt reactors (MSRs)

These systems offer interesting prospects for the future of nuclear energy, both in terms of efficiency and resource management.

2.1.2.5 Uranium Enrichment

Uranium enrichment increases the concentration of the fissile isotope, uranium-235, which is the only uranium isotope capable of sustaining a nuclear fission chain reaction.

Uranium enrichment involves increasing the concentration of uranium-235 by separating uranium isotopes. The most common enrichment methods are gaseous diffusion and gas centrifugation.

The challenge lies in the fact that the atomic masses of these different isotopes are very close.

Uranium enrichment produces high-quality nuclear fuel for nuclear reactors used in electricity production. Enriched fuel can provide greater reaction power and remain in the reactor longer.

Enrichment is also necessary for the manufacture of nuclear weapons, as a higher concentration of uranium-235 is required to produce a nuclear explosion. This is why uranium enrichment is tightly controlled and regulated due to the risks of nuclear proliferation.

2.1.3 Operating Principles of Nuclear Reactors with Uranium Fission

2.1.3.1 Starting the Chain Reaction

In a nuclear reactor, the fission reaction is initiated by bombarding fissile atoms, such as uranium-235 or plutonium-239, with neutrons. These neutrons may come from an external source, such as a particle accelerator, or be produced by the fission reaction itself.

To start the chain reaction, a sufficient amount of fissile fuel is introduced into the reactor core. This fuel is arranged in the form of pellets or rods, which are grouped into fuel assemblies. The assemblies are then placed in the reactor core, where they are subjected to a neutron flux.

When a neutron strikes the nucleus of a fissile atom, it can cause the fission of that nucleus, releasing a large amount of energy in the form of heat and new neutrons. These new neutrons can, in turn, cause the fission of other fissile nuclei, leading to a self-sustaining chain reaction.

2.1.3.2 Controlling the Chain Reaction

To control the chain reaction, control rods made of neutron-absorbing materials, such as boron or cadmium, are used. These rods are inserted into the reactor core and can be moved to adjust the rate of the chain reaction. By modifying the position of the control rods, the reactor's power can be regulated, maintaining the chain reaction at a stable and controllable level.

To generate the initial neutron flux necessary to initiate the chain reaction in a nuclear reactor, an external neutron source is generally used. This neutron source may consist of a neutron generator, which uses a separate nuclear reaction to produce neutrons, or a radioactive source that spontaneously emits neutrons.

In the case of a pressurized water reactor (PWR), for example, the external neutron source is often placed in the reactor core near the fuel assemblies before the chain reaction starts. Once the neutron source is in place, the control rods are gradually withdrawn from the reactor core, allowing neutrons to begin interacting with the fissile nuclei of the fuel.

When neutrons interact with fissile nuclei, they cause the fission of these nuclei, releasing energy in the form of heat and new neutrons. These new neutrons can, in turn, induce further fissions, creating a self-sustaining chain reaction.

Once the chain reaction is initiated, the external neutron source is no longer necessary and can be removed from the reactor core. The chain reaction is then maintained by controlling the neutron flux using the control rods, which absorb some of the neutrons to regulate the reactor's power.

2.1.3.3 Moderation of the Chain Reaction

During uranium fission, the neutrons produced have high kinetic energy, typically on the order of 2 MeV (mega-electronvolts). This kinetic energy translates into a high speed, on the order of 10,000–20,000 km/s. These neutrons are called "fast neutrons," as we have seen.

However, to maintain a controlled chain reaction in a nuclear reactor, it is necessary to slow down these fast neutrons so that they can be more easily absorbed by fissile nuclei and induce further fissions. This is achieved through a process called "moderation."

This process involves having the fast neutron interact with light nuclei, such as those of hydrogen, deuterium, or carbon. When the neutron collides with one of these light nuclei, it loses some of its kinetic energy, reducing its speed.

Note that moderators are used to slow down fast neutrons produced by nuclear fission, enabling a controlled chain reaction in a nuclear reactor, while neutron absorbers, such as boron and cadmium, are used to absorb neutrons, whether fast or thermal, to control reactivity and stop the chain reaction when necessary.

After being slowed down by the moderator, neutrons have a lower kinetic energy, on the order of 0.025 eV (electronvolts), and a speed on the order of a few km/s. These neutrons are called "thermal neutrons," as their kinetic energy is comparable to the thermal energy of the atoms in the surrounding medium. Since thermal neutrons are more likely to be absorbed by fissile nuclei, a controlled chain reaction can be maintained in the reactor.

The moderation process is fundamental in nuclear reactors, as it enables the control of the nuclear fission chain reaction.

2.1.3.4 Operation of a Uranium Fission Nuclear Reactor

Uranium fission nuclear reactors are facilities that produce energy by harnessing the fission of uranium nuclei. This nuclear reaction releases a large amount of energy in the form of heat, which is then converted into electricity.

The basic principle of uranium fission nuclear reactors is as follows:

1. **Fuel Arrangement**: The nuclear fuel, generally enriched uranium, is placed in the reactor core.
2. **Fission Reaction**: Neutrons released by the fission of uranium nuclei are slowed down by a moderator, such as water or graphite, to maintain a controlled chain reaction.

3. **Heat Transfer**: The heat produced by the chain reaction is transferred to a coolant, such as water or steam, which circulates in the reactor.
4. **Electricity Conversion**: The heated coolant is then used to power a thermodynamic cycle, for example, by producing steam that drives a turbine to generate electricity.
5. **Fuel Management**: At regular intervals, spent fuel is removed from the reactor and replaced with fresh fuel to maintain the chain reaction.

Uranium fission nuclear reactors are designed to operate safely and with high efficiency. However, they require strict safety and security measures to prevent accidents and radioactive leaks. Nuclear waste produced by the chain reaction must also be managed appropriately to protect the environment and public health.

2.1.4 Conclusion

This introductory presentation has allowed us to revisit:

1. The atomic basics, particularly the structure of the atom, the nature, and the role of protons, neutrons, electrons, and isotopes.
2. The characteristics of neutrons, particularly the difference between fast, thermal, and epithermal neutrons and their importance in fission.
3. The mechanism of the uranium fission reaction, the distinctions between fissile and fertile nuclei, and the process of enrichment.
4. The operation of nuclear reactors, from the initiation of the chain reaction to moderation, including control by control rods and the conversion of heat into electricity.
5. Breeding, i.e., the possibility of producing additional fissile fuel, offering prospects for closed-cycle reactors.

2.2 OVERVIEW OF NUCLEAR FISSION REACTORS

2.2.1 Introduction

In this overview, we present the main types of nuclear reactors without claiming to be exhaustive. These reactors differ by fuel type, neutron spectrum, moderator, coolant, cycle (direct/indirect), and thermodynamic fluid used, among other characteristics.

First, we present the main reactors used for electricity production that employ steam as the thermodynamic fluid. Then, we examine those that use gas as the working fluid.

Note that we limit ourselves here to a brief presentation of the corresponding thermodynamic cycles, indicating only their type and capacity, the pressure and temperature levels reached, and the efficiencies achieved.

As mentioned in the introduction, we had to limit the number of illustrations for editorial reasons. Therefore, in this chapter, we provide only schematic diagrams, sufficient for understanding the cycles. If you wish to obtain more information, the G4NR Handbook we have already discussed includes many depictions, including cutaways and photographs.

2.2.2 Generations of Nuclear Reactors

In 2001, the United States Department of Energy initiated what is called the *Generation IV Nuclear Energy Systems Initiative*, which has managed to unite a group of nuclear research organizations from 10 countries in a loose cooperation called the *Generation IV International Forum.*(GIF). Its purpose is to study systems that could constitute the fourth generation of reactors, after those in use or planned in the short term (Evolutionary Power Reactor), and that could be deployable by 2030.

The notion of reactor generations has been introduced to distinguish the main developments that have marked the history of civilian nuclear reactors:

- The first generation corresponds to the first prototypes built mainly in the United States until the late 1950s. In France, it saw the development of natural uranium graphite gas systems (Gas-Cooled Reactor), with completion of nine reactors achieving efficiency close to 29%.
- The second generation, from 1960 to 1995, marks the first phase of commercialization of nuclear power plants, with three main systems: light water reactors (BWR – Boiling Water Reactors), PWRs (efficiency approximately 33%), and heavy water reactors (in Canada). Most operational plants are currently second generation.
- The third generation has the main objective of increasing plant safety; accidents at Three Mile Island in the United States, Chernobyl in the USSR, and Fukushima in Japan have shown the major risks associated with certain second-generation units. The Evolutionary Power Reactor is a third-generation reactor. Its performance is slightly higher: 35%.
- It is increased efficiencies that primarily justify the value of work on fourth-generation reactors, with the great innovation being to design the reactor together with its associated cycle and optimize the whole, which allows sustainability issues to be properly addressed.

After studying in detail approximately one hundred reactor types, the Generation IV International Forum has selected six concepts on which the efforts of its members focus:

- supercritical water reactors (SCWR);
- very high temperature reactors (HTRs and VHTRs);
- gas-cooled fast neutron reactors (FNRs);
- sodium-cooled fast neutron reactors (SFR);
- lead-cooled fast neutron reactors (LFR);
- molten salt reactors (MSRs).

Among these six options, a major interest is given to gas-cooled systems.

2.2.3 Reactor Containment

In all reactors, the core is placed in a containment vessel designed to limit the release of radioactive materials resulting from the fission reaction (Figure 2.4.1). The thermodynamic cycle is mostly located outside of this vessel, with the coolant transferring the heat.

FIGURE 2.4.1 Containment vessel under construction, nuclear regulatory commission, public domain, via wikimedia commons.

The containment vessel plays an essential role in the safety and security of nuclear reactors by limiting the consequences of accidents and protecting the environment and public health. It is a fundamental component of the safety systems of nuclear installations, ensuring the containment of radioactive substances, protection against external threats, control of pressure and temperature, acoustic and thermal insulation, and facilitating maintenance and waste management operations.

This photograph shows a containment vessel under construction.

2.2.4 Pressurized Water Reactors (PWRs)

2.2.4.1 Principle of Operation

PWRs, which are used in the majority of nuclear power plants around the world, use high-pressure water (approximately 155 bar) as both a coolant and a moderator. The water circulates in the reactor core at this pressure and at a temperature of approximately 315°C, transferring the heat produced by nuclear fission to a secondary circuit via a steam generator.

This secondary circuit (Figure 2.4.2) includes a turbine, a condenser often cooled by a cooling tower, and a pump. The heat produced is used to heat pressurized water and generate steam, which is then expanded in a turbine coupled to an electrical alternator. This cycle is known as a Rankine or Hirn cycle.

The cycles considered for PWRs are studied in Chapters 5 and 8.

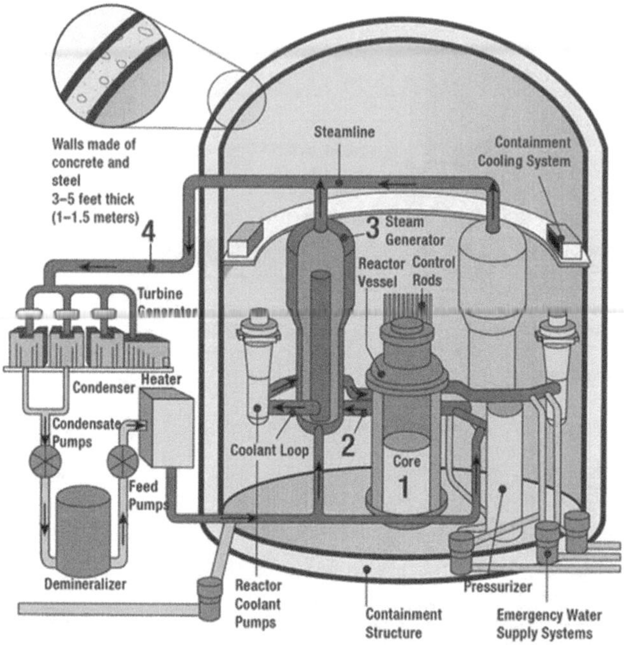

FIGURE 2.4.2 PWR reactor schematic, nuclear regulatory commission from the US, public domain, via wikimedia commons.

2.2.4.2 Main Characteristics

Moderator	The water itself serves as a moderator by slowing down the neutrons emitted during fission
Coolant	High-pressure water also serves as a coolant to transfer the heat
Fuel	Enriched uranium is generally used as fuel
Electrical power	PWRs can vary in capacity, ranging from a few hundred megawatts electric (MWe) to over 1,600 MWe
Pressure and temperature levels	PWRs operate at high pressures, typically around 150–160 bar, with primary circuit temperatures reaching around 300°C
Efficiency	PWRs typically have thermal efficiencies of around 33%–36%
Countries concerned	Many countries use PWRs, including the United States, France, China, the United Kingdom, Russia, and Japan

2.2.5 VVER Reactors

2.2.5.1 Principle of Operation

VVER (Vodo-Vodyanoi Energetichesky Reactor) reactors, also known as Russian PWRs, are used in several countries, including Russia and other former Soviet Union countries, as well as in some countries in Eastern Europe and Asia.

The cycles considered for VVERs are studied in Chapters 5 and 8.

2.2.5.2 Main Characteristics

Moderator	Light water
Coolant	VVER reactors use water as a coolant to transfer the heat produced by nuclear fission
Fuel	VVER reactors typically use enriched uranium as fuel, in the form of uranium dioxide (UO_2) contained in fuel rods
Electrical power	The capacity of VVER reactors can range from a few hundred MWe to over 1 GWe
Pressure and temperature levels	VVER reactors operate at pressures between 100 and 160 bar, with primary circuit temperatures typically reaching between 290°C and 330°C
Efficiency	The efficiencies of VVER reactors are between 32% and 37%
Countries concerned	Russia and other countries of the former Soviet Union, as well as in some countries in Eastern Europe and Asia

2.2.6 Boiling Water Reactors (BWRs)

2.2.6.1 Principle of Operation

Like PWRs, BWRs use water as both a coolant and a moderator. However, in this technology, the water boils directly in the reactor core at a pressure of approximately 70 bar and a temperature close to 285°C, producing steam that drives a turbine without the need for a secondary circuit. A condenser and a pump complete the cycle (Figure 2.4.3).

As shown in the figure, the steam generator is absent in this type of reactor.

The cycles considered for BWRs are studied in Chapters 5 and 8.

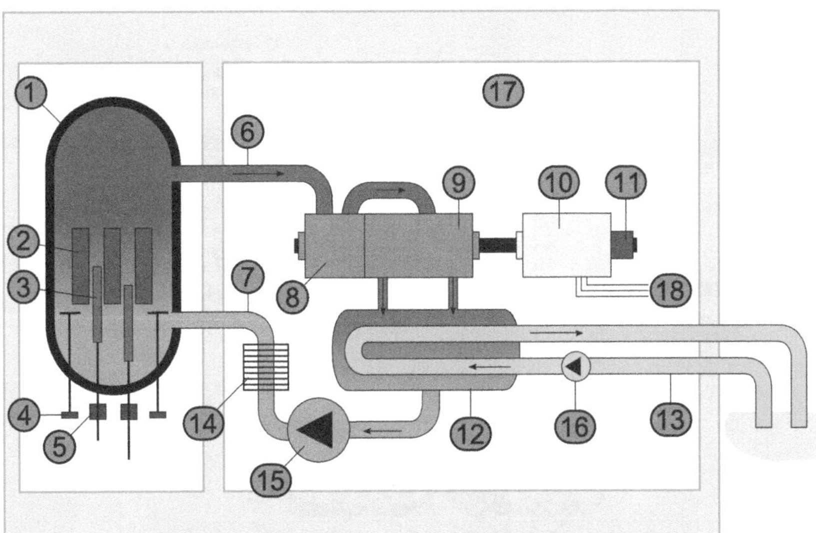

FIGURE 2.4.3 BWR reactor schematic, Robert Steffens (alias RobbyBer 8 November 2004), SVG: Marlus_Gancher, Antonsusi using a file from Marlus Gancher. See File talk:Schema Siedewasserreaktor.svg#License history, GFDL <http://www.gnu.org/copyleft/fdl.html>, via wikimedia commons.

The containment vessel is located on the left, housing the core cooled by pressurized water. On the right, we see the thermodynamic cycle, with the turbine (8-9), below which are the condenser (12) and the pump (15), and the alternator (10).

2.2.6.2 Main Characteristics

Moderator	The water serves as a moderator by slowing down the neutrons
Coolant	The boiling water transfers the heat produced to the turbine
Fuel	Enriched uranium is generally used as fuel
Electrical power	BWRs have a capacity range similar to that of PWRs, from a few hundred MWe to over 1,600 MWe
Pressure and temperature levels	BWRs operate at slightly lower pressures than PWRs, typically around 70–80 bar, with primary circuit temperatures reaching around 285°C–290°C
Efficiency	BWRs have thermal efficiencies comparable to those of PWRs, typically around 33%–36%
Countries concerned	BWRs are also widely used around the world, including in the United States, Sweden, Germany, Finland, Japan, and Switzerland

2.2.7 RBMK Reactors

2.2.7.1 Principle of Operation

RBMK (Russian: Reaktor Bolshoy Moshchnosti Kanalny, "high-power channel-type reactor") reactors are graphite-moderated BWRs primarily used in Russia.

Figure 2.4.4 shows the containment vessel, with the core in the central part, traversed by pressurized water which vaporizes, and the steam exiting at the upper part, both on the right and left.

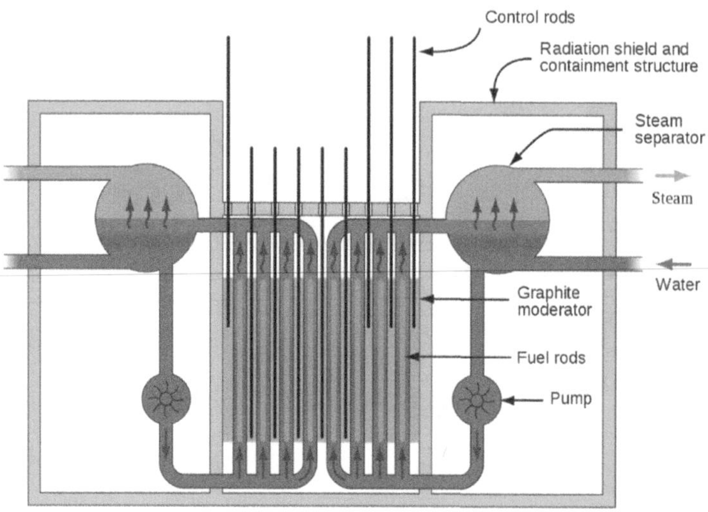

FIGURE 2.4.4 RBMK reactor schematic, Fireice~commonswiki, Sakurambo, Emoscopes, CC BY-SA 3.0 <https://creativecommons.org/licenses/by-sa/3.0>, via wikimedia commons.

The cycles considered for RBMKs are studied in Chapters 5 and 8.

It should be noted that RBMKs had serious safety and design flaws, which were highlighted by the Chernobyl accident in 1986. Since that accident, many improvements have been made to enhance their safety and reliability.

One of the characteristics of these reactors is the ability to extract or add fuel without shutting them down.

2.2.7.2 Main Characteristics

Moderator	RBMKs use graphite as a moderator to slow down neutrons
Coolant	Ordinary water is used as a coolant to transfer the heat produced by nuclear fission
Fuel	RBMKs typically use natural uranium as fuel, in the form of uranium dioxide (UO_2) contained in fuel rods
Electrical power	The first RBMK reactors had a nominal capacity of around 1,000 MWe. However, later versions were designed for a higher nominal capacity, up to around 1,500 MWe
Pressure and temperature levels	RBMKs operate at moderate pressures, with typical values around 70 bar. The temperatures in the reactor core can vary, but they are generally maintained at high levels to ensure optimal operating efficiency
Efficiency	The efficiencies of RBMK reactors can vary, but they are generally comparable to those of other types of BWRs, with typical thermal efficiencies between 30% and 35%
Countries concerned	Russia and other countries of the former Soviet Union, as well as in some countries in Eastern Europe and Asia

2.2.8 Supercritical Water-Cooled Reactors (SCWRs)

2.2.8.1 Principle of Operation

To significantly increase the efficiency of steam-water cycles of the PWR or BWR type, it is possible to use supercritical cycles (SCWR), where the pressure of the water exceeds the critical pressure of 221.2 bar.

While in PWR reactors and those with fast neutrons cooled by sodium, a primary cooling circuit is essential due to the change of state of the steam, this could be eliminated in this type of supercritical cycle, similar to BWR reactors, where the thermodynamic fluid also serves as the reactor cooling fluid, simplifying the system significantly.

On the left of Figure 2.4.5 is the core, traversed by pressurized water which vaporizes. The steam exits the core on the right at the top and is directed toward the turbine coupled to the alternator, underneath which are the condenser and the pump.

The envisaged cycle would achieve an efficiency of 45%, with a turbine inlet temperature ranging from 510°C to 550°C and a pressure of 250 bar. Two versions of the reactor are envisaged, one with thermal neutrons and the other with fast neutrons.

The cycles considered for SCWRs are studied in Chapters 4 and 8.

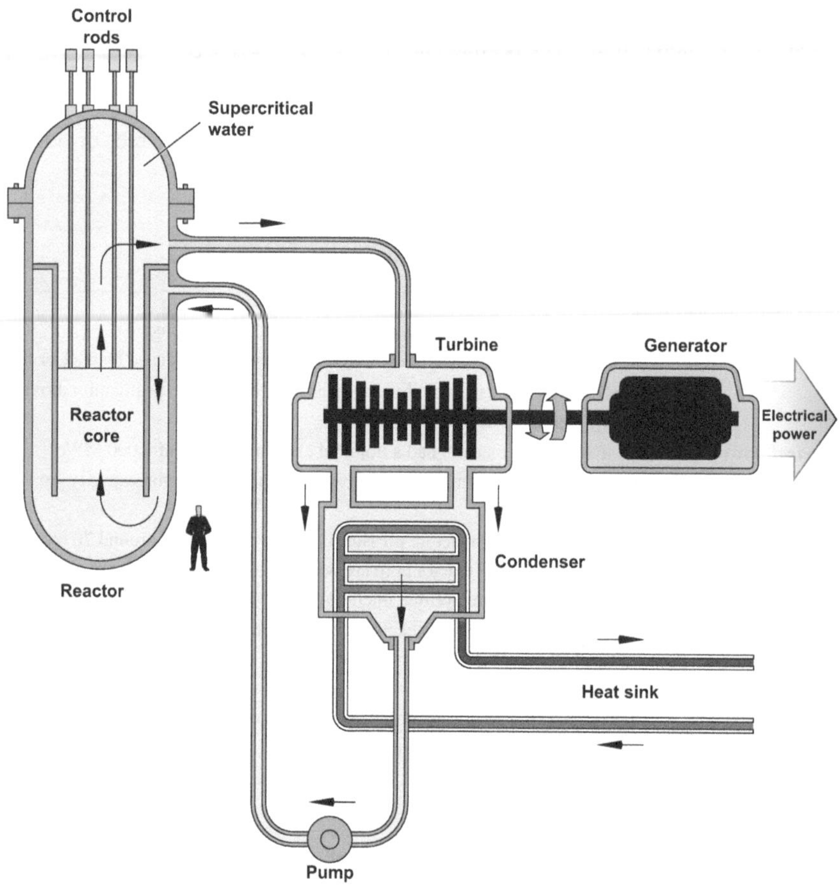

FIGURE 2.4.5 SCWR reactor schematic, US department of energy nuclear energy research advisory committee, public domain, via wikimedia commons.

2.2.8.2 Main Characteristics

Moderator	Water if it uses thermal neutrons
Coolant	Supercritical water (between the liquid and gaseous states)
Fuel	Enriched uranium oxide
Electrical power	1,000–2,000 MWe
Pressure and temperature levels	Pressure of 250 bar and turbine inlet temperature of 510°C–550°C
Efficiency	40%–45%
Countries concerned	Canada, China, South Korea, United States, Russia

2.2.9 Advanced Gas-Cooled Reactors (AGRs)

The first nuclear reactors cooled by gas were known as Uranium Naturel Gaz et Graphite (UNGG) in France and Magnox in the United Kingdom. They used natural uranium metal as fuel, graphite as a moderator, and pressurized carbon dioxide as a coolant.

The Magnox design was later improved by using slightly enriched uranium as fuel and stainless steel as cladding. This new reactor design was named the AGR. AGRs have been the backbone of the UK's nuclear power generation since the 1980s.

2.2.9.1 Principle of Operation

In this type of reactor, the heat produced by nuclear fission is utilized to heat carbon dioxide, which is then used to produce steam and drive a turbine.

AGRs utilize carbon dioxide as a coolant and graphite as a moderator.

The fuels used are natural uranium for the old Magnox reactors and enriched uranium for the AGR reactors.

Figure 2.4.6 illustrates the containment vessel (5) with the core located centrally and cooled by CO_2, which transfers heat to an exchanger traversed by pressurized water, causing it to vaporize.(9) The steam exits from the upper part (10) on both the right and left sides.

In the primary circuit of an AGR, the pressure of the CO_2 generally ranges between 6 and 42 bar, and the temperature is maintained around 650°C.

The cycles considered for AGRs are studied in Chapters 4 and 8.

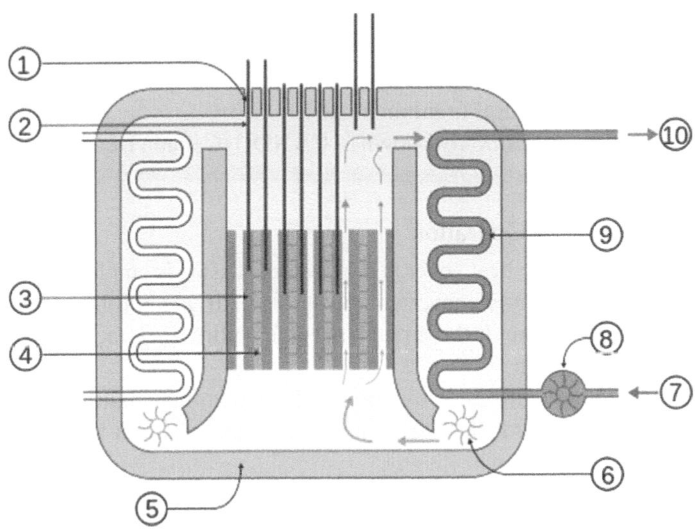

FIGURE 2.4.6 AGR reactor schematic, MesserWoland, CC BY-SA 3.0 <http://creativecommons.org/licenses/by-sa/3.0/>, via wikimedia commons.

2.2.9.2 Main Characteristics

Moderator	Graphite is used as a moderator to slow down neutrons
Coolant	Carbon dioxide is used as a coolant to transfer the heat produced
Fuel	Natural or slightly enriched uranium is generally used as fuel
Electrical power	AGRs are generally designed to produce electricity on a large scale for the power grid, with an electrical power output ranging from several hundred MWe to over 1 GWe
Pressure and temperature levels	The pressure of the steam in the secondary circuit of an AGR is around 170 bar, and the temperature is around 550°C
Efficiency	The efficiency is around 40%
Countries concerned	Reactors using this principle are implemented in several countries, including the United Kingdom, France, Germany, Italy, and Japan

2.2.10 Pressurized Heavy Water Reactors (PHWRs) or CANDUs

2.2.10.1 Heavy Water

Heavy water, also known as deuterium oxide (D_2O), is a variation of ordinary water (H_2O) in which almost all of the hydrogen atoms are replaced by deuterium, an isotope of hydrogen with two neutrons instead of one.

Deuterium serves as an excellent moderator because it effectively slows down neutrons, similar to hydrogen, but without absorbing them in the process. However, deuterium occurs naturally in water in only low proportions (0.015%), and its concentration is costly.

Pressurized Heavy Water Reactors (PHWRs) utilize heavy water as both a moderator and a coolant. Thanks to the efficient moderation provided by heavy water, these reactors can use natural uranium as fuel.

This technology is commonly known as CANDU (Canada Deuterium Uranium).

2.2.10.2 Principle of Operation

In this type of reactor, the heat generated by nuclear fission is employed to heat heavy water, which is subsequently converted into steam to drive a turbine.

One noteworthy feature of these reactors is the ability to extract or add fuel without the need for shutdown.

On the left side of Figure 2.4.7, we observe the containment building with the core cooled by heavy water, transferring heat to a secondary circuit (not depicted here) via a steam generator.

The fuel is housed within a calandria (2) containment building containing heavy water, and a specialized device facilitates its handling.

The cycles considered for PHWRs are studied in Chapters 5 and 8.

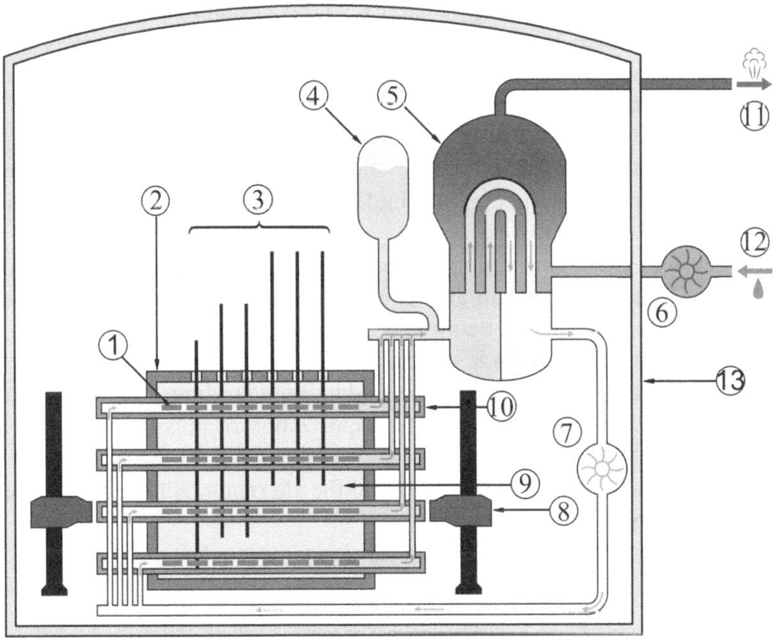

FIGURE 2.4.7 PHWR/CANDU reactor schematic, Inductiveload, CC BY-SA 2.5 <https://creativecommons.org/licenses/by-sa/2.5>, via wikimedia commons.

2.2.10.3 Main Characteristics

Moderator	Heavy water is used as a moderator to slow down neutrons
Coolant	Heavy water is also used as a coolant to transfer the heat produced
Fuel	Natural or slightly enriched uranium is generally used as fuel
Electrical power	They have a power output that generally ranges from 600 to 2,100 MWe
Pressure and temperature levels	These reactors operate at a pressure of around 100 bar and a primary circuit temperature of around 310°C
Efficiency	The typical efficiency is around 30%
Countries concerned	Reactors using this principle are implemented in several countries, including Canada, India, Argentina, China, South Korea, and Romania

2.2.11 Sodium-Cooled Fast Reactors (SFRs)

2.2.11.1 Principle of Operation

SFRs represent a category of FNRs specifically utilizing liquid sodium as cooling fluid, operating at temperatures ranging from 520°C to 550°C and pressures close to atmospheric. The selection of this coolant imposes significant technological constraints, particularly regarding reactivity with water and air: sodium is a highly reactive metal that reacts vigorously with water and air, posing risks of fires or explosions. Therefore, sodium cooling systems must be meticulously designed to prevent any sodium leakage and contact with water or air.

Various variants of SFRs are under consideration, differing in terms of capacity and fuel. In all cases, a primary cooling loop is incorporated, with the thermodynamic fluid distinct from the cooling fluid. The core outlet temperature typically reaches approximately 550°C.

These reactors are engineered to utilize high-level waste, particularly plutonium and other actinides, but they can also employ fissile and fertile materials, yielding significantly higher efficiencies than PWR reactors.

On the left side of Figure 2.4.8, the containment building depicts the core cooled by liquid sodium, transferring heat to a secondary liquid sodium circuit via a primary exchanger. This exchanger facilitates the vaporization of pressurized water in the steam generator. On the right side, the remaining portion of the thermodynamic cycle is illustrated, featuring the turbine coupled to the alternator, with the condenser and pump positioned beneath.

The cycles considered for SFRs are studied in Chapters 6 and 8.

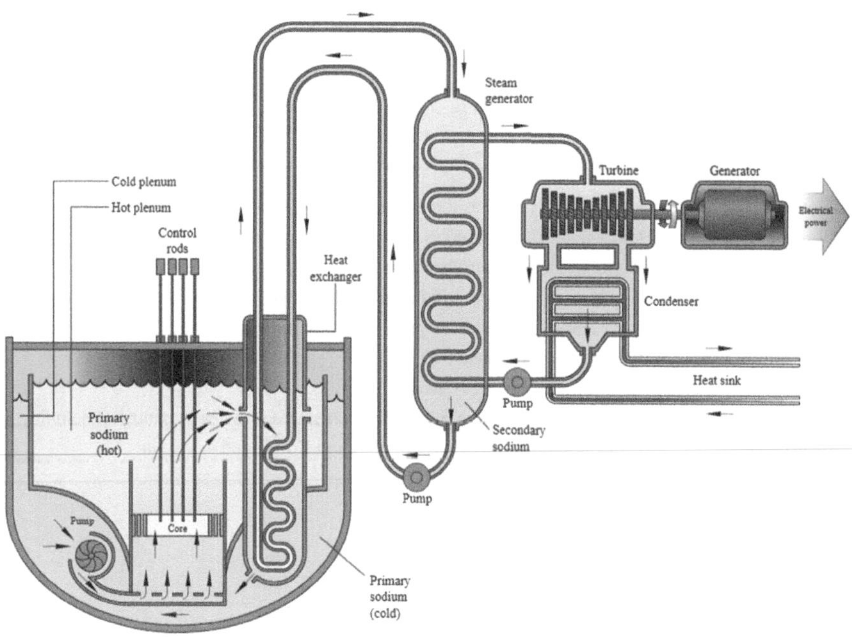

FIGURE 2.4.8 SFR reactor schematic, Sfr.gifderivative work: Beao, Public domain, via wikimedia commons.

2.2.11.2 Main Characteristics

Coolant	Liquid sodium is used as a coolant
Fuel	Mixed oxides of uranium and plutonium, or uranium nitride
Electrical power	600–1,500 MWe
Pressure and temperature levels	In general, the pressure of the water vapor in the secondary circuit of an SFR is between 40 and 180 bar, while the temperature varies between 400°C and 500°C
Efficiency	35%–40%
Countries concerned	France, Russia, China, India, Japan, United States

2.2.12 Fast Neutron Reactors (FNRs)

2.2.12.1 Principle of Operation

FNRs utilize fast neutrons without a moderator to induce fission. This term encompasses a broad range of reactors.

Two FNR concepts are considered. The first, which uses sodium as a coolant, was discussed in the previous section. The second, currently under development as part of the Generation IV program, uses helium as the coolant.

One of the advantages of reactors utilizing gas as the working fluid is their capability to surpass temperature limitations imposed by water-based systems, enabling high conversion efficiencies.

The thermodynamic cycles employed are typically variations of the Brayton cycle utilized in gas turbines, comprising four main components: a gas compressor, an exchanger where the coolant transfers its heat to the thermodynamic fluid, an expansion turbine, and a heat exchanger that cools the expanded fluid prior to compression.

In Figure 2.4.9, the left-hand side illustrates an FNR in which helium serves both as the reactor coolant and as the working fluid of the thermodynamic cycle. The containment building houses the reactor core, cooled by helium. On the right-hand side, the rest of the thermodynamic cycle is shown, featuring the turbine coupled to the alternator (above) and two compressor stages with intermediate cooling (below). A regenerator and a precooler are located to the right.

FNRs operate without a moderator, since the neutrons remain fast throughout the process.

The cycles considered for FNRs are studied in Chapters 4 and 6.

As we shall see throughout this book, the variety of reactor types is much greater than that of thermodynamic cycles, especially among the medium- and high-temperature concepts currently under development. In the chapters where these systems are discussed, we will therefore avoid listing all reactor types exhaustively so as not to burden the reader with unnecessary repetition.

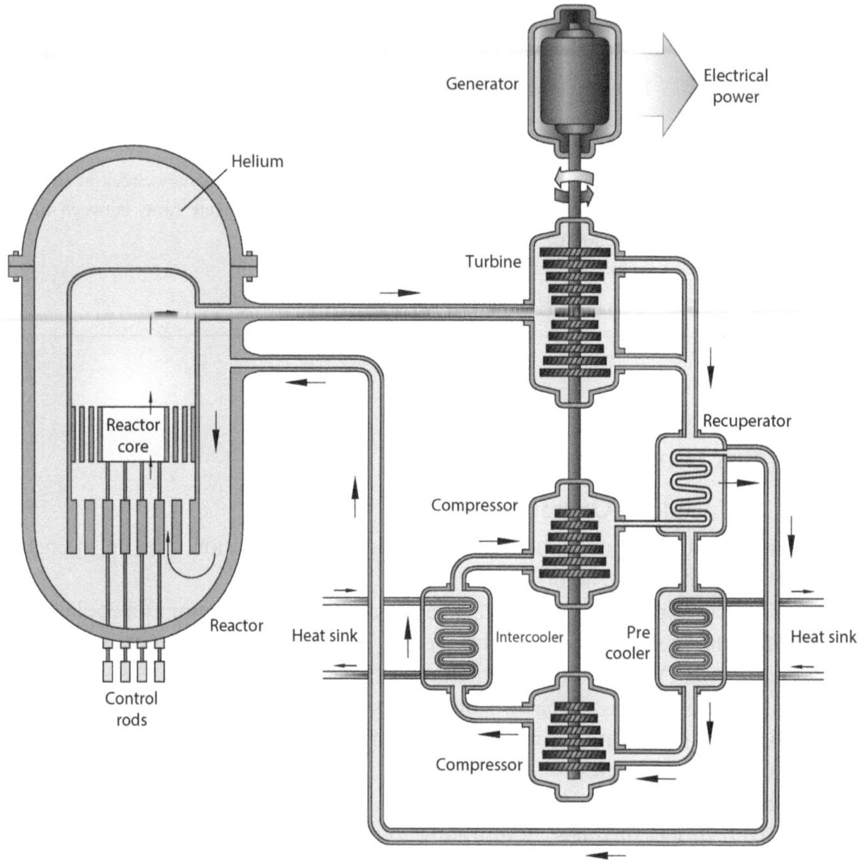

FIGURE 2.4.9 FNR reactor schematic, Gfr.gifderivative work: Beao, Public domain, via wikimedia commons.

2.2.12.2 Main Characteristics

Moderator	There is no moderator in FNRs, as stated above
Coolant	Helium or liquid sodium is used as a coolant
Fuel	Uranium or plutonium can be used as fuel, sometimes in the form of mixed oxide fuel
Electrical power	FNRs can have power outputs ranging from a few MW to several hundred MW
Pressure and temperature levels	In general, the pressure of the water vapor in the secondary circuit of a FNR with liquid sodium is between 40 and 180 bar, while the temperature varies between 350°C and 550°C. Helium pressure is around 70 bar and temperature around 850°C
Efficiency	The thermal efficiency of FNRs depends on many factors and can vary depending on the specific design and application. It is between 38% and 42%
Countries concerned	FNRs are less common than PWRs and BWRs, but some countries such as Russia, China, India, and Japan are pursuing development programs in this area. France was at the forefront of this technology until the late 1990s, and then relaunched studies on the subject between 2010 and 2019

2.2.13 Lead-Cooled Fast Reactors (LFRs)

2.2.13.1 Principle of Operation

LFRs are derived from highly efficient Russian submarine reactors, which were discontinued due to maintenance issues. Their power output typically ranged between 120 and 400 MWe. Future developments are targeting even higher power outputs.

The core outlet temperature of LFRs falls within the range of 550°C–800°C, with consideration for utilizing a gas-based thermodynamic cycle (such as He or CO_2), although in submarines the working fluid was water/steam.

In Figure 2.4.10, the containment vessel is depicted on the left, housing the core cooled by lead, which transfers heat to a secondary circuit traversed by gas similar to the one we presented for the FNRs. This gas undergoes expansion in a turbine coupled to an alternator, and then passes through a regeneration exchanger before being cooled by an external source. Subsequently, it undergoes recompression in a two-stage compressor before being redirected toward the lead exchanger.

The cycles considered for LFRs are studied in Chapter 6.

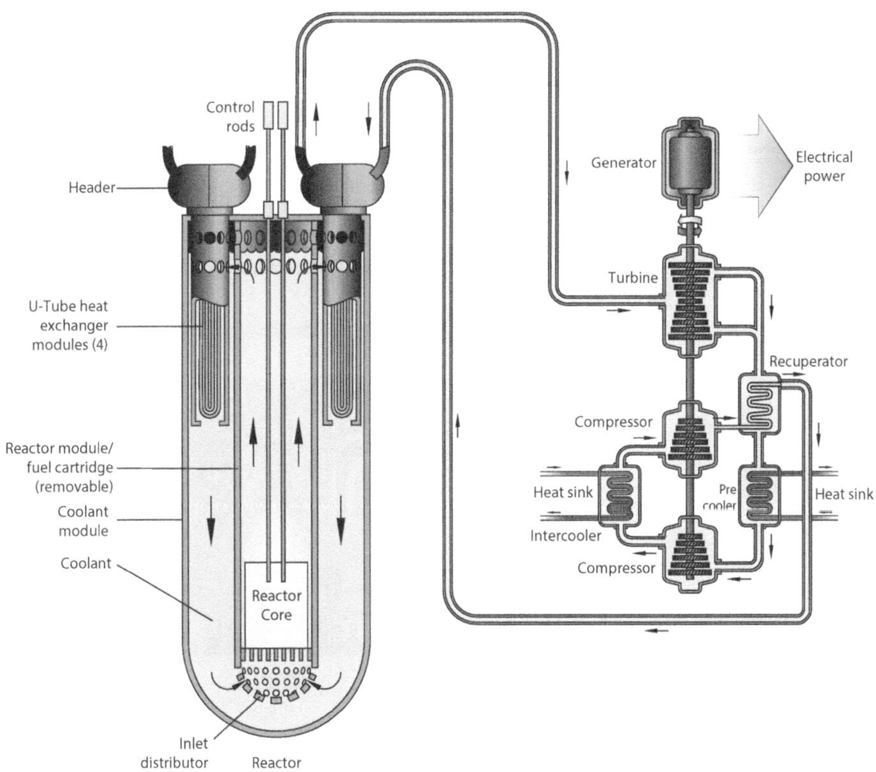

FIGURE 2.4.10 LFR reactor schematic, Idaho National Library derivative work: Beao, Public domain, via wikimedia commons.

2.2.13.2 Main Characteristics

Coolant	Lead or lead-bismuth alloy
Fuel	Mixed oxides of uranium and plutonium, or uranium nitride
Electrical power	600–1,500 MWe
Pressure and temperature levels	The temperature varies between 400°C and 800°C
Efficiency	35%–40%
Countries concerned	Russia, China, United States, South Korea

2.2.14 High-Temperature Reactors (HTRs)

2.2.14.1 Principle of Operation

HTRs cooled by gas are thermal neutron reactors designed to operate at elevated temperatures.

In helium-cooled HTRs, the helium pressure is typically maintained at a relatively low level, between 5 and 10 bar. Helium's low density and low viscosity, coupled with its high thermal capacity, help to minimize pressure losses in the cooling circuit.

In HTRs, the secondary circuit can be configured to generate steam for driving a turbine and producing electricity.

Although this section deals with HTRs, Figure 2.4.11 illustrates a VHTR, which will be discussed later. The two concepts are similar, differing mainly in their

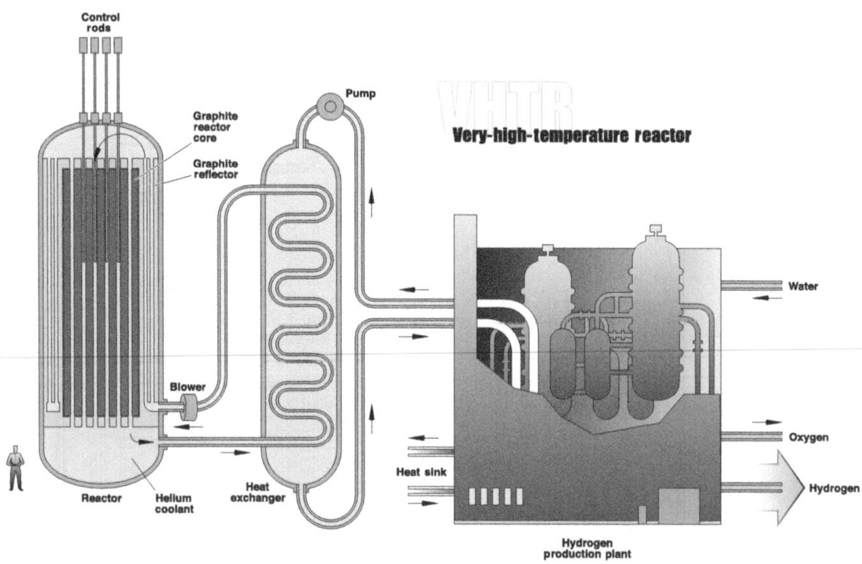

FIGURE 2.4.11 VHTR reactor schematic, US department of energy nuclear energy research advisory committee, public domain, via wikimedia commons.

temperature levels. On the left-hand side of the figure, we see the reactor core cooled by helium. The helium transfers heat in the central heat exchanger to the secondary circuit, which in this example corresponds to a hydrogen production plant, the details of which are not specified.

The cycles considered for HTRs are studied in Chapters 6 and 7.

2.2.14.2 Main Characteristics

Moderator	HTRs generally use graphite as a neutron moderator
Coolant	Helium is used as the coolant in HTRs
Fuel	HTRs use TRISO fuel, composed of carbon-coated fuel particles, optimized for high temperatures
Electric Power	25–250 MW
Pressure and Temperature Levels	In general, the pressure in the secondary circuit of an HTR is between 7 and 10 bar, while the temperature ranges from 700°C to 950°C
Efficiency	Their efficiency ranges between 40% and 50%
Countries Involved	HTRs are less widespread than PWRs and BWRs, but some countries like Russia, China, India, and Japan are pursuing development programs in this field

2.2.15 Very High Temperature Reactors (VHTRs)

2.2.15.1 Principle of Operation

VHTRs are High-Temperature Gas-Cooled Reactors designed to operate at even higher temperatures than standard High-Temperature Gas-Cooled Reactors, offering greater versatility for industrial applications.

In Figure 2.4.11, the core cooled by helium is shown on the left, which transfers heat to the secondary circuit associated with a hydrogen production plant, the details of which are not illustrated.

VHTRs can utilize a variety of working fluids in their secondary circuits to achieve even higher temperatures and meet specific industrial requirements. For example, some VHTRs can use helium as the working fluid in the secondary circuit.

Their high temperature allows them to be used for various industrial processes, such as cement, glass, and steel production, coal gasification, and thermochemistry. However, given the inertia of industrial processes and the hopes placed in the development of combustibles for transportation, the main application of VHTRs could be hydrogen production.

The cycles considered for VHTRs are studied in Chapters 6 and 7.

2.2.15.2 Main Characteristics

Moderator	VHTRs generally use graphite as a neutron moderator. The choice of graphite is due to its ability to withstand high temperatures (up to 1,600°C)
Coolant	Helium is used as a coolant in VHTRs
Fuel	VHTRs use TRISO fuel particles, which are optimized for high-temperature operation
Electrical power	The power output of VHTRs could potentially reach several GW
Pressure and temperature levels	VHTRs operate at moderate pressures, generally below 100 bar, and at very high temperatures, exceeding 1,000°C
Efficiency	VHTRs aim for a thermal efficiency of over 50%
Countries concerned	Research on VHTRs is being conducted in several countries, including the United States, China, Germany, South Africa, and Poland, among others

2.2.16 Molten Salt Reactors (MSRs)

2.2.16.1 Principle of Operation

In an MSR, the fuel is in liquid form, mixed with molten salts at temperatures ranging from 600°C to 900°C. The nuclear fuel is dissolved in these salts, which also serve as the coolant.

The heat produced by nuclear fission is used to heat the molten salts, which then transfer the heat to a secondary circuit. This secondary circuit can use either water or gas as the working fluid, with the pressure and temperature varying based on the chosen fluid and reactor design.

The salts used in MSRs are generally fluorides or chlorides, capable of dissolving large quantities of actinides, fission products, and other nuclear elements.

While adjustments to fuel composition are possible in various types of nuclear reactors, the high solubility of fuels in MSRs provides unique flexibility. This allows for more efficient optimization of neutron reactivity using epithermal neutrons.

Favoring epithermal neutrons offers several advantages, including the possibility of transmuting long-lived nuclear waste, reducing material fatigue in the reactor, conserving fissile material, and producing medical isotopes. These benefits are particularly relevant for advanced nuclear reactors and specific nuclear applications.

The pressure of the salts in MSRs is generally maintained close to atmospheric pressure, i.e., around 1 bar. This is because molten salts are liquids at high temperatures with low volatility and do not easily vaporize.

On the left of Figure 2.4.12, we see the core cooled by molten salts, which transfer heat to a secondary salt circuit through a primary heat exchanger. The heat is then transferred to the secondary circuit, traversed by a gas, through a secondary heat exchanger. This gas is expanded in a turbine coupled to an alternator, and then passes through a regeneration heat exchanger before being cooled by an external source, recompressed in a two-stage compressor, and redirected to the secondary heat exchanger with the molten salts.

The cycles considered for MSRs are studied in Chapters 6 and 7.

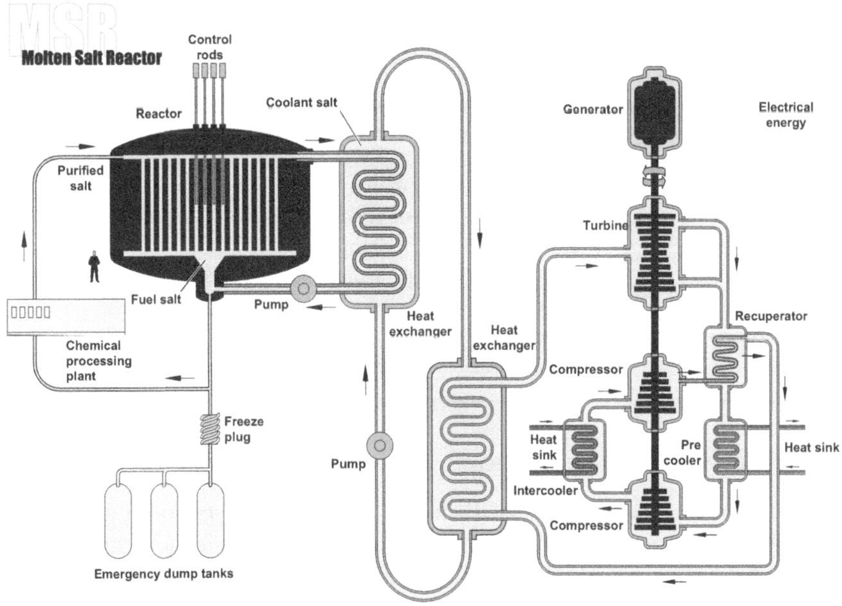

FIGURE 2.4.12 MSR reactor schematic, US department of energy nuclear energy research advisory committee, public domain, via wikimedia commons.

2.2.16.2 Main Characteristics

Moderator	The moderator is graphite or the molten salts themselves
Coolant	Molten salts
Fuel	Thorium or enriched uranium
Electrical power	MSRs can be designed to produce a wide range of power outputs, from a few MW to several GW
Pressure and temperature levels	Atmospheric pressure, 700°C
Efficiency	Around 45%
Countries concerned	Many countries are developing MSRs, including the United States, China, Canada, the United Kingdom, France, Russia, India, Japan, and South Korea

2.2.17 Fast Molten Salt Reactors (FMSRs)

2.2.17.1 Principle of Operation

In Fast Molten Salt Reactors (FMSRs), there is no moderator.

Their secondary circuit generally uses a gaseous coolant, such as helium or supercritical carbon dioxide, rather than water.

They differ from MSRs by their ability to incinerate actinides

The cycles considered for FMSRs are studied in Chapters 6 and 7.

2.2.17.2 Main Characteristics

Coolant	Molten salts, such as fluorides and chlorides, are used as a coolant
Fuel	The fuel can be composed of thorium, uranium, plutonium, and minor actinides
Electrical power	Their capacity could vary from a few MW to a GW or so
Pressure and temperature levels	They operate at atmospheric pressures and at high temperatures, which can exceed 700°C–800°C in the primary circuit
Efficiency	FMSRs would aim for high thermal efficiencies, similar to or higher than MSRs, around 40%–50%
Countries concerned	Research on FMSRs is less advanced than on MSRs, but work is being done in the United States and China, among other countries

2.2.18 Relative Sizes of Reactors

It should be noted that the size of AGRs is much larger than that of other technologies. This is due to the properties of their moderator, which is carbon in the form of graphite.

Graphite, like diamond, is a crystalline form of carbon. It consists of layers of carbon atoms arranged in hexagons, with the layers stacked on top of each other (Figure 2.4.13). Each carbon atom is bonded to three other carbon atoms in the same layer, forming strong covalent bonds, while the bonds between the layers are weaker, giving graphite its layered structure.

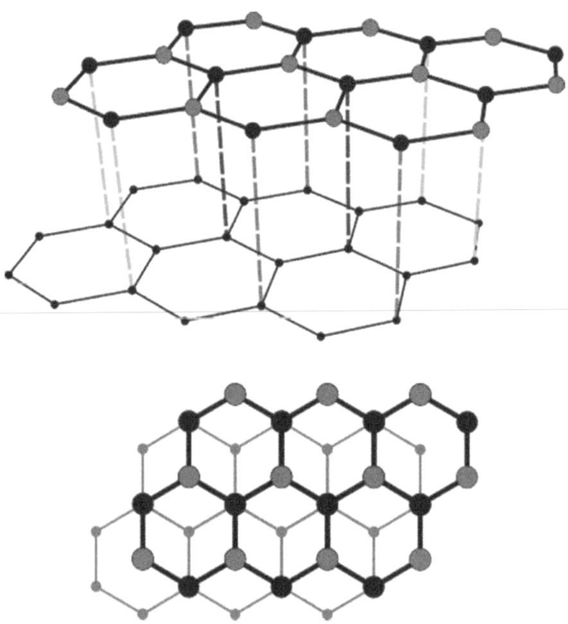

FIGURE 2.4.13 Graphite structure, Persia, CC BY-SA 4.0 <https://creativecommons.org/licenses/by-sa/4.0>, via wikimedia commons.

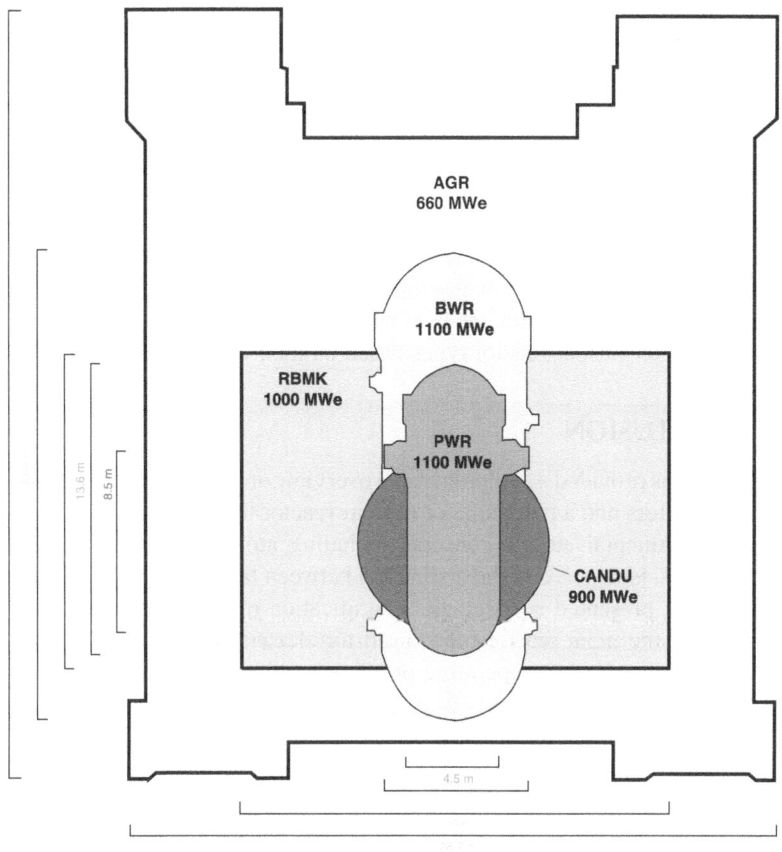

FIGURE 2.4.14 Summary of main nuclear reactor characteristics, Original: antoine fleitz? Vector: Doanri, CC BY-SA 4.0 <https://creativecommons.org/licenses/by-sa/4.0>, via wikimedia commons.

Graphite is an abundant and inexpensive element with the unique characteristic of moderating neutrons without absorbing them. However, its moderating power is limited, requiring the use of large quantities to achieve the desired effect.

In nuclear reactors, graphite can be used as a moderator. However, it poses a potential safety risk because it can burn in the presence of air and produce radioactive gases.

This is indeed what happened during the accident at the Chernobyl nuclear power plant in 1986.

The Chernobyl accident was triggered by a series of design errors and poor operating procedures that led to an overheating of the reactor. When the reactor reached excessive temperatures, the nuclear fuel melted and the graphite of the moderator caught fire, causing an explosion and releasing massive amounts of radioactive material into the environment.

Figure 2.4.14 compares the sizes of various reactors: an AGR of 660 MW, an RBMK of 1 GW, a PWR and a BWR of 1.1 GW each, and a CANDU of 900 MW. It clearly illustrates the impact of the type of moderator used on the size of the reactors.

The maximum height is 36 m, and the width is 28.7 m.

2.2.19 Summary

Figure 2.4.15 summarizes the main characteristics of the nuclear reactors that have just been presented.

2.2.20 Mind Maps

These four mind maps present the different types of existing reactors in graphical form, categorized by moderator, coolant, fuel, and thermodynamic fluid (Figures 2.4.16–2.4.19).

These graphical representations help visualize the complex relationships and distinctions between various reactor types based on their key characteristics.

2.3 CONCLUSION

This chapter has provided a comprehensive overview of physical phenomena involved in nuclear reactors and a panorama of nuclear reactor technologies. The first section revisited fundamental atomic concepts including atomic structure, protons, neutrons, electrons, isotopes, and the distinction between fast and thermal neutrons. The second section presented a systematic classification of nuclear reactor generations and examined the main reactor types used for electricity production. Each reactor type was described with its operating principles and main characteristics including

Type	Neutron spectrum	Moderator	Coolant	Fuel	Thermo. cycle
PWR	thermal	water	water	enriched uranium	Rankine
VVER	thermal	water	water	enriched uranium	Rankine
BWR	thermal	water	water	enriched uranium	Rankine
RBMK	thermal	graphite	water	natural uranium	Rankine
SCWR	thermal or fast	water	water	enriched uranium	Rankine
AGR	thermal	graphite	CO2	enriched uranium	Rankine
CANDU	thermal	heavy water	heavy water	natural uranium	Rankine
FNR	fast	without	liquid sodium	uranium or plutonium	Rankine
SFR	fast	without	liquid sodium	MOX or uranium nitrure	Rankine
LFR	fast	without	lead or Pb-Bi alloy	MOX	Brayton
HTR	thermal	graphite	helium	TRISO (enriched uranium)	Brayton
VHTR	fast	without	helium	optimized TRISO	Brayton
MSR	epitherm. fast	graphite or molten salts	molten salts	thorium or enriched uranium	Rankine
FMSR	epitherm. fast	without	molten salts	thorium or enriched uranium	Brayton

Type	Working fluid	Pressure (bar)	Temperature (°C)	Capacity (Mwe)	Efficiency
PWR	water	160	300	100 1600	33 - 36 %
VVER	water	100 - 160	290-330	1000-1500	32 - 37 %
BWR	water	70	285	100-1600	33 - 36 %
RBMK	water	70	300-330	1000-1600	30 - 35 %
SCWR	super-crit. steam	250	510 - 550	1000 - 2000	40 - 45 %
AGR	water	170	550	600-1500	40%
CANDU	water	100	310	600-2100	30%
FNR	water	30 - 50	550 - 600	10-500	40%
SFR	water/super-crit. steam	30 - 50	600 - 650	150 - 1500	40 - 45 %
LFR	Gas, sCO2	70	550 - 800	600 - 1500	35 - 40%
HTR	Gas, sCO2	70	700 - 1000	25-250	40-50%
VHTR	Gas, sCO2	70	> 1000	> 1000	> 50%
MSR	water	100 - 160	700	10-500	45%
FMSR	water	100 - 160	700 - 800	1000	40 - 50%

FIGURE 2.4.15 Main characteristics of nuclear reactors.

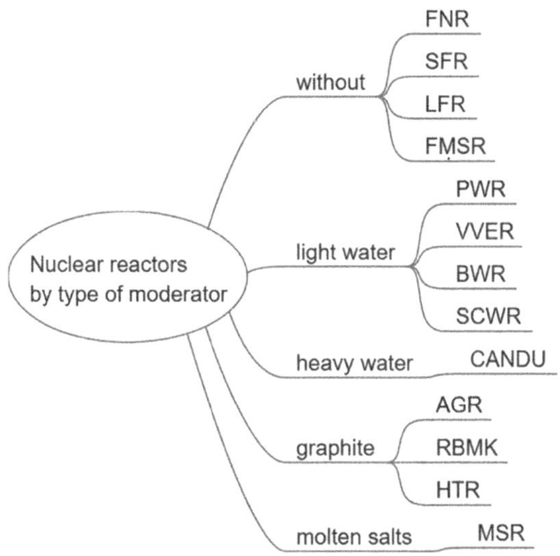

FIGURE 2.4.16 Reactor classification by moderator.

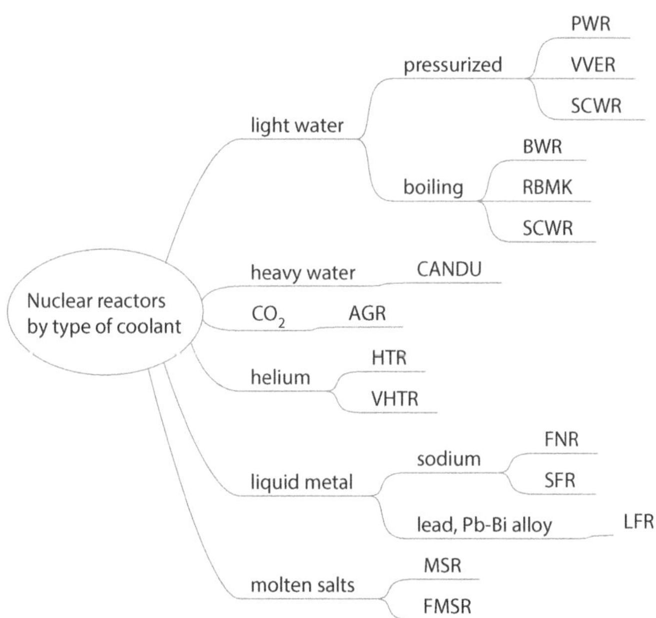

FIGURE 2.4.17 Reactor classification by coolant.

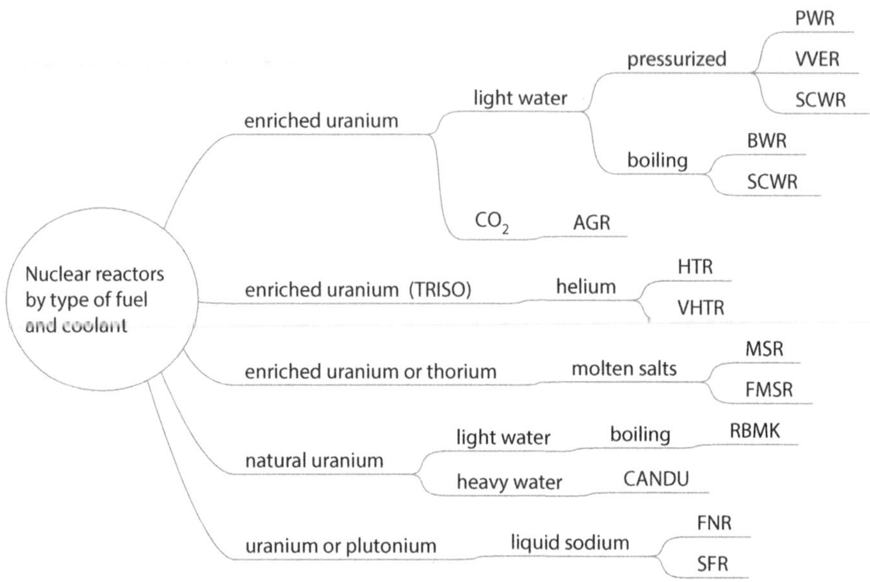

FIGURE 2.4.18 Reactor classification by fuel.

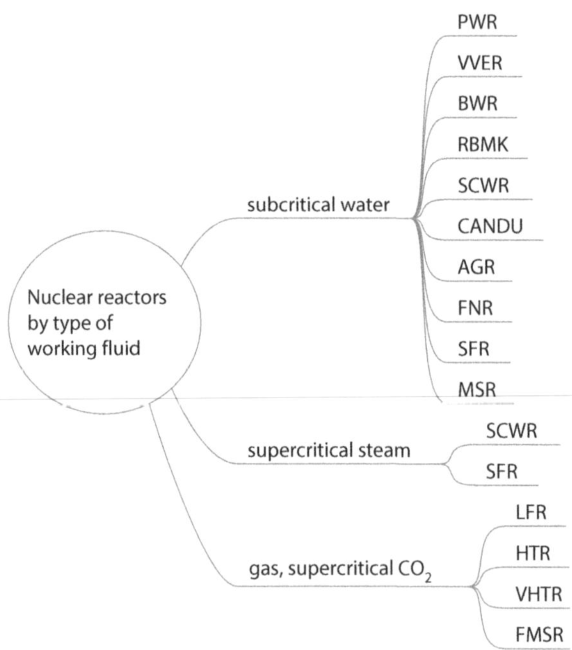

FIGURE 2.4.19 Reactor classification by thermodynamic fluid.

moderator, coolant, fuel type, electrical power, pressure and temperature levels, efficiency, and countries of deployment.

Understanding the physical mechanisms of nuclear reactors—from atomic interactions to system-level architectures—is indeed essential for grasping both their technological diversity and their role in energy production.

BIBLIOGRAPHY

Lamarsh, J.R. & Baratta, A.J. (2001). *Introduction to Nuclear Engineering*, 3rd ed. Prentice Hall, New York. ISBN 978-0201824988.

Lee, J.C., et al. (2020). *Nuclear Reactor Physics and Engineering*. Wiley, New York. ISBN 978-1119582328. DOI: 10.1002/9781119582342.

Masterson, R.E. (2018). *Introduction to Nuclear Reactor physics*. CRC Press, Boca Raton, FL. ISBN 978-1498751483. DOI: 10.1201/9781315375144.

World Nuclear Association. (2025). "Nuclear power reactors." Available at: https://world-nuclear.org/information-library/nuclear-fuel-cycle/nuclear-power-reactors/nuclear-power-reactors (consulted November 2025).

CREDITS

Containment Building

Source: NRC

Pressurized Water Reactors

Source: NRC Source Steffen Kuntoff; translation by Enr-vDerivative work: Cjp24

Boiling Water Reactors

Source: Robert Steffens, SVG: Marlus Gancher, Antonsusi, GFDL, via Wikimedia Commons

RBMK Reactors

Source: Leovilok (talk), Fireice~commonswiki, Sakurambo, Emoscopes, CC BY-SA 3.0

Supercritical Water Reactors

Source: US Department of Energy Nuclear Energy Research Advisory Committee, Public domain, via Wikimedia Commons

Advanced Gas-Cooled Reactors

Source: Scheme-of-Advanced-Gas-cooled-Reactor-AGR-Wikimedia-2012-Heat-exchanger-is

Heavy Water Reactors

Source: Inductiveload, CC BY-SA 2.5, via Wikimedia Commons

Sodium-Cooled Fast Reactors

Source: Sfr.gifderivative work: Beao, Public domain, via Wikimedia Commons

Fast Neutron Reactors

By Gfr.gifderivative work: Beao - Gfr.gif, Public Domain, https://commons.wikimedia.org/w/index.php?curid=8030370

Lead-Cooled Fast Reactors

Source: Idaho National Laboratory derivative work: Beao, Public domain, via Wikimedia Commons

High-Temperature Reactors

Source: US Department of Energy Nuclear Energy Research Advisory Committee, Public domain, via Wikimedia Commons

Very High Temperature Reactors

Source: US Department of Energy Nuclear Energy Research Advisory Committee, Public domain, via Wikimedia Commons

Molten Salt Reactors

Source: US Department of Energy Nuclear Energy Research Advisory Committee, Public domain

AGR Size

Source: DeepKling, CC BY-SA 3.0, via Wikimedia Commons Source: antoine fleitz, Vector: Doanri, via Wikimedia Commons

3 Steam and Gas Power Cycles

Understanding, Analysis, and Improvement Methodologies

INTRODUCTION

The panorama of nuclear reactors presented in Chapter 2 has shown that most use water as a working fluid, at temperatures varying from 550°C–800°C for the highest (MSR, SFR, FNR, AGR, SCWR) to 285°C–330°C for water-cooled reactors (PWR, BWR, RBMK, VVER, CANDU). Since the steam temperatures of the first group are close to those used in conventional hydrocarbon-burning steam thermal power plants, the cycles used and envisaged are similar. As the temperatures in the second group are much lower, specific cycles are employed.

The other four reactor types generally intend to use so-called Brayton gas cycles (LFR,[1] HTR, VHTR, FMSR), with temperatures varying from 550 to more than 1,000°C. The cycles considered are therefore similar to those using combustion gas turbines, except that they are closed rather than open cycles, due to the absence of a combustion chamber.

The design and optimization of thermal power cycles—particularly steam and gas cycles—require thorough understanding of both their thermodynamic principles and the methodologies used to analyze and improve them. While many reference texts concentrate on specific technologies or components, this chapter adopts a broader, more structured perspective.

It offers a general introduction to steam and gas power cycles, beginning with their physical and thermodynamic foundations, followed by typical configurations of steam and closed-cycle gas turbine plants. This chapter concludes with an overview of methodological tools for performance analysis and improvement. The goal is to provide readers with a solid foundation before exploring more detailed applications in subsequent chapters.

This chapter serves as an entry point, introducing key elements relied upon later: the two basic cycles, the Thermoptim simulator and its integrated browser, and tools used to improve thermodynamic cycles. The next three chapters demonstrate how to apply these tools to each of the two basic cycles.

Across all these chapters, we continue to focus on these two reference cycles and explore different variants that emerge as we attempt to improve them. They serve as benchmarks for analyzing the various cycle configurations associated with the reactor categories presented in the panorama.

DOI: 10.1201/9781003606024-4

3.1 INTRODUCTION TO STEAM AND GAS POWER CYCLES

This section presents the basic principles of thermodynamic power cycles used in electricity generation. It outlines the main components and operating parameters of these systems and introduces key concepts required to understand how energy is transformed and transferred in practical installations.

3.1.1 Conversion of Heat into Work

The question of how to convert heat into mechanical energy has been pressing since the 17th century and the beginning of the Industrial Revolution. When a high-temperature heat source is available, several points are now established:

- Only a fraction of the available heat can be converted into work, and this fraction is higher when the temperature of the heat source is higher.
- At least one cold source is required to transfer the heat that is not converted into work.
- To perform this conversion of heat into work, devices called heat engines (such as steam engines, internal combustion engines, or gas turbines) are used.
- These machines operate according to a thermodynamic cycle that uses a working fluid (usually steam or gas), which undergoes several processes.
- The theoretical thermodynamic cycle that allows for maximum work output is called the Carnot cycle, named after its inventor.

Power plants coupled with nuclear reactors use this principle. The energy released by nuclear reactions in the reactor core is used to heat a thermodynamic fluid, which undergoes a cycle that produces work. For safety reasons, especially radiological safety, the reactor is commonly cooled by another fluid, called a coolant, which transfers the heat it receives to the thermodynamic fluid.

As indicated, the higher the reactor temperature, the higher the conversion efficiency. Therefore, efforts are always made to increase this temperature as much as possible, recognizing limits related to several factors, such as the mechanical strength of materials used and, especially for nuclear power plants, safety criteria. Consequently, the maximum temperature of the conversion cycle is often much lower than theory would suggest. We examine the consequences of these constraints on cycles implemented in practice.

Since the Carnot cycle allows for the best conversion efficiency of heat into work, it serves as a reference. It is therefore important to understand it well and to know why real cycles deviate from it. This is the subject of the following section.

3.1.2 Carnot Cycle

The various engines used to convert heat into mechanical energy mainly differ in:

- the type of thermodynamic cycle they use;
- the nature of the thermodynamic fluid that passes through them;

- the types of hot source;
- the compression and expansion components used.

Different typologies can thus be established, but the categories partially overlap.

Generally, there are two main categories of engines:

The first category corresponds to compressible fluid engines, where the fluid remains in the gaseous or vapor state throughout the cycle. Depending on the case, a boiler or combustion chamber is used as a hot source for an open cycle if the working fluid contains oxygen (usually air).

The second category corresponds to condensable fluid engines, in which the fluid changes state. At the condenser outlet, it is liquid and is compressed by a pump, then heated and transformed in a boiler into vapor which is then expanded in a turbine or positive displacement engine. The compression work, proportional to the specific volume of the liquid, is much lower in these engines than in the previous ones.

3.1.2.1 Carnot's Effectiveness

Let us now examine Carnot's effectiveness, which is very important in practice.

Sadi Carnot demonstrated in 1824 that the effectiveness of an ideal thermal machine describing a cycle between two heat sources depends only on the temperatures T_h and T_c of the hot and cold sources with which it is in contact, and is given by this formula, where T_h and T_c are expressed in degrees Celsius:

$$\eta = 1 - \frac{T_c + 273.15}{T_h + 273.15} = \frac{T_h - T_c}{T_h + 273.15} \tag{3.1.1}$$

A Carnot cycle consists of two isentropes, corresponding to the compression and expansion of the fluid, and two isotherms corresponding to heat exchanges with the hot and cold sources (Figure 3.1.1).

The expression for Carnot effectiveness shows that its value is greater the larger the difference ($T_h - T_c$).

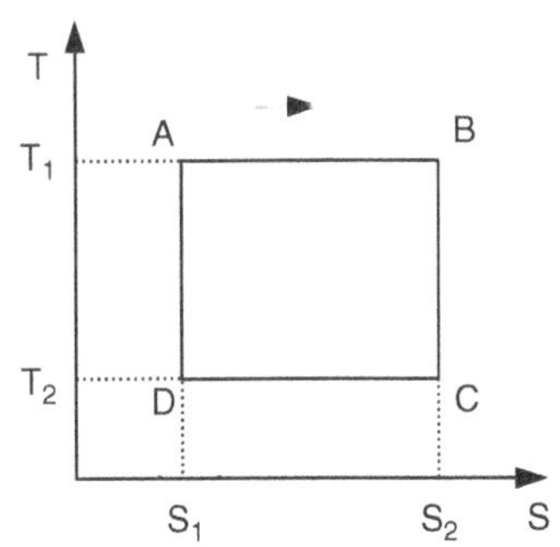

FIGURE 3.1.1 Carnot cycle.

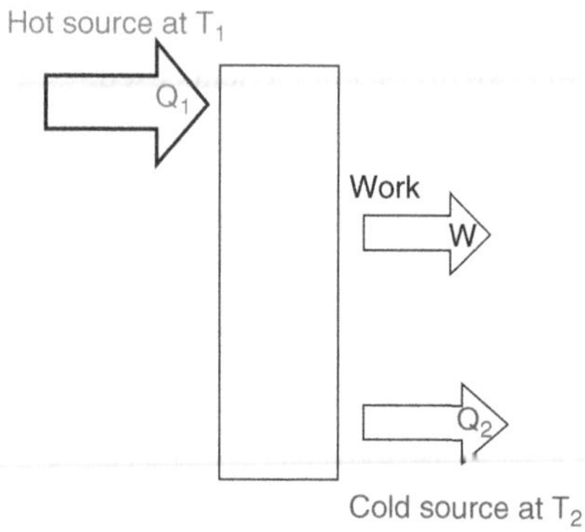

FIGURE 3.1.2 Conversion of heat into work.

The existence of a wide variety of technological solutions is mainly due to the multiplicity of existing heat sources, as there are relatively few cold sources that can be used in practice, usually ambient air or a river. Since T_c is generally close to ambient air temperature, the higher the hot source temperature T_h, the higher the Carnot effectiveness.

Recall that the First Law of thermodynamics states that the energy of the system is conserved, while Carnot's theorem states that only the fraction η of the heat provided by the hot source is converted into mechanical energy.

As these two statements may seem contradictory, some additional explanation is preferable.

Consider a hot source providing a machine with a quantity of heat Q_l at temperature T_1.

The conversion of this heat in a perfect machine exchanging with a cold source at temperature T_2 produces (Figure 3.1.2):

- On the one hand, work W equal to $\eta\ Q_l$, with η being equal to the Carnot effectiveness;
- On the other hand, heat Q_2 at low temperature equal to $(1 - \eta)\ Q_l$.

The First Law is duly respected, since Q_l equals $W + Q_2$, and Carnot's theorem has made it possible to determine η, which represents the fraction of the heat at high temperature that is converted into work.

Table 3.1.1 and Figure 3.1.3 show how Carnot's effectiveness changes when the temperature of the hot source varies from 100°C to 1,300°C for a cold source temperature of 15°C.

The values reached for different types of cycles converting heat into mechanical power or electricity are indicated:

- 80% for a high-temperature gas turbine
- 65% for a flame-fired power plant
- 50% for a PWR-type nuclear power plant
- 20% for a solar thermal power plant with vacuum tubes

TABLE 3.1.1
Carnot Efficiency, $T_0 = 15°C$

	T_1 (°C)	Carnot eff. (%)
High-temperature gas turbine	1,300	82
Flame steam power plant	560	65
PWR nuclear power plant	300	50
Vacuum tube collector solar thermal power plant	100	23

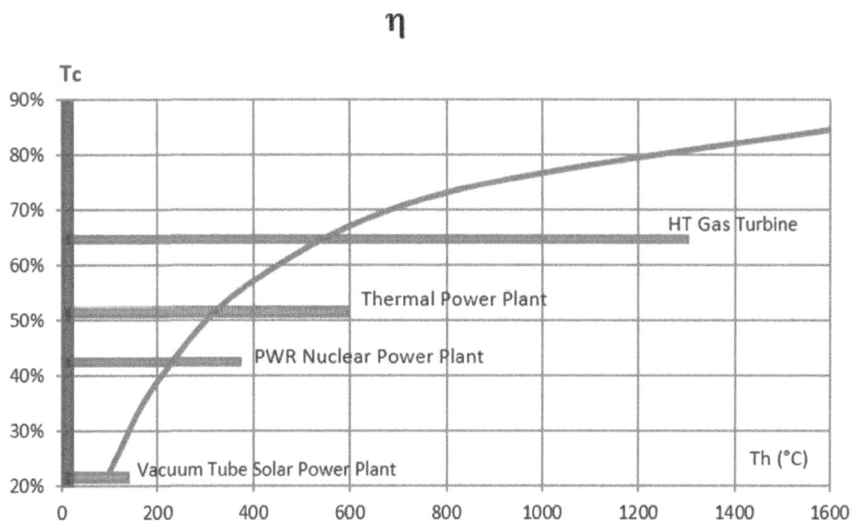

FIGURE 3.1.3 Carnot effectiveness.

Generally speaking, the objective in developing power cycles is to produce as much mechanical power as possible from the available heat source, taking into account existing technological constraints.

In practice, the actual efficiency is much lower than that of the Carnot cycle, on the one hand because the hypotheses that establish equation (3.1.1) are almost never met, and on the other hand due to the imperfections of the machines used, which generate what are called irreversibilities.

The differences between a real cycle and the Carnot cycle arise from the following points, among others:

- First, in practice there must be a certain temperature difference between the machine and the hot and cold sources, which is a primary cause of irreversibility.
- Heat exchange with the hot and cold sources is isobaric and not isothermal, because for technological reasons, we generally cannot achieve isothermal heat exchange.

- Then, it is exceptional that the hot and cold sources can themselves be considered isothermal: most often, it is a fluid that exchanges heat between two temperature levels.
- Finally, when compression and expansion are adiabatic, they are not reversible due to mechanical irreversibilities.

Actual power cycles therefore deviate significantly from the Carnot cycle, with isotherms AB and CD most often replaced by isobars or isochores. However, attempts are of course made for expansion BC and compression AD to approach isentropic processes.

3.1.2.2 Regenerative Cycles

We can imagine cycles where the isentropic compression AD and expansion BC are replaced by other processes, each deduced from the other by a translation in the entropy diagram (Figure 3.1.4). This is called regeneration.

In these circumstances, it is theoretically possible to make all internal and external heat exchanges at constant temperature, and the ideal cycle with regeneration reaches the same efficiency as the Carnot cycle.

In practice, the internal heat exchange is of course not at constant temperature, and we introduce the notion of regenerator effectiveness to characterize performance.

3.1.2.3 Compression and Expansion Work

For various technological reasons, we do not know how to industrially manufacture components capable of both transferring heat and achieving efficient compression or expansion.

Machines performing the compression or expansion of a fluid have a very compact design for reasons of weight, size and cost. For similar reasons, they rotate very fast (several thousand revolutions per minute). Each parcel of fluid remains there very briefly. Moreover, fluids brought into play in compressors and turbines are often gases whose heat exchange coefficients have low values.

Short residence time, small areas of fluid-wall contact, and low exchange coefficients imply that the heat exchange is minimal and that the operation of these machines is nearly adiabatic.

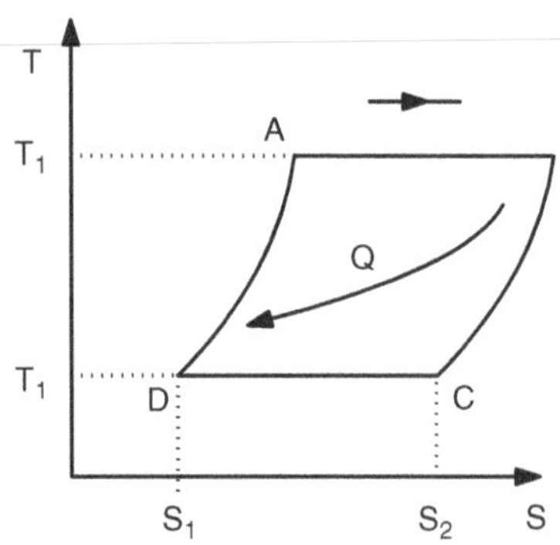

FIGURE 3.1.4 Regeneration cycle.

Moreover, as we will see in the analytical developments relating to the gas turbine (Section 3.2), the work involved in reversible adiabatic compression or expansion is, for an ideal gas,[2] proportional to the absolute temperature of the fluid at the suction of the machine.

$$h_2 - h_1 = c_p T_1 (r^{\beta} - 1)$$

This relationship in which r is the compression or expansion ratio, c_p the specific heat T the absolute temperature, h the enthalpy, and β a positive parameter, shows that the work is lower the colder the fluid.

It is therefore always valuable to cool a gas before compressing it, and if the compression ratio is large and it is technologically possible, it may be necessary to split the compression and cool the gas between two compression stages using a heat exchanger.

We will see examples of this when we study variants of gas turbine cycles.

Similarly, the expansion work is greater the higher the temperature of the fluid.

It is therefore always advantageous to heat a gas or steam before expansion.

Therefore, if the expansion ratio is large and technologically feasible, the expansion is split and the gas is heated between two expansion stages.

We will see examples of this when we study variants of steam power plant and gas turbine cycles.

3.1.2.3.1 Remark

Carnot's theorem is probably rightly one of the best known in thermodynamics. It is of great practical importance, given its implications:

- First, with the assumptions made (reversible cycle without friction and heat transfer without temperature differences), it is easy to show that this effectiveness is the highest that can be achieved by a simple fluid heat engine operating between two sources at T_h and T_c. It is therefore a maximum limit, and real cycles generally have much lower efficiencies.
- Second, the Carnot effectiveness of all reversible machines operating between two sources at T_h and T_c is the same and depends only on their temperatures, not on the working fluid used.

3.2 STEAM POWER PLANT AND GAS TURBINE CYCLES

Throughout this and subsequent chapters, we work on variants of nuclear heat power generation cycles, the first corresponding to the steam cycle of the Hartlepool AGR plant in County Durham, North East England, and the second to a helium cycle of the type envisaged for HTRs.

3.2.1 Simple Steam Power Plant

This section introduces the general structure and operation of steam power plants. It covers both conventional fossil-fuel-based systems and water-cooled nuclear power plants, emphasizing their similarities and distinctive characteristics. The objective is to provide a broad overview of technologies that rely on steam as the working fluid.

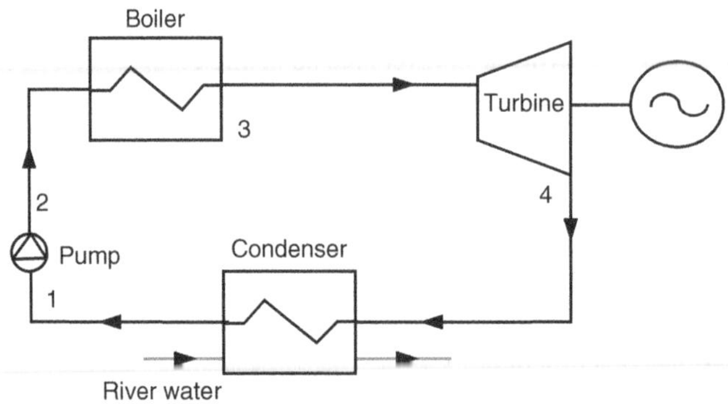

FIGURE 3.2.1 Sketch of a steam power plant.

A steam power plant (Figure 3.2.1) is an energy system that converts the heat provided in a boiler into electricity. In most countries, a significant proportion of the power plant fleet consists of such facilities, often referred to as thermal power plants.

The boiler can be of different types. In our case, it is the core of a nuclear reactor, for example an AGR type plant. In this case, the reactor core is cooled by a flow of gaseous CO_2, which transfers its heat to the working fluid (water) in a steam generator.

As we have already mentioned in Chapter 1, the plant operation can be broken down into four successive steps.

- The first step takes place in the steam generator: the transfer of heat to pressurized water which turns into steam.
- The second step is electricity production: the steam is expanded in a turbine that drives an alternator, which produces electric current that is transported to places of consumption by high-voltage lines.
- The third step is the condensation of steam exiting the turbine into liquid water. This water is then repressurized in a fourth step before being redirected to the steam generator.

The Hirn (or Rankine with superheating) cycle uses a condensable fluid, which is cooled at a temperature and pressure sufficient for it to be fully liquefied before compression. Under these conditions, the compression work is almost negligible compared to the expansion work (although it represents about 60% in a gas turbine as we will see). The compressed liquid is vaporized and superheated in the boiler by heat exchange with the hot source, then expanded, and condensed. The fluid two-phase state during condensation and vaporization phases is very favorable for heat transfer.

The block diagram of a steam power plant operating on the ideal Hirn cycle is given in Figure 3.2.1. Note that AGRs have been the backbone of the UK's nuclear power generation since the 1980s.

We start by using a very simplified model, which will be gradually refined by making successive improvements, to arrive at a model very close to the cycle actually used in this plant.

Such a plant includes four basic components: a pump, a boiler, a turbine and a condenser, through which the same water flow rate passes.

As the sea temperature in this area varies from 7°C to 15°C between winter and summer, we consider a condensation temperature of 15°C, which corresponds to a pressure of 0.017 bar. The numerical values chosen in Figure 3.2.2 correspond to a simplified model of such an AGR cycle.[3]

As indicated before, the pump and turbine can be assumed adiabatic. As for the boiler and condenser, at first approximation we can assume they are isobaric.

At point 1, a flow rate of 0.55 t/s of water enters the pump in the liquid state and at very low pressure (LP) (0.04 bar).

It is compressed and exits at point 2 at a pressure of 163 bar, still in the liquid state.

Compression (1–2) of the liquid can be legitimately considered isentropic, and temperature T remains approximately constant. As the liquid isobars are virtually merged with the beginning of the vaporization curve, point 2 is virtually identical with point 1 on the entropy diagram of Figure 3.2.3.

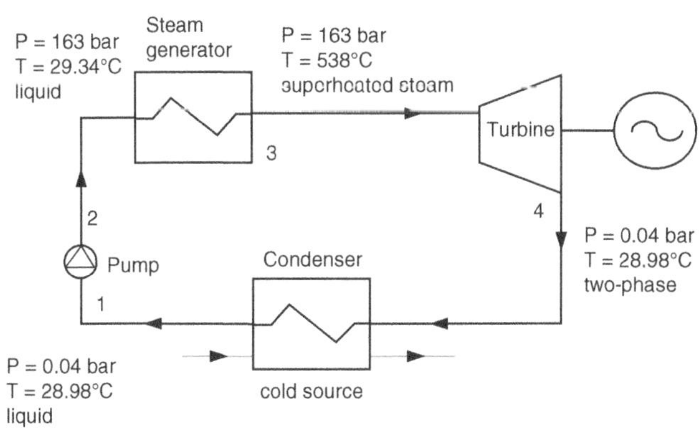

FIGURE 3.2.2 Steam cycle parameters.

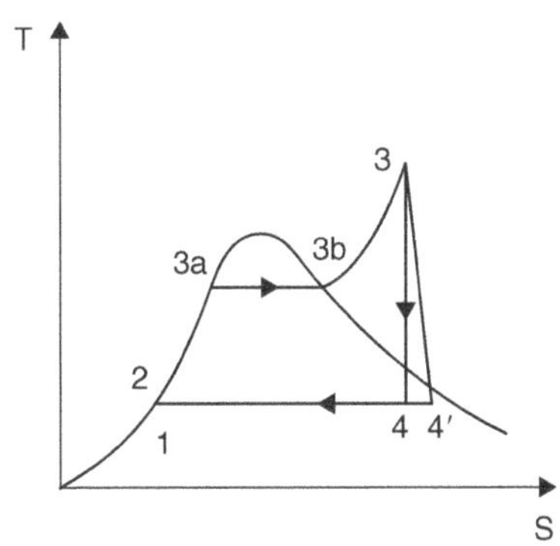

FIGURE 3.2.3 Steam cycle in an entropy diagram.

The nuclear reactor core is cooled by the coolant, namely CO_2, which transfers that heat to water, the working fluid, in the steam generator.

Pressurized water is thus heated to high temperature in the boiler, with heating comprising the following three steps, clearly visible in Figure 3.2.3:

- liquid water heating in the economizer from 29°C to approximately 349°C, the saturation temperature at 163 bar: process (2–3a) in the entropy diagram. Point 3a is on the vaporization curve at ordinate 349°C on the same isobar as point 2.
- vaporization at constant temperature 349°C in the evaporator: process (3a–3b). Since vaporization is carried out at constant pressure and temperature, it results in a horizontal segment 3a–3b in the diagram. Point 3b is therefore on the descending branch of the vaporization curve, or dew point curve, at its intersection with the horizontal line of temperature 349°C, still at 163 bar pressure P_2.
- superheating from 349°C to 538°C in the superheater: process (3b–3). Point 3 is still assumed to be at the same pressure, but at temperature $T_3 = 538$°C. It is thus at the intersection of isobar P = 163 bar and the horizontal of ordinate T = 538°C.

The superheated steam at point 3 is then expanded in a steam turbine, which produces mechanical power on its shaft.

It exits at point 4 at the LP of the cycle, and its temperature again becomes equal to 28.98°C. Its state corresponds to a mixture of liquid and vapor, called the two-phase state.

Process (3–4) is an adiabatic expansion from 163 bar to 0.04 bar. In the ideal cycle, obtained without irreversibility and hence constant entropy, point 4 is at the intersection of the vertical passing through point 3 and isobar 0.04 bar. The point being in the mixed zone, the latter coincides with the horizontal $T \approx 29$°C, with quality x = 0.746. The real point 4′ is at the same pressure as point 4, but its entropy is different because of irreversibilities (larger by the Second Law). Its enthalpy can be determined if one knows the turbine isentropic or polytropic efficiency as we will see below.[4] Two cases may arise:

- either point 4′ is in the mixed zone, and it is also on isotherm $T \approx 29$°C, closer to the dew point curve;
- or it is in the vapor zone, on isobar P = 0.04 bar.

In this example, with a polytropic efficiency of 0.8, the point at the end of expansion is 4′, to the right of 4, and quality reaches the value of 0.816.

It is important that the quality is not lower than a value of around 0.7, mainly because the presence of water droplets has a destructive effect on turbine blades.

The steam exiting the turbine at point 4 is then completely liquefied at point 1 by cooling in the condenser. The cycle is thus closed.

The pump then restores the water to boiler pressure at point 2.

Note that the condenser is cooled by an external cold source, here the sea.

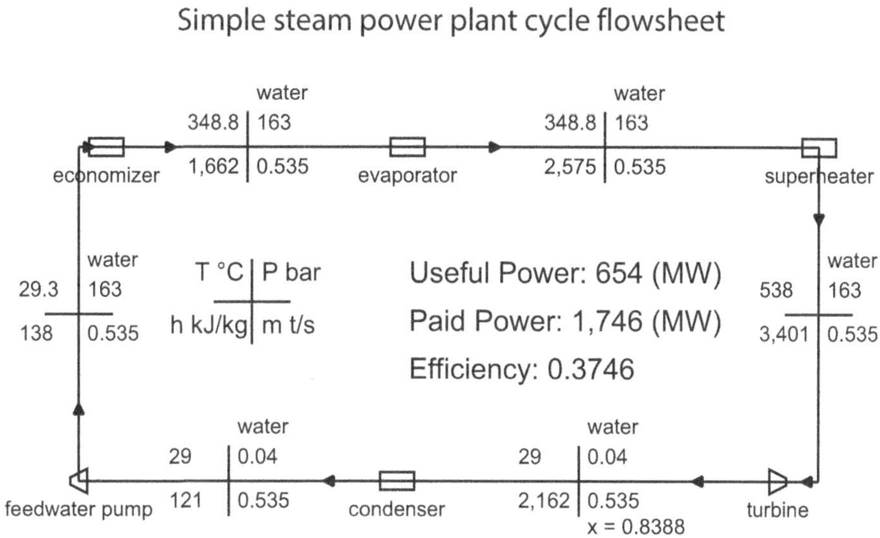

FIGURE 3.2.4 Flowsheet of the cycle.

The working fluid circuit has **two parts at different pressures**: **the high pressure (HP) at 163 bar** in the boiler and the **LP at 0.04 bar** in the condenser.

This is a characteristic common to most cycles we will study: they implement at least two pressure levels, with pumps and compressors passing the working fluid from LP to HP, and turbines and valves from HP to LP.

This cycle can be represented by the flowsheet in Figure 3.2.4, with four values of the state of each point displayed around a cross: pressure in bar at the top right, flow rate, here in t/s at the bottom right, enthalpy in kJ/kg at the bottom left, and temperature in °C at the top left, with the name of the substance indicated above the pressure. In the rest of this book, we will frequently use this kind of representation automatically generated by Thermoptim.

3.2.1.1 Plot of the Cycle in the Thermodynamic Diagrams

The cycle plot in the (h, ln(P)) enthalpy diagram is shown in Figure 3.2.5. The one in the (T, s) entropy diagram will be presented in Section 3.3.1.

The isobaric processes in the steam generator and in the condenser, represented by horizontal segments, are clearly recognizable.

3.2.1.2 Technological Aspects

Let us now briefly present the technologies implemented in the components of this steam plant.

3.2.1.2.1 Pump

The pump is generally of the centrifugal type, multistage because it must achieve a high compression ratio (around 4,000 with our settings).

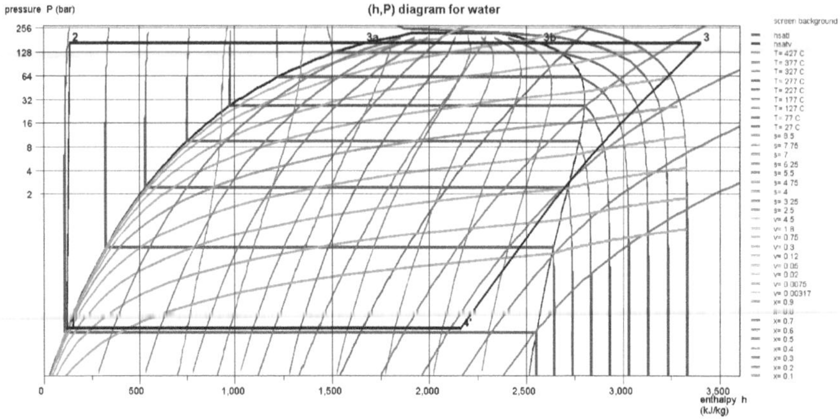

FIGURE 3.2.5 Cycle plot in the (h, ln(P)) enthalpy diagram.

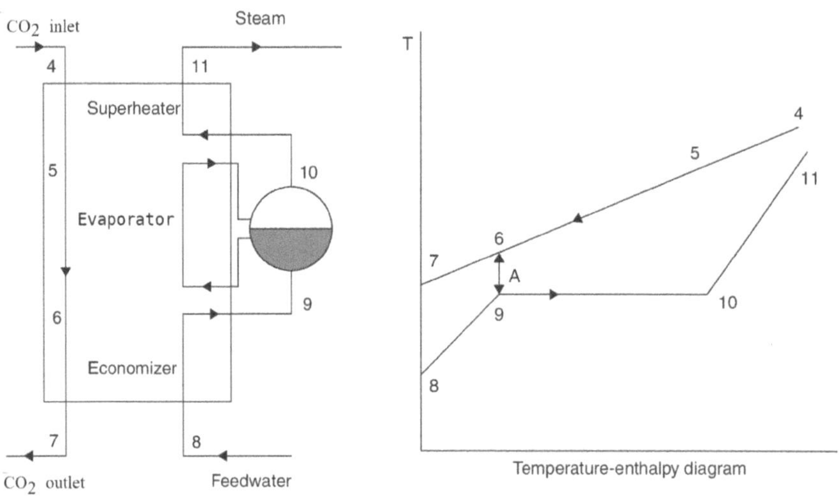

FIGURE 3.2.6 Steam generator exchange configuration.

3.2.1.2.2 Steam Generator

A steam generator has three successive functions:

- heat pressurized feedwater (in the economizer) to the vaporization temperature corresponding to the HP;
- vaporize steam in the evaporator;
- finally superheat steam to the desired temperature.

It behaves like a triple heat exchanger, and may be represented in terms of heat exchange by the diagram in Figure 3.2.6.

At point 6 on the diagram, you can see a minimum in the temperature difference between the two fluids, called the pinch. This point plays a fundamental role in the design of the steam generator, as it represents the smallest temperature difference in the facility. To reduce equipment cost, it is preferable that the pinch is not too low. However, thermodynamic optimization of the complete system requires it to be as small as possible. Generally, the pinch point is the most constrained zone in the thermal system.

Several types of steam generators are used in nuclear reactors. In AGRs, this is a technology called once-through, where water is successively heated to a liquid state, vaporized, and superheated inside tubes placed in parallel, without recirculation.

3.2.1.2.3 Steam Turbines

Steam turbines are always multistage given the very large expansion ratio. We are content to provide here some indications on the technologies implemented in steam power plants, knowing that Appendix 2 develops this subject in much more detail. You can therefore refer to it if you want further explanations.

3.2.1.2.4 Condenser

The condenser is a heat exchanger whose particularity is to work at a pressure lower than atmospheric, given the low vapor pressure of water at ambient temperature.

The performance of steam systems operating in closed circuit is very sensitive to conditions in which steam may be condensed. Indeed, a condenser is a heat exchanger of a particular type, phase change, whose thermal equilibrium determines the condensation temperature of steam, and thus its condensation pressure.

The cycle LP, which directly determines the power that the turbine can produce, is thus determined by the condenser.

Heat transferred to the condenser is typically removed by cooling water flowing inside the tubes, which must be renewed (open-loop cooling) or cooled by contact with ambient air (air cooling).

For systems that operate in open loop, cooling water is drawn and released directly in the sea or a river, after passing through the condenser, while in others, condenser cooling water circulates in closed circuit cooled with air, usually in a cooling tower.

In both cases, the amount of water required is very important, which is a burden, especially in the context of global warming and depletion of water resources. Section 5.11 introduces innovative cycles today proposed to reduce this constraint.

Due to the properties of water, the condenser pressure is very low, so air leakage is inevitable. We must therefore always extract air from the condenser, which results in significant energy consumption (up to 0.5% of plant capacity), especially since, for reasons of simplicity and cost, low efficiency steam ejectors are generally used. Their influence on the power plant's performance remaining limited, we will not address ejectors here.

3.2.2 Simple Closed Gas Turbine Cycle

This section explores the fundamentals of closed-cycle gas turbines, which offer an alternative to traditional steam-based systems. It highlights their thermodynamic advantages, configurations, and potential applications in advanced or high-temperature contexts. The goal is to familiarize the reader with an important class of power systems that is less commonly discussed but increasingly relevant.

Let us clarify right away that, for reasons we will explain later, there is currently no commercially operating power plant utilizing closed cycles of gas turbines, unlike the widely used open cycles, especially in combined cycle. The results we present are therefore related to projects that have not yet been realized.

In its simplest and most common form (Figure 3.2.7), a closed-cycle gas turbine is composed of four elements:

- a compressor, usually centrifugal or axial, which is used to compress the working fluid to a pressure 10–30 times the intake pressure in modern machines;
- a heat exchanger or heater, in which the fluid is brought to high temperature;
- a turbine, usually axial, in which the high-temperature gas exiting the heater is expanded. A significant proportion (60%–70%) of the work recovered on the turbine shaft is used to drive the compressor;
- a cooler in which the fluid is brought to low temperature before being compressed again.

For the reasons presented above, the compressor and turbine can be assumed to be adiabatic. As to the heater and cooler, at first approximation we can assume they are isobaric.

The sketch of a closed-cycle gas turbine coupled to a nuclear reactor of the HTR type is shown in Figure 3.2.7, for the case where cooling of the reactor is directly provided by the thermodynamic fluid.

In this cycle, helium is used, which plays the role of both the heat transfer fluid in the nuclear reactor and the working fluid in the secondary circuit.

If we disregard the change in thermodynamic fluid, the cycle is very similar to that of the steam plant. The heater is used to transfer heat to helium and heat it to high temperature. The turbine produces mechanical power; the cooler allows the

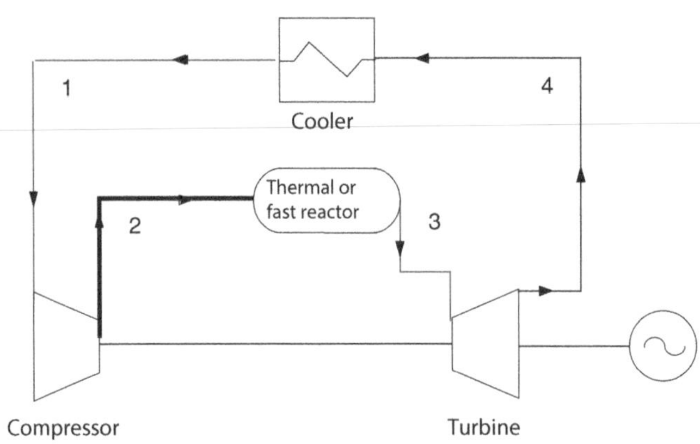

FIGURE 3.2.7 Gas cooled HT reactor.

temperature of the helium to be lowered before compressing it, which greatly reduces the compression work. The pump is replaced by a compressor, the working fluid being gaseous.

Let us call τ_c the compressor work, τ_t the work done by the turbine, $\dot{m}$ the flow rate passing through them, and Q_h the heat provided by the nuclear reactor.

The enthalpy balances of the three components give:

$$\tau_c = \dot{m} \quad (h_2 - h_1) > 0$$

$$\tau_t = \dot{m} \quad (h_4 - h_3) < 0$$

$$Q_h = \dot{m} \quad h_3 - \dot{m}\, h_2 > 0$$

The First Law writes

$$Q_h + \tau_c + \tau_t = 0$$

The cycle efficiency is written:

$$\eta = \frac{|\tau_t| - |\tau_c|}{Q_h}$$

This model based on the component enthalpy balances is very general and rigorous, but incomplete as models of fluids have not been selected. In what follows, we will show the very significant differences in results which are obtained depending on the assumptions made at that level.

3.2.2.1 Perfect Gas Cycle without Irreversibilities

The basic thermodynamic cycle used in gas turbines is the Brayton cycle. Let us first assume that the working fluid passing through the gas turbine is perfect and there is no irreversibility. Its plot in an entropy diagram is given in Figure 3.2.8.

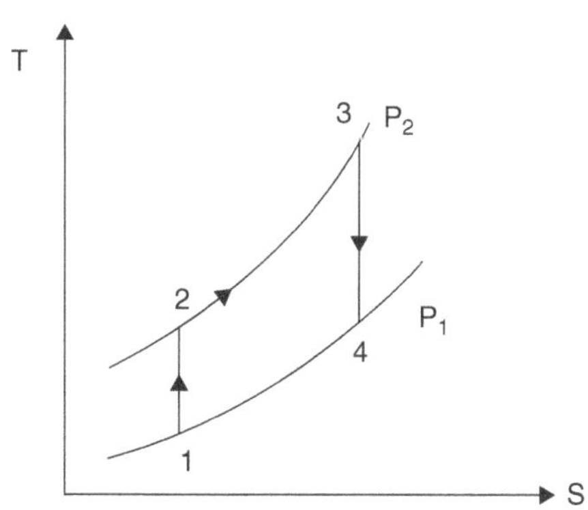

FIGURE 3.2.8 Theoretical Brayton cycle.

In a closed cycle, the introduction of two exchangers has the effect of increasing costs and lowering heat exchange effectiveness. However, it allows a significant increase in the average pressure of the cycle, and thus its specific power. All things considered, it is justified only in special cases such as power production from nuclear reactors.

Its efficiency is:

$$\eta = \frac{(h_3 - h_4) - (h_2 - h_1)}{h_3 - h_2}$$

The working fluid being assumed perfect, we have: $h = c_p\, T + \text{Const}$, or $\Delta h = c_p\, \Delta T$.

Moreover, on isentropes 1–2 and 3–4 of Figure 3.2.8, we have:

$$\frac{P_2}{P_1} = \frac{P_3}{P_4} = \left(\frac{T_1}{T_2}\right)^{\gamma/(\gamma-1)} = \left(\frac{T_3}{T_4}\right)^{\gamma/(\gamma-1)}$$

$$\eta = 1 - \frac{T_1 - T_4}{T_2 - T_3}$$

$$\eta = 1 - \left(\frac{P_2}{P_1}\right)^{(1-\gamma)/\gamma} \tag{3.1.1}$$

We see that, assuming first that the fluid follows the perfect gas law, and second there is no irreversibility, the efficiency is an increasing function of the compression ratio.

3.2.2.2 Perfect Gas Cycle with Irreversibilities

In reality, we know that the compression and expansion are not isentropic but rather irreversible adiabatic processes (Figure 3.2.9).

Even if we keep the assumption of perfect gas, it is desirable to refine the model by taking into account irreversibilities.

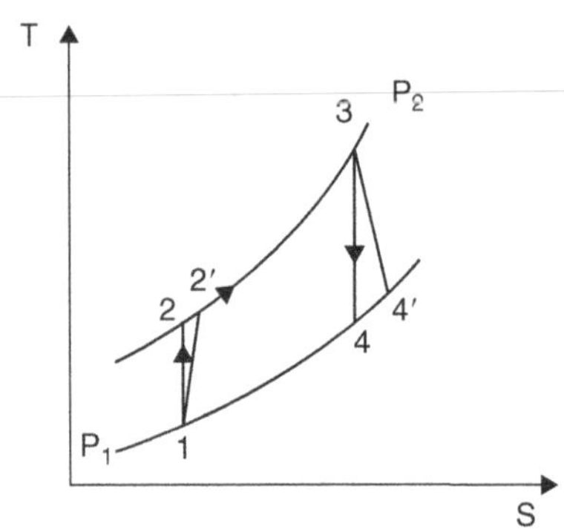

FIGURE 3.2.9 Actual open Brayton cycle.

3.2.2.2.1 Isentropic Formulation

Taking into account the isentropic efficiencies of the compressor η_c (between 0.88 and 0.92), and of the turbine η_t (near 0.86), the cycle efficiency is given by equation 3.1.2.

$$\eta = \frac{\eta_t(h_3 - h_4) - \left(\dfrac{h_2 - h_1}{\eta_c}\right)}{h_3 - h_{2'}} \tag{3.1.2}$$

Introducing ratios:

$$\theta = \frac{T_3}{T_1}, \quad \beta = \frac{\gamma - 1}{\gamma}, \text{ and } r = \frac{P_2}{P_1}, \text{ we get :}$$

$$h_2 - h_1 = c_p T_1 (r^{\beta} - 1)$$

$$h_3 - h_4 = c_p T_3 (1 - r^{-\beta})$$

$$h_3 - h_4 = c_p T_1 \theta (1 - r^{\beta})$$

$$h_3 - h_{2'} = c_p (T_3 - T_{2'})$$

$$h_3 - h_{2'} = c_p T_1 \left(\theta - 1 - \frac{r^{\beta} - 1}{\eta_c}\right)$$

All calculations done, we get:

$$\eta = \frac{(r^{\beta} - 1)(\eta_c \eta_t \theta - r^{\beta})}{r^{\beta}(\eta_c(\theta - 1) + 1 - r^{\beta})} \tag{3.1.3}$$

The graph in Figure 3.2.10, established for helium, shows that the efficiency drops significantly compared to the theoretical cycle, and there is an optimum compression ratio, which is an increasing function of the ratio θ of the inlet temperatures in the turbine and compressor, and depends on the isentropic efficiencies of these devices and γ, ratio of heat capacities of the working fluid.

3.2.2.2.2 Polytropic Formulation

This way of working is however not the most relevant, especially if one examines the influence of compression ratio r on machine performance, as isentropic compression and expansion efficiencies η_c and η_t are then a function of r. It is preferable to use a polytropic formulation, and this is all the more justified as compressors and turbines are most often multistage axial machines, whose polytropic efficiency is substantially equal to the isentropic efficiency of a stage.

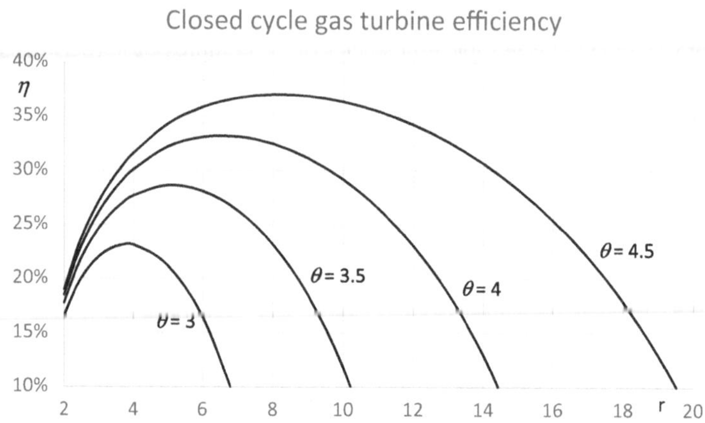

FIGURE 3.2.10 Efficiency of the closed Brayton cycle.

Introducing ratios βc for the compressor and βt for the turbine:

$$\beta_c = \frac{k_c - 1}{k_c} = \frac{1}{\eta_{pc}} \cdot \frac{\gamma - 1}{\gamma}$$

$$\beta_t = \frac{k_t - 1}{k_t} = \eta_{pt} \frac{\gamma - 1}{\gamma}$$

We get:

$$T_{2'} = T_1 \left(\frac{P_2}{P_1} \right)^{\beta c} = T_1 r^{\beta c}$$

$$T_{4'} = T_3 \left(\frac{P_1}{P_2} \right)^{\beta t} = T_1 \theta r^{-\beta t}$$

The efficiency of the Brayton cycle with irreversibilities is equal to:

$$\eta = \frac{(T_3 - T_{4'}) - (T_{2'} - T_1)}{T_3 - T_{2'}}$$

$$\eta = \frac{\theta(1 - r^{-\beta t}) + 1 - r^{\beta c}}{\theta - r^{\beta c}} \tag{3.1.4}$$

The output power can be expressed in dimensionless form by dividing its value by the product cp (T3 − T1) = cp T1 (θ − 1).

We get:

$$W_0 - \frac{\theta(1 - r^{-\beta t}) + 1 - r^{\beta c}}{\theta - 1} \tag{3.1.5}$$

From these two expressions, it is possible to plot the diagram of Figure 3.2.11, which shows for various values of θ and r curves binding η and the reduced power Wo. This diagram shows that the compression ratio value leading to maximum power output is not that which leads to best efficiency.

The maximum efficiency is obtained for compression ratios greater than those which result in maximum power, because for a given turbine inlet temperature, the temperature of expanded gas is lower the higher the compression ratio.

As the upper curves are relatively flat, manufacturers generally prefer to optimize their engines for maximum power, even if efficiency slightly drops.

These curves also show that the maximum efficiency is an increasing function of θ. To obtain high efficiencies, it is thus necessary to increase as much as possible the turbine inlet temperature.

These theoretical developments are very interesting, especially for helium gas turbines for which the ideal gas hypothesis is verified, but they find their limits when the cycles are more complex, in which case it becomes preferable to use a simulation environment.

In the example of interest, which may correspond to a single gas turbine cycle coupled to an HTR (Figure 3.2.12), a flow rate of 140 kg/s of helium enters at point 1 into a compressor with a polytropic efficiency of 0.9, a pressure of 24.7 bar, and a temperature of 30°C. It is compressed to 70 bar and then directed to the reactor core, from which it exits at 900°C at point 3, and at 66.5 bar.

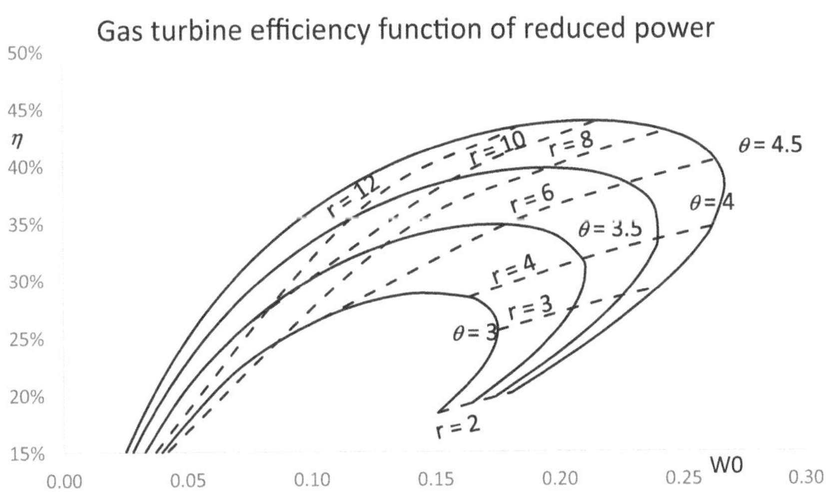

FIGURE 3.2.11 Efficiency and reduced power of a gas turbine.

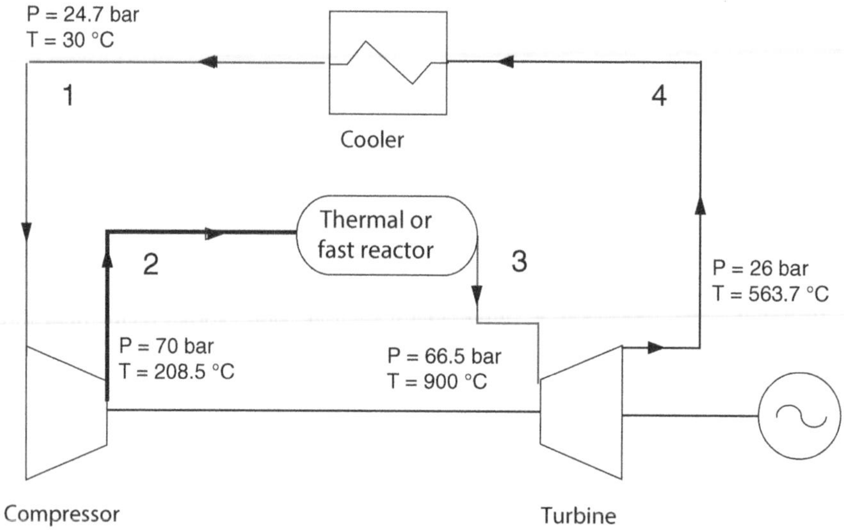

FIGURE 3.2.12 Gas turbine cycle parameters.

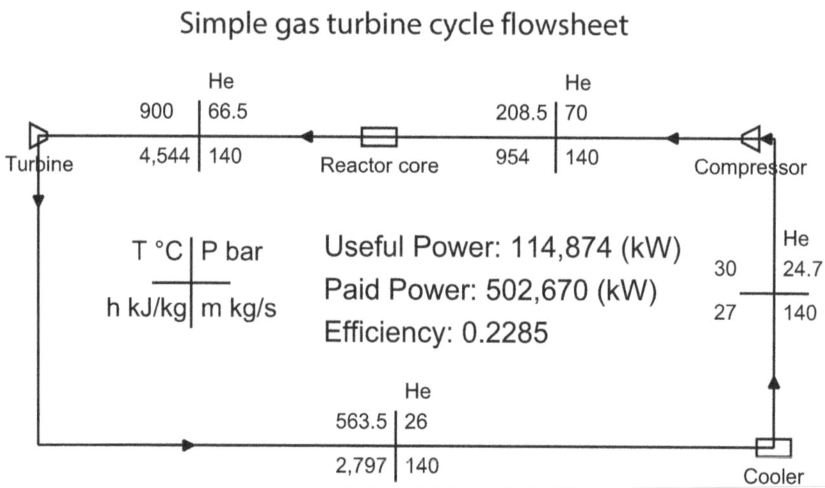

FIGURE 3.2.13 Gas turbine cycle flowsheet.

This flow is expanded to 26 bar in a turbine (3–4) coupled to the compressor, and then cooled to 30°C at point 1, which closes the cycle.

As with the steam cycle, this one can be represented by the flowsheet in Figure 3.2.13.

3.2.2.3 Plot of the Cycle in the (h, P) Thermodynamic Diagram

The cycle plot in the (h, P) enthalpy diagram is shown in Figure 3.2.14. The one in the entropy diagram will be presented in Section 3.3.1.

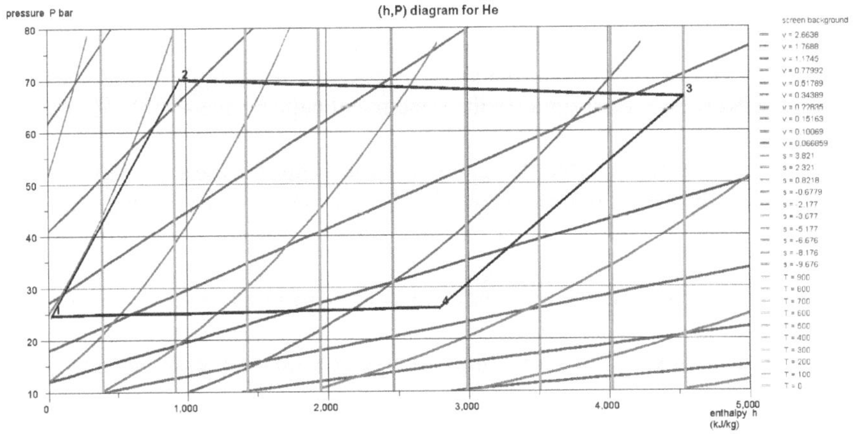

FIGURE 3.2.14 Cycle plot in the (h, P) enthalpy diagram.

The pressure drops in the reactor and the cooler, which are not negligible here, are clearly visible.

3.2.2.4 Technological Aspects

As with steam power plants, we are content to provide here some indications on the technologies implemented in closed-cycle gas turbines.

The development of gas turbines is much more recent than that of steam power plants. It started around the 1930s but only really accelerated during the Second World War, with the development of turbojets, which are machines derived from gas turbines. The main reason for this slow and late start was that the isentropic or polytropic efficiencies of the first turbocompressors and turbines were poor, so the overall performance of these machines was low.

Over the past 70 years, considerable progress has been made in developing high-performance turbomachinery and turbines capable of withstanding increasingly high temperatures.

For open-cycle gas turbines, a common distinction is made between industrial machines and aeroderivative ones. This classification does not apply to closed-cycle gas turbines, which are rather characterized by the working fluid (helium, nitrogen, supercritical CO_2, etc.) and by their field of application (nuclear, solar, or space power systems).

One distinctive feature of aeroderivative open-cycle gas turbines is their compact design with concentric shafts. In contrast, closed-cycle gas turbines are usually built around a single-shaft configuration, prioritizing robustness and thermomechanical reliability over extreme compactness, except in space applications where compact integration of turbomachinery and generator becomes a major design driver.

The goal is to simplify the machine, reduce the number of moving parts, and minimize leaks, which is crucial for closed systems. The design of bearings and sealing systems is also simplified.

In summary, while both types of machines can be very compact, they achieve this result through different design choices: concentric shafts for aeroderivatives to optimize efficiency at different rotational speeds, and generally a single shaft for a closed-cycle gas turbine to simplify the system and better manage leakage and pressure constraints.

In addition, there is another technological constraint for these machines, which limits the possibilities of innovation for closed-cycle gas turbines. We have indicated that it has taken several decades to achieve high performance in combustion gas turbines.

Turbomachinery manufacturers now know how to build high-performance gas turbines using gases that have thermodynamic and transport properties close to those of air and flue gases, but they have little experience with other gases such as helium or CO_2, which are now considered potentially interesting.

It will therefore take many years, perhaps decades, to develop machines as efficient as open-cycle machines, even if the modeling and calculation capabilities are now much more powerful than in the past.

We will study in Chapter 6 the improvements that can be made to the simple gas turbine cycle and the corresponding specific technological constraints.

3.3 COMPARISON WITH THE CARNOT CYCLE IN THE ENTROPY DIAGRAM

Since the Carnot cycle achieves the best efficiency, comparing actual cycles to it is nearly always instructive. This section performs such a comparison for the two simple cycles studied.

3.3.1 Steam Power Plant

In the water entropy diagram (Figure 3.3.1), points 1 and 2, representing compression in the liquid state, are nearly superimposed, and heating in the liquid state almost coincides with the liquid saturation curve (2–3a). Vaporization occurs along a horizontal line segment (3a–3b).

The isobaric superheating (3a–3) leads to the cycle's maximum temperature, and the irreversible expansion (3–4′) is reflected by an increase in entropy; point 4 is located within the vapor-liquid equilibrium zone (quality equal to 0.816).

To plot the Carnot cycle (A–B–C–D), we assume $T_1 = 600°C$ and $T_2 = 15°C$. Comparing both cycles yields the following observations:

The condenser (process 4′–1) is a finite-size heat exchanger; condensing water cannot reach the same temperature as the cold source, which represents a first difference from the Carnot cycle. However, the heat exchange can be considered nearly isothermal with minimal irreversibilities.

Compression (process 1–2) in the pump may be approximated as isentropic at first; its irreversibilities are low. However, it deviates significantly from the Carnot cycle because the compression end temperature approximates that of the cold source rather than the hot source.

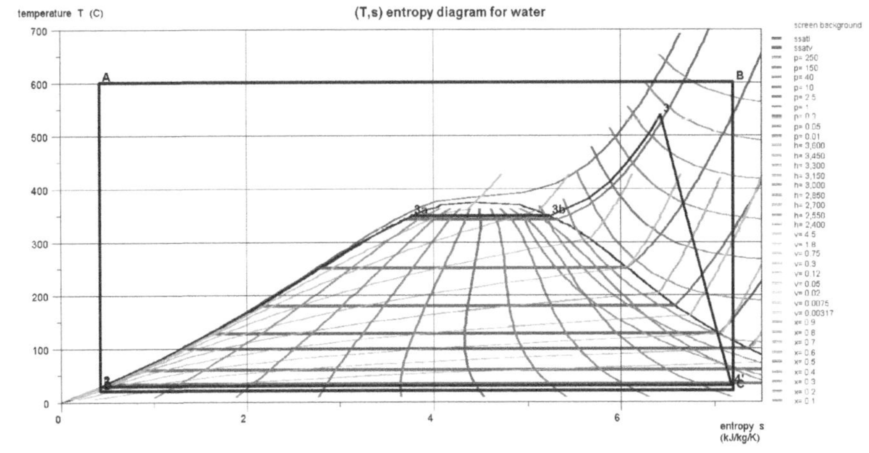

FIGURE 3.3.1 Comparison of the steam cycle with the Carnot cycle.

Isobaric pressurized water heating occurs in three steps:

- Liquid heating in the economizer to the saturation temperature at the operating pressure (process 2–3a);
- Vaporization at constant temperature in the evaporator (process 3a–3b);
- Superheating to the maximum cycle temperature in the boiler (process 3b–3).

Given the shape of isobars on the entropy diagram, this isobaric heating process deviates considerably from the Carnot cycle, which requires that the heat engine exchange heat at constant temperature with the hot source. In particular, the temperature difference with the hot source is maximum in the economizer: high-temperature heat is used to warm water below 100°C.

The turbine (process 3–4′) has a polytropic efficiency of 0.853. Again, the difference from the Carnot cycle is significant.

As illustrated in Figure 3.3.1, the shape of the steam power plant cycle differs markedly from the Carnot cycle: it resembles a triangle rather than a rectangle, and its area is much smaller. There is substantial room for improvement, as will be shown later.

This cycle can be improved through reheats that enable so-called para-isothermal expansion, and extractions that allow partial regeneration.

3.3.2 Gas Turbine

In the helium entropy diagram (Figure 3.3.2), irreversible compression and expansion result in entropy increases (1–2 and 3–4). Heating in the reactor is isobaric, except for the pressure drops.

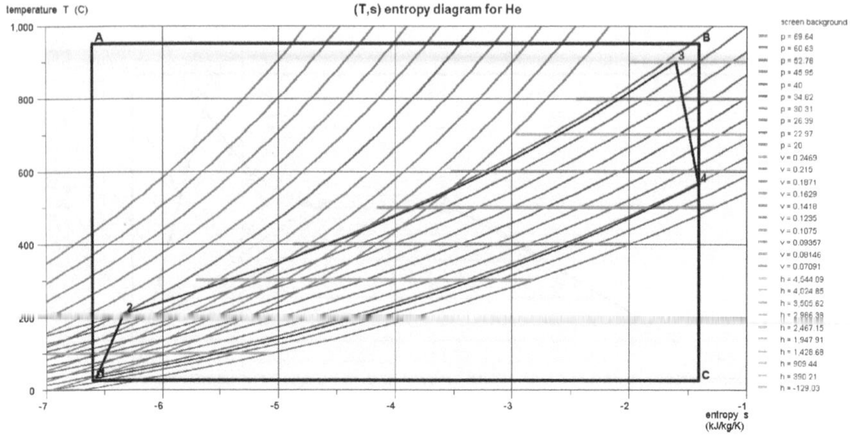

FIGURE 3.3.2 Comparison of the gas turbine cycle with the Carnot cycle.

To plot the Carnot cycle (A–B–C–D), we assume T_1 = 950°C and T_2 = 25°C. Comparing both cycles yields the following observations:

1. Compression (process 1–2) cannot be assumed isentropic due to irreversibilities in the compressor. It departs from the Carnot cycle.
2. The heating (process 2–3) occurs at almost constant pressure. Given the exponential-like shape of isobars on an entropy diagram, the difference between this process and the Carnot cycle is significant, as the ideal cycle requires heat exchange at constant temperature with the hot source.
3. Expansion (process 3–4): the turbine has a polytropic efficiency of 0.9. Again, the difference from the Carnot cycle is substantial.
4. The cooling (process 4–1) occurs at almost constant pressure. Given the shape of isobars on an entropy diagram, the difference between this process and the Carnot cycle is substantial.

As illustrated in Figure 3.3.2, the shape of the gas turbine cycle differs markedly from the Carnot cycle: it resembles a skewed parallelogram rather than a rectangle, and its area is much smaller. Improvements can be achieved through cooled staged compression, staged expansion with reheats, regeneration, and combined cycle configurations.

Thus far, we have studied two simple power cycles: the steam power plant and the gas turbine. We will now examine how they can be improved, with the objective of minimizing irreversibility.

In practice, cycle modifications essentially relate to:

- Reducing temperature differences both with the system surroundings and internally
- Splitting compressions and expansions

3.4 METHODOLOGICAL TOOLS FOR ANALYSIS AND IMPROVEMENT

Improving the efficiency of steam and gas power cycles requires not only a strong grasp of thermodynamic principles but also effective methods and tools to analyze and optimize system performance. This section introduces the core methodological approaches used throughout this book, establishing the foundation for the detailed applications that follow.

While many references focus either on theoretical concepts or specific technologies, our approach aims to provide readers with practical and operational tools directly applicable to real-world systems. We concentrate on three main tools included in the Thermoptim software:

- The functional structure environment, which supports thermodynamic modeling and simulation
- The exergy analysis framework using exergy structures, which offers a powerful method to evaluate irreversibilities
- The pinch method, which helps design efficient heat exchanger networks through thermal integration

Other approaches—such as basic energy balances or techno-economic analyses—exist and are sometimes useful, but they are not developed here, as our focus is on thermodynamic performance improvement.

These tools will be applied systematically to the two reference cycles previously introduced. They are presented not in isolation but as part of an integrated methodology combining simulation, analysis, and design. The aim is to provide the reader with both the means to model power cycles and the perspective to improve them, based on rigorous principles and clear performance criteria.

Each method comes with its own assumptions and limitations. The Thermoptim simulator, while powerful and flexible, abstracts away certain physical details such as fluid dynamics. Exergy analysis assumes steady-state operation and relies on appropriate environmental references. The pinch method requires well-defined temperature and flow rate data to construct composite curves.

Using a simulation tool is necessary for several reasons: calculating fluid properties, handling component complexity, and managing the number of components in real cycles.

We emphasize that the objective of this work is educational and that actual installations use far more sophisticated industrial software.

Taken together, however, these tools offer a robust and accessible framework for analyzing and enhancing thermodynamic cycles—one that is both theoretically sound and practically implementable.

We limit ourselves here to a brief presentation of these methodological tools. Further information is available on this book's website.[5]

3.4.1 Introduction to Thermoptim

Applied thermodynamic calculations are often complex for two main reasons:

- Many governing laws are highly nonlinear.
- Substance properties are difficult to model accurately; even the simplest case, the ideal gas law PV = nRT, involves nonlinear hyperbolic relationships, contrasting with proportional laws in mechanics or electricity. Real fluids require far more complex equations of state and tabulated thermodynamic data.

Given these challenges, particularly when conducting in-depth analyses or optimizations, dedicated computational tools are highly advisable.

We provided a brief introduction to Thermoptim in the first chapter, indicating the various resources available to help you become acquainted with the software. Therefore, we will not revisit this here and assume you have sufficient understanding of the software to follow subsequent developments.

We recall that, to help you master this tool, we have developed approximately forty-five guided explorations of models with progressive difficulty. These guided explorations can be viewed in a particular browser called the Thermoptim Browser, which can directly emulate the software package.[6]

3.4.2 Thermoptim Model of the Steam Cycle

Modeling the steam cycle in Thermoptim presents no particular difficulty.

Simply place the four components on the diagram screen and configure each of them by assigning a name, defining their inlet and outlet points, and selecting the working fluid from the available options. The next step is to connect these components together, thereby defining the diagram. By double-clicking on any component, you open its configuration window, where you can specify the flow rates, temperatures, pressures, and isentropic or polytropic efficiencies. In just a few minutes, the entire cycle can be defined and configured.

The steam cycle is a variant of the one presented in guided exploration S-M3-V7: Exploration of a Simple Steam Power Plant.

The plant capacity is approximately 654 MW, slightly less than the information provided by available data. The efficiency is also slightly lower.

These differences are partly explained by cycle simplification. We will discover later where they originate.

Figure 3.4.1 shows how this steam plant cycle appears in the diagram editor.

As discussed in the next chapter, this cycle can be improved by approximately 20% through various methods, including reheat and extraction. Note that the cycle maximum temperature and pressure, set here at 538°C and 163 bar, are limited by nuclear reactor technological constraints and the durability of the boiler materials, typically steel for cost reasons.

3.4.3 Thermoptim Model of the Gas Turbine Cycle

Modeling the simple helium gas turbine cycle in Thermoptim also presents no difficulty: operate in a manner similar to that used for the steam cycle. The model yields the synoptic view presented in Figure 3.4.2.

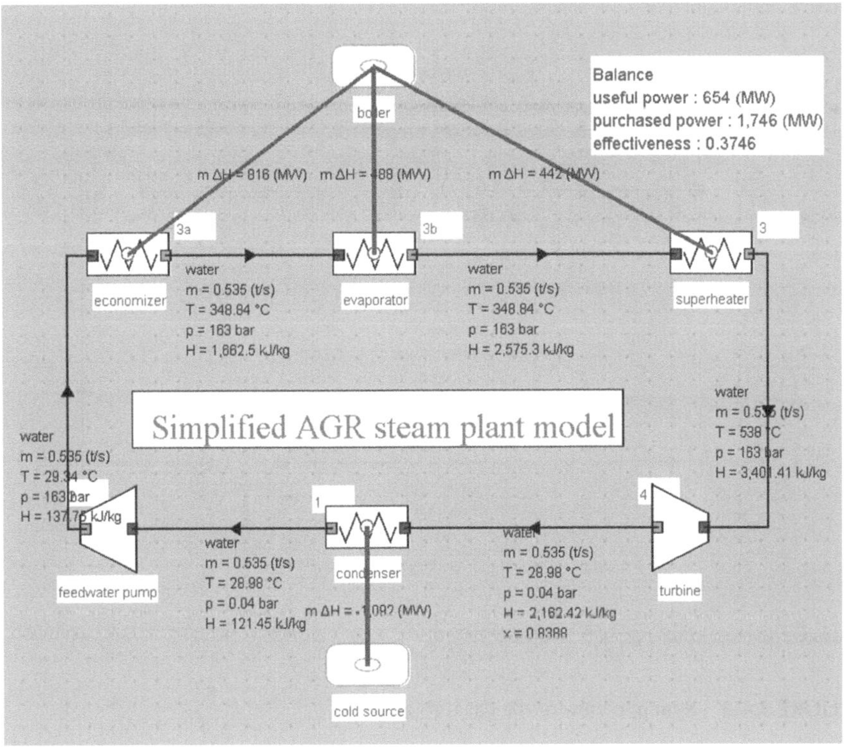

FIGURE 3.4.1 Synoptic view of the power plant.

The gas turbine cycle is the subject of guided exploration GT_He: Helium Gas Turbine.

In this cycle, pressure drops in the two heat exchange processes were considered non-negligible and assumed equal to approximately 5% of the inlet pressure.

The overall efficiency is quite low: 22.85%. We will examine how it can be improved.

The plant capacity is approximately 115 MW.

3.4.4 Exergy Balances, Exergy Structures

To thoroughly analyze and improve steam and gas power cycles, engineers must go beyond basic thermodynamic principles and apply specialized methodologies and tools. This section introduces two key approaches that serve as the foundation for analyses developed throughout this book: exergy-based analysis through exergy structures and the pinch method for thermal integration.

Each tool addresses specific challenges in energy system design and optimization: exergy analysis provides a rigorous framework to assess energy quality and system efficiency, while the pinch method supports optimal heat exchanger network design.

Together with the Thermoptim diagram editor and simulator, they form a powerful methodological toolkit enabling the identification and implementation of meaningful improvements in power cycle performance.

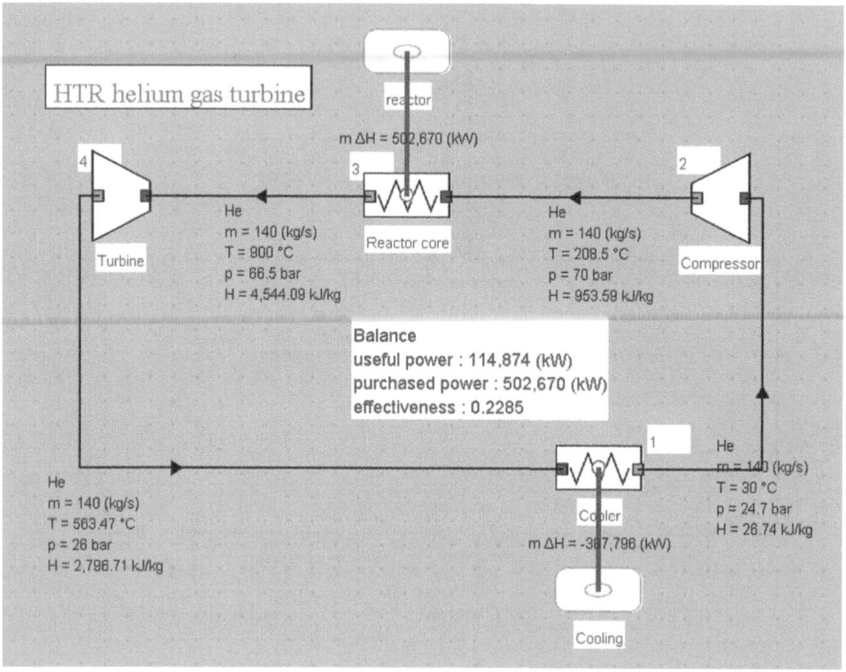

FIGURE 3.4.2 Synoptic view of the gas turbine cycle.

3.4.4.1 Exergy Balances

Exergy theory provides an integrated framework encompassing both the First and Second Laws of thermodynamics, enabling consideration not only of the quantity of energy involved but also its quality (something the First Law alone cannot account for). Its main advantage lies in offering a rigorous thermodynamic foundation for evaluating energy quality within any system, whether open or closed, and whether operating under steady-state or transient conditions.

Exergy depends on both the system state and the chemical composition of its elements. Unlike energy, which is conserved, exergy is destroyed whenever irreversibilities exist. Since it is not a conservative quantity, a specific approach is needed to establish its balance.

Using the notations introduced in Volume 1 of this set, we obtain the following: for an open system, the function $x_h = h - T_0 s$ is commonly referred to as the *exergy*, where T_0 is the temperature at the system boundary, h the enthalpy, and s the entropy.

The term Δx_{h_i} represents the *exergy dissipation* due to irreversibilities.

We define the Carnot factor as $\theta = 1 - T_0 / T$. It expresses the fraction by which a quantity of heat Q, available at temperature T, must be multiplied to obtain its exergy value. The *heat exergy* (or *useful heat*) is thus given by $x_q = \theta Q$.

We recall some results:

$$-\tau = \sum_{k=1}^{n} x_{qk} - \Delta x_h - \Delta x_{hi} \tag{3.4.1}$$

The maximum work that an open system can provide equals the sum of heat exergies from sources with which it exchanges heat, minus the change in exergy of the fluid passing through it and the exergy destroyed due to irreversibilities.

For an open multitherm system in the steady state traversed by a constant fluid flow, which exchanges work and heat with n external sources at constant temperatures T_k, and the environment at temperature T_0, the exergy equation (3.4.1) can be rewritten as:

$$\Delta x_{hi} = \tau - \Delta x_h + \sum_{k=1}^{n} x_{qk} \tag{3.4.2}$$

Δx_{hi} represents the exergy dissipation from irreversibilities, τ the work received by the system, Δx_h the exergy variation of the fluid passing through it, and x_{qk} is the heat exergy received from the source at temperature T_k.

We specify that what we call irreversibilities includes both the exergy destroyed and the exergy lost by the system.

When heat sources are no longer at constant temperature, when multiple shaft systems must be considered, and when the system is traversed by multiple fluids, equation (3.4.2) must be replaced by a more complex one.

To calculate irreversibilities of each component j, it becomes (3.4.3):

$$\Delta x_{hij} = \sum_{k=1}^{n} x_{qjk} - m_j \Delta x_{hj} + \tau_j \tag{3.4.3}$$

If heat exchange occurs at variable temperature:

$$x_{qk} = \int_{T_a}^{T_b} \left(1 - \frac{T_0}{T_k}\right) \delta Q_k \tag{3.4.4}$$

Assuming a linear variation of heat exchange as a function of T:

$$x_{qk} = \left(1 - \frac{T_0}{T_b - T_a}\right) \ln\left(\frac{T_b}{T_a}\right) Q_k$$

To calculate irreversibilities in a heat exchanger, simply add equations (3.4.3) for each fluid flowing through it. If it is adiabatic (no heat exchange with surroundings), heat exergies are zero. Since useful work is zero, the irreversibility equals the sum of exergy variations of both fluids.

Exergy allows rigorous definition of system efficiency and therefore quantification of its thermodynamic quality: it is the ratio of exergy uses to exergy resources.

It is always between 0 and 1, and is higher when irreversibilities are lower. Exergy resources are the sum of all exergies provided to the cycle from outside. Exergy uses represent the net cycle balance, i.e., the algebraic sum of exergy produced and consumed within it.

$$\eta = \frac{\text{exergy uses}}{\text{exergy resources}}$$

If the system consists of components satisfying equation (3.4.3), this general definition is mathematically translated by equation (3.4.5):

$$\eta_x = 1 - \frac{\Sigma \Delta x_{hij}}{\Sigma \Delta x_{qi}^{+} + \Sigma \tau_j^{+}} \tag{3.4.5}$$

The denominator represents the exergy supplied, i.e., the sum of positive heat exergies and useful work provided to the cycle. The numerator is the sum of cycle irreversibilities.

3.4.4.2 Exergy Structures

Using a spreadsheet, constructing the exergy balance of a complex system can prove challenging in practice.

To address this difficulty, exergy balances can be automated, even for intricate systems, by employing a graphical representation known as a productive structure, introduced by Valero and collaborators (Valero et al., 2000), which corresponds to the exergy structure introduced in Chapter 1.[7]

An exergy structure is a graph representing the system's exergy production and consumption pathways. It can be derived from the functional structure of the system, with the addition of a few key details: mechanical power connections, potential useful output recovery, and temperatures of external heat sources and sinks.

The structure comprises Productive and Dissipative Units (PDUs), as well as junctions and branches:

- Each PDU is associated with a specific product flow but may also generate secondary outputs or byproducts.
- These units consume either external resources or intermediate products supplied by other units.
- At each junction, a product flow results from the combination of similar incoming flows from various sources.

This graph-based approach provides a systematic way to visualize and quantify exergy interactions within a system and to identify inefficiencies.

The exergy structure of the AGR steam plant is given in Figure 3.4.3.

It can be interpreted as follows: the steam power plant is a machine receiving an external exergy input in the boiler, corresponding to the three productive units at the top left of the screen, and through internal recycling, exergy is provided to the pump at the bottom left.

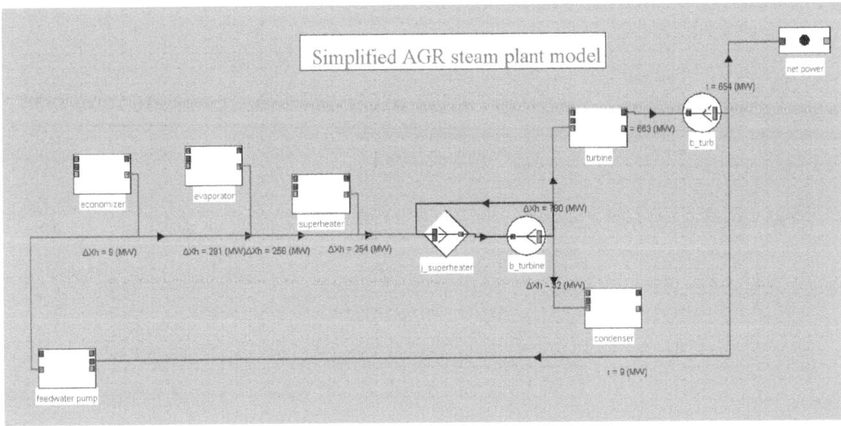

FIGURE 3.4.3 Exergy structure of the AGR steam plant.

This exergy is partly converted into mechanical form in the turbine and partly dissipated in the condenser. The net power is the fraction of mechanical power not recycled.

Note that exergy flow values can be displayed on the exergy structure.

Figure 3.4.4 shows the correspondence between the functional and exergy diagrams.

Once the exergy structure is established, the exergy balance can be automatically deduced.

The only information that must be added to the Thermoptim functional and exergy structures is the temperature of the sources with which the power plant exchanges heat and the mechanical coupling between the turbine and the pump. We have assumed it is 20°C for the condenser and 700°C for the nuclear reactor core, as the peak coolant (CO_2) temperature is limited to approximately 660°C.

Figure 3.4.5 shows the exergy balance of the steam plant.

This balance confirms what the comparison with the Carnot cycle demonstrated, namely that in such a cycle, the greatest irreversibility occurs in the economizer due to temperature differences with the hot source.

Analyzing the irreversibility distribution in this balance reveals that nearly half (51.96%) occurs at the economizer level.

As previously indicated, one great advantage of exergy balances is that they quantify irreversibilities, whereas comparisons with the Carnot cycle in the entropy diagram provided only qualitative information.

In particular, the penultimate column of this balance provides the irreversibility distribution among different system components.

Guided explorations BESP1 and BESP2 explain practical operations when using exergy structures to build exergy balances.

3.4.4.2.1 Implementation in Thermoptim

A specific editor with the various components used in exergy structures has been developed in Thermoptim versions 2.6 and 2.8 (Figure 3.4.6).

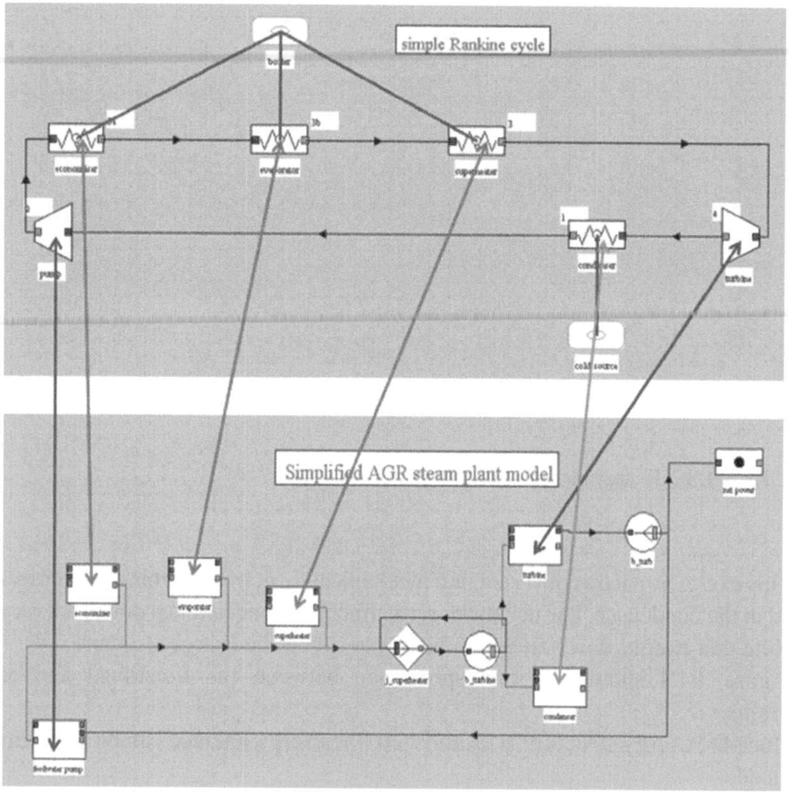

FIGURE 3.4.4 Correspondence between the functional and the exergy diagrams.

component	Resource	Product	Exergy efficiency	Irreversibilities	% total	settings
feedwater pump	8.72	8.722	100.0000%	0	0	
turbine	780.182	662.86	84.9622%	117.322	20.7304%	
condenser	32.472	0	0	32.472	5.7377%	Tk = 20.00 °C
economizer	570.008	291.426	51.1267%	278.582	49.2244%	Tk = 700.00 °C
superheater	308.832	254.356	82.3606%	54.476	9.6257%	Tk = 700.00 °C
evaporator	341.24	258.15	75.6504%	83.091	14.6819%	Tk = 700.00 °C
global	1,220.08	654.137	53.6143%	565.943	1	

FIGURE 3.4.5 Exergy balance of the steam plant.

It is available from the Special menu of the simulator (Ctrl+B). The palette includes process-point icons to represent inlet and outlet flows, PDUs, junctions, and branches.

You construct an exergy structure either manually by selecting palette icons, placing them on the workspace, and connecting them with links—analogous to diagram construction—or by automatically generating the exergy structure from the simulator and physical diagram.

Each diagram component corresponds to a PDU. These are connected to links representing exergy variation fluxes in the studied system.

Figure 3.4.7 shows the exergy structure of the helium gas turbine studied earlier.

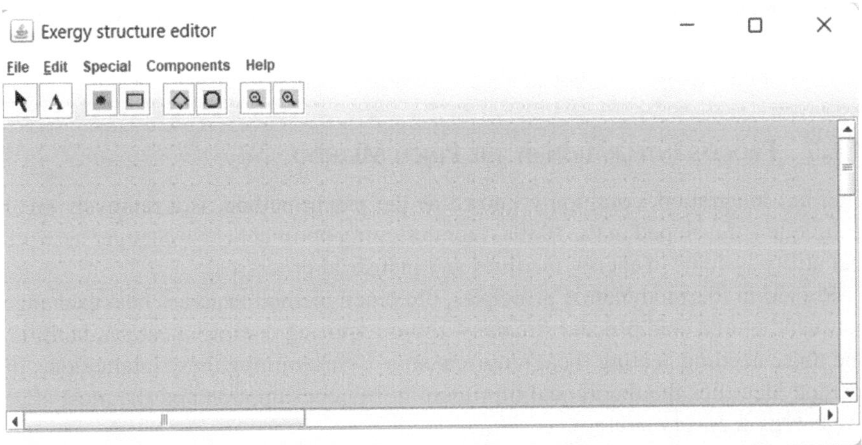

FIGURE 3.4.6 Exergy diagram editor.

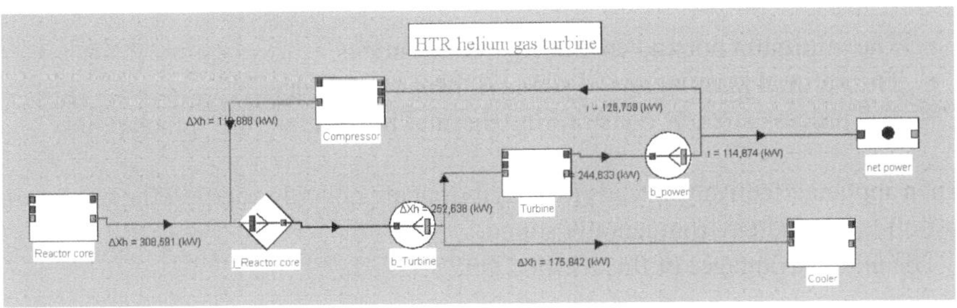

FIGURE 3.4.7 Helium cycle exergy structure.

component	Resource	Product	Exergy efficiency	Irreversibilities	% total	settings
Compressor	129,759.03	119,888.462	92.3932%	9,870.568	3.6924%	
Turbine	252,638.126	244,632.89	96.8313%	8,005.236	2.9946%	
Cooler	175,841.714	0	0	175,841.714	65.7790%	Tk = 20.00 °C
Reactor core	382,195.975	308,591.379	80.7417%	73,604.596	27.5340%	Tk = 950.00 °C
global	382,195.975	114,873.861	30.0563%	267,322.114	1	

FIGURE 3.4.8 Helium cycle exergy balance.

This exergy structure can be interpreted as follows: the gas turbine receives a heat contribution from the nuclear reactor core and, through internal recycling, an exergy contribution from the compressor. This exergy is converted into mechanical form in the turbine, with the remainder dissipated through discharge via the cooler. Net power corresponds to the fraction of mechanical power not recycled.

Note that the compressor is by default counted as an external resource. To correct this setting, we must connect it downstream of the turbine as shown.

The project exergy balance is given in Figure 3.4.8. Note that the nuclear reactor is assumed to be a heat source at 950°C.

Heat rejected from the cooler accounts for more than 65% of irreversibilities. These could be reduced by using a regenerative cycle.

Additional explanations are provided in the fourth volume of the reference manual and on the Thermoptim-Unit portal.[8]

A dedicated getting started guide is also available.[9]

3.4.5 Process Integration by the Pinch Method

Thermal integration, commonly known as the pinch method, is a relatively recent technique—developed in the 1980s—for designing optimal heat exchanger networks and utility systems in energy facilities and industrial processes.

Rooted in thermodynamic principles, the pinch method analyzes heat exchanges between hot and cold process streams—those requiring cooling (heat availabilities) and those needing heating (heat requirements). By examining these interactions, the method identifies the theoretical minimum utility consumption and the most effective energy recovery strategy.

The pinch method allows engineers to minimize internal irreversibilities within the heat exchanger network, thereby improving its overall thermodynamic performance. It also helps determine:

- The minimum hot and cold utility requirements
- The optimal placement and sizing of heat exchangers
- The process streams constraining thermal integration (the "pinch points")

When applied effectively, it leads to designs that are not only energy-efficient but also cost-effective and environmentally sound.

The main advantages of this method are:

- It is a visual and graphic method, allowing the engineer to maintain a physical understanding of the phenomena, whereas optimization methods are often purely numerical.
- Most importantly, optimization is performed without any *a priori* assumptions about heat exchanger network configuration, which is defined only later.
- It has become widespread over the past 50 years because it has proven capable of reducing **both capital costs and operating expenses**, which is generally not the case.

We limit ourselves here to a few generalities about this method, referring the reader to the Thermoptim-Unit portal.[10]

3.4.5.1 Basic Principles

Consider a system in which matter is transformed by various processes, some exothermic, others endothermic. Each process receives or provides heat. To optimize overall energy expenditure and reduce external inputs, all internally available heat must be utilized optimally. Thermal or heat integration provides a rigorous method for this, optimizing the overall facility configuration and ensuring greater consistency between needs and energy availability.

For simplicity, we present the basic principles of heat integration without using exergy analysis, but this approach can be generalized.

The method is based on identifying a critical point in the system known as the pinch point, where the temperature difference between hot and cold streams is at its minimum (usually defined as ΔT_{pinch}). This point:

- Divides the system into two regions with distinct energy behavior (above and below the pinch)
- Dictates design rules: no external heating should be applied below the pinch, and no external cooling above the pinch

3.4.5.1.1 Pinch Point

The basic principle of heat integration is to classify all heat needs according to their temperature levels to find the best match between hot and cold streams. For this we use a graph with temperature on the ordinate and enthalpy on the abscissa (enthalpy diagram of Figure 3.4.9). Note that for purely thermal exchanges, enthalpies equal heat exchanged. The evaporating fluid corresponds to the lower curve (1–2–3), which presents an angular point 2 corresponding to the boiling onset. The hot fluid is cooled from 4 to 5, decreasing its sensible heat.

On segments (1–2) and (4–5) we have $\Delta H = \dot{m}c_p\Delta T$, and thus $\Delta T = \frac{1}{\dot{m}c_p}\Delta H$.

The slopes of segments (1–2) and (4–5) equal the inverse of the capacity rates of the corresponding fluids. On segment (2–3), $\Delta T = 0$, as vaporization occurs at constant temperature for a pure component or azeotrope.

We observe at point 2 a minimum temperature difference P between both fluids, called the pinch. This point plays a fundamental role in heat exchanger network design, representing the smallest temperature difference in the facility.

To reduce equipment costs, it is preferable that the pinch not be too low. However, thermodynamic optimization of the complete system often requires it to be as small as possible. Generally, the pinch point is the most constrained zone in the thermal system.

On this graph it is clear that below the pinch, the cold fluid segment slope is greater than that of the hot fluid:

$$\frac{1}{\dot{m}c_{pf}} > \frac{1}{\dot{m}c_{pc}} \quad \dot{m}c_{pc} > \dot{m}c_{pf}$$

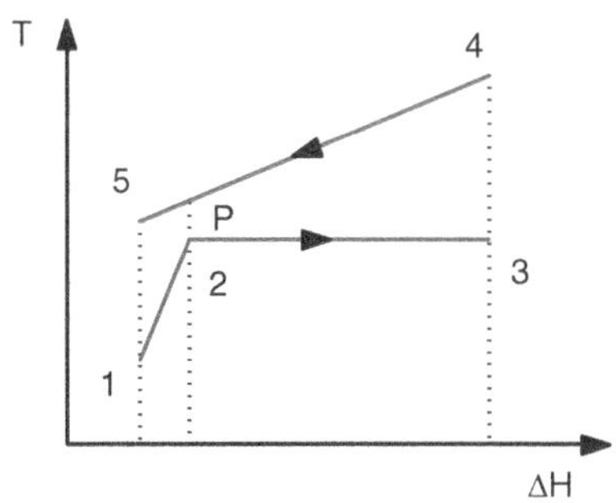

FIGURE 3.4.9 Pinch point.

Above the pinch, the reverse is true: $\dot{m}c_{pc} < \dot{m}c_{pf}$.

The heat exchanger, sized according to the pinch, operates outside this point with larger temperature differences between fluids, indicating increased irreversibility without compensation, which is undesirable. However, given the peculiarity of the exchanger considered here (an evaporator with a vaporization plateau), these irreversibilities cannot be reduced except by using a condensable vapor as the hot fluid; this is precisely what mechanical vapor compression achieves, when possible (see Section 7.2.7).

As discussed below, the pinch point represents a break in the thermal system and leads to an independent study of each zone. In most cases, facilities are designed so that such points are located between elements.

For a more complex system, heat integration involves ranking needs and energy availability as a function of temperature and then plotting them on a (T, h) diagram.

3.4.5.1.2 Composite Curve

The temperature level at which heat may be exchanged being a major constraint in heat exchanger network design, we begin by sorting heat according to this parameter.

For this, we typically assume linear enthalpy variation for sensible heat transfer (assuming constant stream heat capacities). We perform a combination of availability and needs for different system temperature intervals, whose bounds are determined by the original specifications.

We can plot these energies in a (T, h) diagram seeking to maximize enthalpy values according to their respective temperature levels. We overlay the needs and availability curves, shifting them so the former is below the latter.

This yields the diagram in Figure 3.4.10, the so-called composite curves. The top line represents cumulative availability as a function of temperature; the bottom line represents requirements. In the example shown, the availability temperature levels are insufficient to meet the requirements. There is a need for additional heat at high temperature, identified as the "hot target," and a surplus heat termed "cold target," which must be removed by cold utilities, for example through a cooling tower exchanging with ambient air.

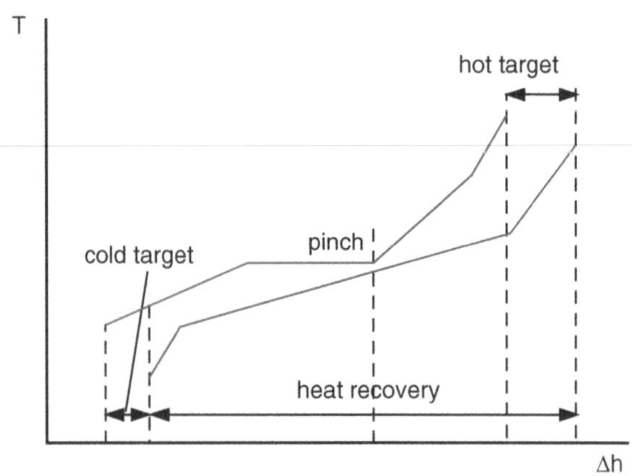

FIGURE 3.4.10 Composite curves.

3.4.5.2 Design of Heat Exchanger Networks

This brief overview has shown the usefulness of thermal integration to visualize irreversibilities in a complex system. It allowed us to highlight temperature differences within heat exchangers but not to define the overall network architecture. That is what we address in this section. For reasons just explained, we use a classical enthalpy approach.

On the composite curve (Figure 3.4.10), the pinch distinguishes two zones:

- The lower zone, where "cold" streams (needs) cannot absorb all available heat in "hot" streams (availability), with surplus heat discharged through complementary coolant. This area, with excess heat, generally behaves as a heat source and is called the exothermic zone or cold-end.
- The upper zone, where the situation is opposite: the heat requirement of "cold" streams exceeds the enthalpy available in "hot" streams, requiring additional heating. This area, lacking heat, behaves as a heat sink and is called the endothermic or hot-end.

As shown in Figure 3.4.11, by shifting the cold composite from right to left, we vary the gap with the hot composite, representing thermal irreversibilities. When the pinch value is zero, no heat transfer exists between these zones: the pinch separates the system into two thermally independent parts. This is obviously a borderline case corresponding to minimum heat absorbed by the endothermic zone and minimum heat rejected by the exothermic zone.

When the pinch value is not zero (Figure 3.4.12), heat absorbed at high temperature and released at low temperature both increase by the same amount α, corresponding to heat passing through the pinch between zones (called cross-pinch heat transfer). Similarly, heat provided in the exothermic zone (β) or cooling in the endothermic zone (γ) increases hot and cold utility requirements. We see an additional advantage of thermal integration: by highlighting the pinch point, it reduces α to minimum

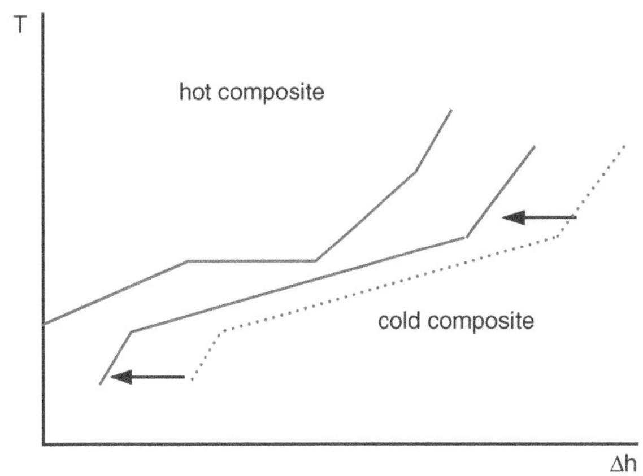

FIGURE 3.4.11 Shifting composite curves.

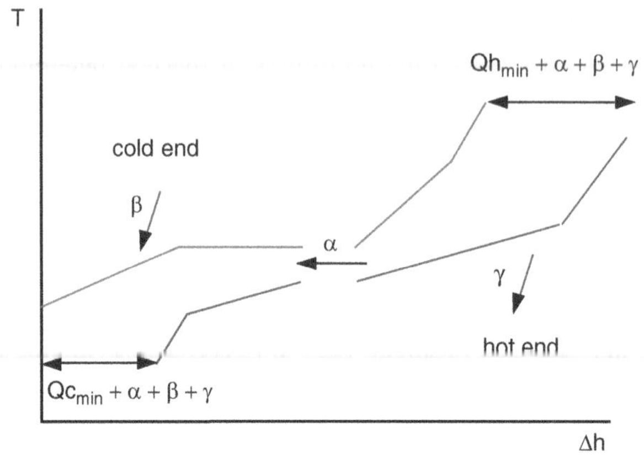

FIGURE 3.4.12 Plus-minus principle.

economic size and therefore allows optimal heat exchanger network design in upper and lower areas. Indeed, an exchanger design error in one area, resulting in increased enthalpy exchanged, inevitably leads to oversizing in the other zone.

The pinch method enables determining an exchange network configuration minimizing heat gain by utilities in the hot endothermic zone and heat extraction by cold utilities in the exothermic area. This configuration, optimal at least according to this criterion, may often require more heat exchangers than other arrangements. It may therefore be necessary to consider alternatives, which is done by relaxing the above criteria.

3.4.5.3 Minimizing the Pinch

Since searching for the pinch by shifting composite curves is neither easy nor precise, Professor B. Linnhoff proposed a method whose main steps we present here. It works directly on enthalpy balances by temperature levels, whereas when constructing the composite curves, we separated the availability curve from the requirements curve.

The pinch minimization algorithm determines the minimum energy consumption corresponding to the minimum pinch, i.e., the hot target. The original Linnhoff algorithm, also called the Problem Table Algorithm, forms the basis of Thermoptim's method.

The algorithm comprises four steps:

- Shift all minimum and maximum temperatures by $\Delta T_{pinch}/2$, subtracting it for hot streams and adding it for cold streams (this ensures that heat exchange between streams at the same shifted temperature can occur)
- Determine problem temperature intervals by sorting upper and lower temperature limits
- Build enthalpy balances (by temperature interval) of hot and cold composites and their difference
- Build the overall summary table and determine the minimum hot utility value (the hot target)

3.4.5.4 Establishment of Actual Composite Curves

Once this target is known, we can return to actual temperatures and recalculate the actual requirements and availability composite curves, reusing part of the pinch minimization algorithm:

- Determine the problem's actual temperature intervals by sorting actual upper and lower temperature bounds.
- Build hot and cold composite enthalpy balances (by temperature interval).
- Plot composite curves. Plotting the hot composite curve poses no problem: we know its origin, which corresponds to the lowest hot stream temperatures and zero enthalpy. Interval by interval, it is constructed without difficulty.
- Moreover, the cold composite curve's maximum enthalpy equals the sum of the hot composite's maximum enthalpy and the minimum target. Its maximum temperature is the highest temperature of cold streams. Interval by interval, the cold composite is built. Its minimum corresponds to the enthalpy value to be evacuated by external cooling (the cold target).

The method will be detailed in Section 7.1 for studying a dual-pressure combined cycle.

You will find in the last section of the first volume of the Thermoptim reference manual a complete presentation of this method and how it is implemented in the software.

3.5 CONCLUSION

This chapter provided an introduction to steam and gas thermodynamic cycles for nuclear power generation. The two reference cycles that we will use subsequently were presented: a steam power plant based on the Hartlepool AGR reactor and a closed-cycle helium gas turbine for HTRs. Comparison with the Carnot cycle revealed the main sources of irreversibility, notably during heat exchanges with the hot source and the compression–expansion processes. Three essential methodological tools were introduced: the Thermoptim simulator for thermodynamic modeling, exergy analysis for quantifying irreversibilities, and the pinch method for heat exchanger network optimization.

NOTES

1 The first LFR-type reactors used water cycles, but those planned for the future consider gas cycles.
2 In this book, we distinguish between an **ideal gas**, which obeys the equation of state PV = nRT, and a **perfect gas**, defined as an ideal gas with constant specific heat. It should be noted that in much of the English-language literature, particularly in North America, the term "ideal gas" is often used for both concepts.
3 The data used to model AGR power plants in this work come from two documents: Nonbel (1996) and Shengqi Zhou and Alan Turnbull (2002).
4 We assume these notions are known by the reader; if not, they are explained in Appendix 1.

5 Available at: https://server.s4e2.com/crc/tcnpp/chapters/chapter3/_resources/index.html.
6 https://direns.minesparis.psl.eu/Sites/Thopt/en/co/explor-scenar.html.
7 Given their similarities, we will indiscriminately talk about productive or exergy structures.
8 https://direns.mines-paristech.fr/Sites/Thopt/en/co/structures-productives.html.
9 https://direns.minesparis.psl.eu/Sites/Thopt/en/co/guide-prise-en-mains-struc-prod.html.
10 https://direns.minesparis.psl.eu/Sites/Thopt/en/co/Optimisation_systemique_methode_pincement.html.

BIBLIOGRAPHY

El-Sayed, A.F. (2008). *Aircraft Propulsion and Gas Turbine Engines.* CRC Press, Boca Raton, FL. ISBN 978-0-8493-9196-5.

Guzović, Z. (ed.) (2023). *Gas and Steam Turbine Power Plants: Applications in Sustainable Power.* Cambridge University Press, Cambridge. ISBN 978-1009201308.

Linnhoff, B. (1982). *User Guide on Process Integration for the Efficient Use of Energy.* Pergamon Press, Oxford.

Linnhoff, B. (1994). "Use pinch analysis to knock down capital costs and emissions." *Chemical Engineering Progress*, 90, 32–57.

Linnhoff, B. & Hindmarsh, E. (1983). "The pinch design method for heat exchanger networks." *Chemical Engineering Science*, 38(5), 745–763.

Nonbel, E. (1996). *Description of the Advanced Gas Cooled Type of Reactor (AGR).* Risø National Laboratory, Roskilde, Denmark.

Valero, A., Serra, L., & Uche, J. (2002). Fundamentals of thermoeconomics (Lectures 1–3). EURO Summer Course on Sustainable Assessment of Clean Air Technologies, CIRCE – Centro de Investigación de Recursos y Consumos Energéticos, Universidad de Zaragoza, Zaragoza, Spain.

Zhou, S. & Turnbull, A. (2002). *Steam turbine operating conditions, chemistry of condensates, and environment assisted cracking – a critical review.* NPL Report MATC (A) 95, May.

4 Improvement of Steam Power Cycles

INTRODUCTION

In the previous chapter, we presented the basic steam power plant cycle and demonstrated its limitations. We will now apply the methodology and tools we have introduced to improve the performance of this cycle.

This chapter presents high-temperature steam power plant cycles, which are primarily used today for centralized electricity generation: variants of the Hirn (or Rankine with superheating) cycle studied previously.

Most steam power plants using these cycles are either conventional (i.e., boiler flames burning mainly coal or oil) or nuclear. However, other hot sources may also be used, although on a much smaller scale: solar energy in solar power plants (see Chapter 8 of Part 4 of the 2021 edition), geothermal energy (see Chapter 16 of the 2021 edition), and gas or liquid waste of any kind, provided their temperature level is sufficient.

We remind you that general technological explanations on turbines are available in Appendix 2. We will therefore limit ourselves in this chapter to giving some specific additions relating to certain particular cycles.

The beginnings of civil nuclear power were marked by the development of four technologies, three of which correspond to water-cooled reactors (WCRs): the PWR and BWR in the USA, and the RBMK and VVER, which is a kind of PWR, in Russia. A few years later, the CANDU was developed by Canada, which did not want to depend on foreign countries for the manufacture of its reactors, whether it was the pressurized vessels of the PWRs or enriched uranium.

The fourth technology is that of gas-cooled reactors, which was developed in France and the UK.

It should also be noted that in Russia the development of a sodium-cooled RNR reactor was initiated very early on, with France following later with SuperPhénix.

In this chapter, we will begin to address the last two technologies, starting with the AGRs. At the end of this chapter, we will say a few words about the SFRs.

The WCRs will be studied in Chapter 5.

4.1 HIGH-TEMPERATURE STEAM POWER CYCLES

4.1.1 Basic Hirn or Rankine Cycle with Superheat

In Section 3.2, we studied the simple Hirn cycle corresponding to the Hartlepool AGR nuclear power plant. We will now examine the modifications to this cycle that may increase its performance.

DOI: 10.1201/9781003606024-5

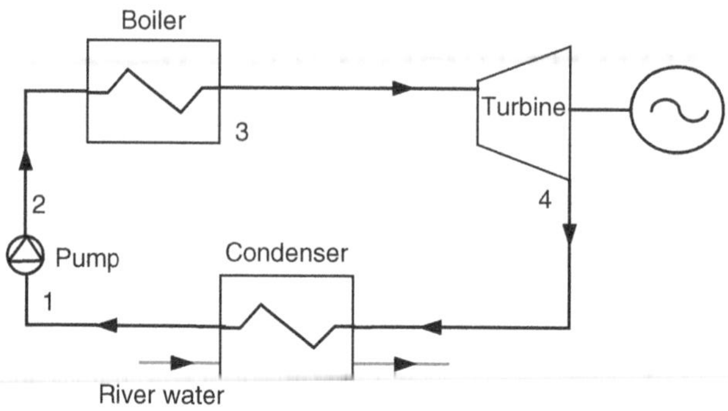

FIGURE 4.1.1 Sketch of a steam power plant.

Let us recall that the Hirn (or Rankine with superheat) cycle uses a condensable fluid, which is cooled at a temperature and pressure sufficient for complete liquefaction before compression. The compressed liquid is vaporized and superheated in the boiler through heat exchange with the hot source, then expanded, and condensed. The fluid's two-phase state during condensation and vaporization phases is very favorable for heat transfer.

The block diagram of a steam power plant operating on the ideal Hirn cycle is given in Figure 4.1.1 (we return to the cycle studied as an example in Section 3.2.1).

Such a plant includes four basic components: a pump, a boiler, a turbine, and a condenser, through which the same water flow rate passes.

Figure 4.1.2 shows the synoptic view of this cycle.

Figure 4.1.3 shows a plot of this cycle in an (h, P) diagram, and Figure 4.1.4 the corresponding flowsheet.

4.1.2 Exergy Balance

We have established the exergy balance of this cycle when we presented the productive structures (Figure 4.1.5).

This balance confirms what the comparison with the Carnot cycle had shown: in such a cycle, the greatest irreversibility is located in the economizer due to the large temperature differences with the hot source.

If you analyze the distribution of irreversibilities given in this balance, you see that nearly half (49.22%) occur at the economizer level, with other losses distributed between the turbine (20.7%), the evaporator (14.7%), the superheater (9.6%), and the condenser (5.7%).

Clearly, the improvement effort must focus mainly on the economizer and the turbine. Much has already been accomplished by manufacturers to achieve the best possible performance from their turbines, and few new gains are now foreseeable.

Methods for improving the steam cycle can be deduced from the exergy balance:

- Increase the average temperature of the cycle
- Reduce the exergy losses in the economizer

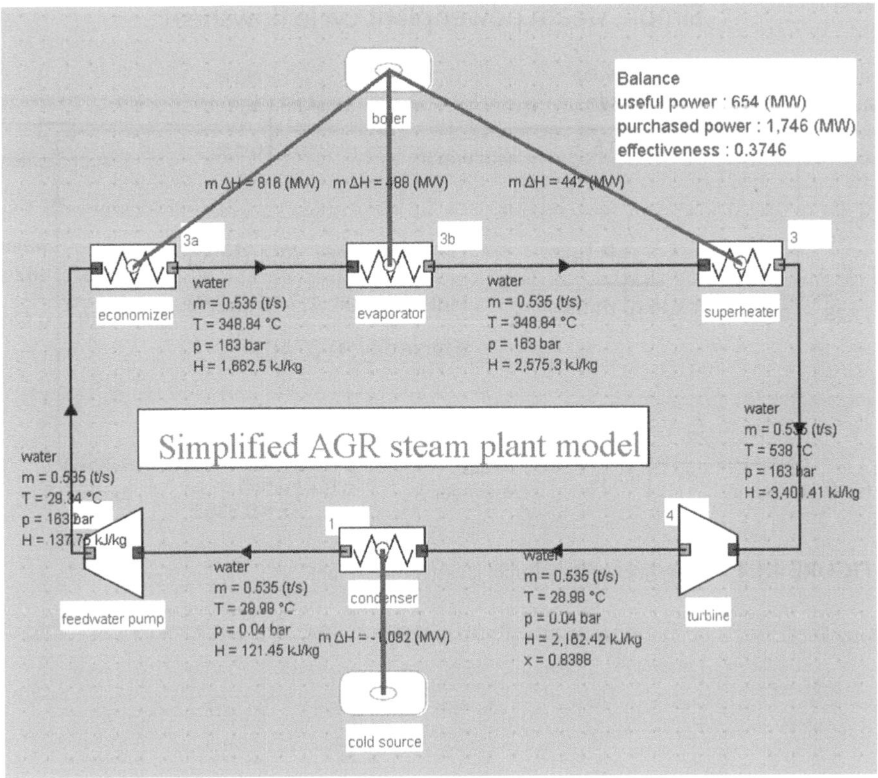

FIGURE 4.1.2 Synoptic view of the simple steam power plant.

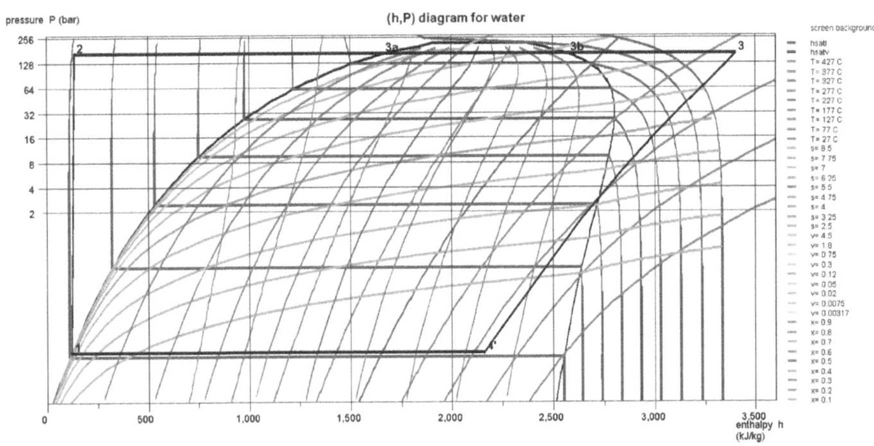

FIGURE 4.1.3 Simple steam cycle in (h, P) diagram.

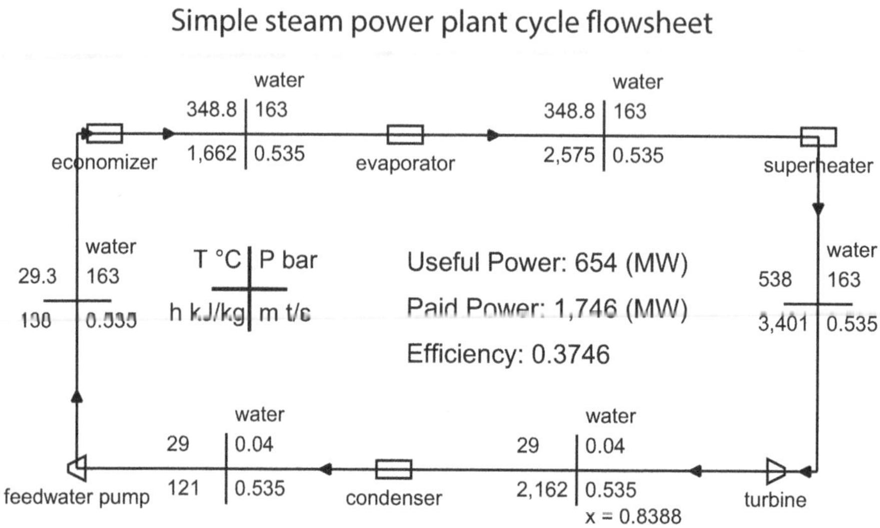

FIGURE 4.1.4 Flowsheet of the simple steam power plant.

component	Resource	Product	Exergy efficiency	Irreversibilities	% total	settings
feedwater pump	8.72	8.722	100.0000%	0	0	
turbine	780.182	662.86	84.9622%	117.322	20.7304%	
condenser	32.472	0	0	32.472	5.7377%	Tk = 20.00 °C
economizer	570.008	291.426	51.1267%	278.582	49.2244%	Tk = 700.00 °C
superheater	308.832	254.356	82.3606%	54.476	9.6257%	Tk = 700.00 °C
evaporator	341.24	258.15	75.6504%	83.091	14.6819%	Tk = 700.00 °C
global	1,220.08	654.137	53.6143%	565.943	1	

FIGURE 4.1.5 Simple steam power plant cycle exergy balance.

We will now examine how these operations are realized in practice by resorting to reheats and by implementing extractions. In order to be able to compare the different models with each other, we will assume that the polytropic efficiencies of the turbines always remain equal to the same value: 0.805.

4.1.3 Thermodynamic Limits of the Simple Hirn Cycle

The relatively low Hirn cycle efficiency compared to that of the Carnot cycle is due to its shape in an entropy diagram, which differs significantly from a rectangle as we discussed it in Section 3.3.1.

In the diagram of Figure 4.1.6, we see that it is theoretically possible to approach the Carnot cycle by modifying the Hirn cycle in several ways. In practice, however, all these ideas are confronted with various impossibilities.

Cycles (1-a-d-e) or (1-b-3-4) are impossible because the compression (1-a) or (1-b) in the liquid phase would require enormous pressures.

Cycle (f-c-d-e) would presuppose that one can achieve isentropic compression and expansion in the two-phase zone, which is currently technologically impossible.

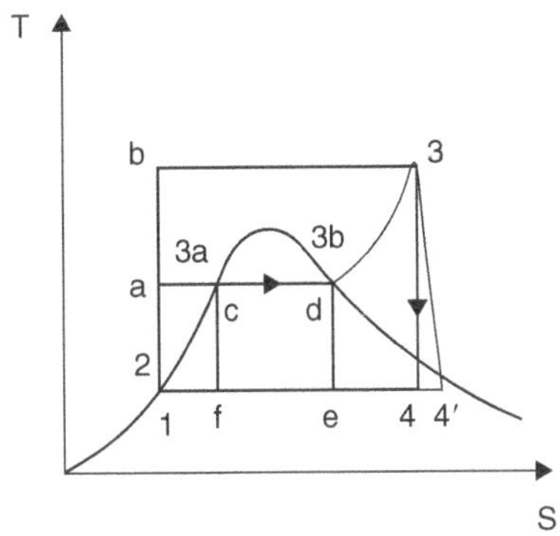

FIGURE 4.1.6 Impossible Carnot cycles.

To improve cycle performance, we first select a fluid having a sharply ascending saturation curve, and we seek to increase the average temperature at which heat is provided, as discussed below.

Water appears to be an excellent working fluid for the Hirn cycle, with its critical temperature of 374°C and high latent heat of vaporization at ambient pressure and temperature. Its low viscosity helps limit auxiliary consumption, and its low cost and non-toxicity position it very favorably relative to its competitors. However, for some applications, other fluids can sometimes be more appropriate as we will discuss in Section 5.10.

Water does not, however, have only benefits: condensation at 15°C requires maintaining a somewhat low vacuum in the condenser (0.017 bar), which necessitates a complex device for extracting air given inevitable leaks. This is especially challenging since it is necessary to use huge low-pressure turbines considering water's very low specific volume at 15°C, which induces very large volume flows, and thus very large flow sections and high velocities. This may lead to leaving losses at the turbine outlet if all available kinetic energy is not converted into work.

Thus, in the example discussed above, the specific volume of steam at the end of expansion, at point 4′, is equal to around 29 m^3/kg, while at point 3, it is 0.02 m^3/kg.

The volume expansion ratio is considerable: it is close to 1,500.

The large variation in volumetric flow during expansion represents a significant technological constraint for steam power plants, especially nuclear ones.

Another constraint is that the quality at the end of expansion must not be too small; otherwise, the droplets present may have a destructive abrasive effect on the turbine blades.

4.1.4 Cycle with Reheat

A first idea to improve the Hirn cycle is to approach the Carnot cycle by conducting reheats. In this case, we begin by partially expanding the fluid, and then it passes again into the boiler, where it is heated at the new pressure to approximately the maximum cycle temperature (Figure 4.1.7). This operation can optionally be repeated several times, generating what is called a quasi-isothermal cycle.

This results in efficiency gains of a few percent and, most importantly, as shown in the diagram, increased quality at the end of expansion, which is always beneficial for extending the life of turbine blades.

The price, however, is higher complexity, but since the expansion must be split anyway, this improvement has no major technological impact on the power plant.

4.1.4.1 Hartlepool Power Plant Cycle with Reheat

With this configuration and an intermediate pressure (IP) of 41 bar, the reheat cycle modeling results provided by Thermoptim show: the efficiency increases from 37.5% for the reference cycle to 39.85%, and the power output from 654 to 791 MW (Figures 4.1.8 and 4.1.9).

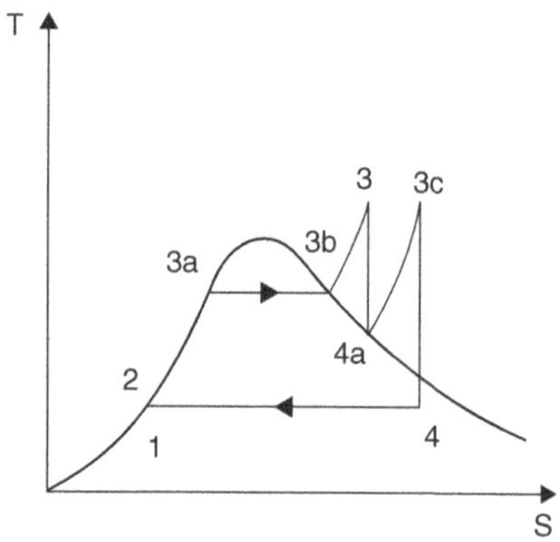

FIGURE 4.1.7 Reheat Hirn cycle.

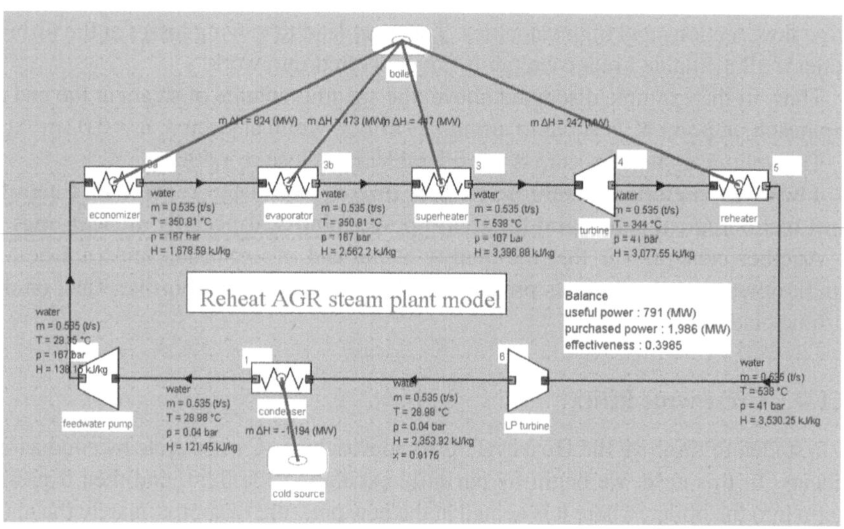

FIGURE 4.1.8 Synoptic view of the reheat cycle.

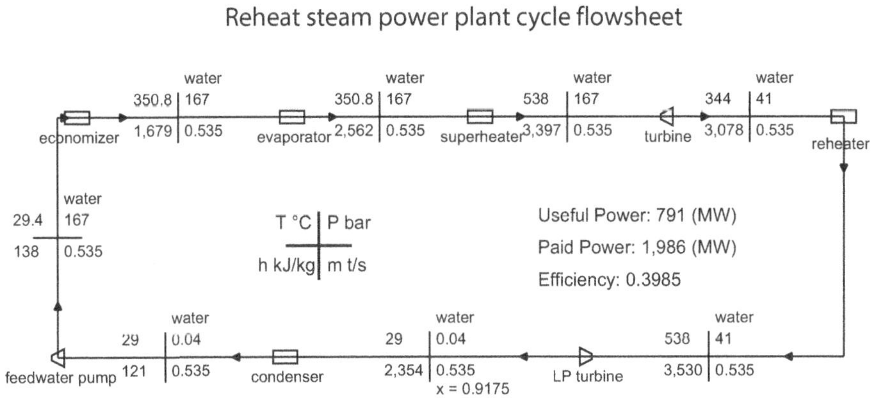

FIGURE 4.1.9 Flowsheet of the reheat cycle.

component	Resource	Product	Exergy efficiency	Irreversibilities	% total	settings
feedwater pump	8.94	8.936	99.9519%	0.004297	0.0007%	
turbine	191.887	170.84	89.0314%	21.047	3.5300%	
LP turbine	727 18	629.34	86.5453%	97.84	16.4096%	
condenser	35.519	0	0	35.519	5.9572%	Tk = 20.00 °C
economizer	575.878	295.946	51.3905%	279.931	46.9498%	Tk = 700.00 °C
superheater	312.032	257.187	82.4231%	54.845	9.1985%	Tk = 700.00 °C
evaporator	330.326	250.595	75.8630%	79.731	13.3724%	Tk = 700.00 °C
reheater	169.24	141.922	83.8586%	27.318	4.5817%	Tk = 700.00 °C
global	1,422.995	826.76	58.1000%	596.235	1	

FIGURE 4.1.10 Reheat steam power plant cycle exergy balance.

To obtain this model, it was sufficient to insert:

- a new "exchange" process at the outlet of the first HP turbine section and to set the outlet point temperature to the desired value for the superheat, here 538°C.
- a second turbine to complete the expansion of the superheated steam.

The exergy efficiency increases slightly, but the exergy balance is not radically changed (Figure 4.1.10).

4.1.5 Cycle with Extraction

The second area of improvement in power cycles consists of reducing irreversibility due to temperature heterogeneity.

When we analyzed the exchanges of the simple Rankine cycle with its external sources by comparing this cycle with the Carnot cycle, we saw that the most significant thermal irreversibilities are located in the economizer.

Remember that on the cold source side, isothermal condensation makes it possible to limit the temperature difference between the working fluid and the cooling fluid, so that the irreversibilities due to temperature heterogeneity are low.

On the hot source side, the nuclear reactor, capable of producing gas at around 700°C, heats the working fluid to much lower temperatures: in the economizer, the water exits the pumps at a few tens of degrees Celsius, while vaporization and superheating take place at several hundred degrees Celsius. In all cases, the temperature differences with the hot source are very large, but it is clearly in the economizer that the greatest irreversibility due to temperature heterogeneity occurs.

It is possible to undertake partial regeneration by using part of the heat rejected during expansion for preheating the pressurized liquid water before it enters the boiler.

Complete regeneration cannot of course be achieved, because processes (2–3a) and (3–4) in Figure 4.1.11 cannot be deduced from each other by translation (see Section 3.1.2.2). However, it is possible to take advantage of regeneration to significantly improve the Hirn cycle.

Consider a cycle with reheat (Figure 4.1.7). If we extract a small quantity of steam, called extraction steam or bleed steam, at the outlet of the first expansion, at point 4a in the figure, its pressure remains high enough to condense it at a temperature that allows preheating the pressurized water leaving the pump at point 2.

The enthalpy of the vapor is much greater than that of the liquid, due to the latent heat of vaporization; it is then possible to preheat the liquid using a small extraction of bleed steam during expansion. This operation is carried out in specific components called feedwater reheaters.

This cycle is called a regenerative Rankine cycle, also known as an extraction and reheat steam cycle.

Of course, for the operation to be possible, the bleed steam extracted at point P must be at a temperature higher than that of the liquid, which means that in practice, we must limit ourselves to partial water preheating to point A (Figure 4.1.11).

It is possible, if the economic study validates it, to combine reheat and extraction by extracting steam just before the reheat (Figures 4.1.12 and 4.1.13).

It should be noted that due to extraction, the mass flow rate of the fluid is not the same in the different parts of the machine. If ε is the steam extracted, the fluid flow

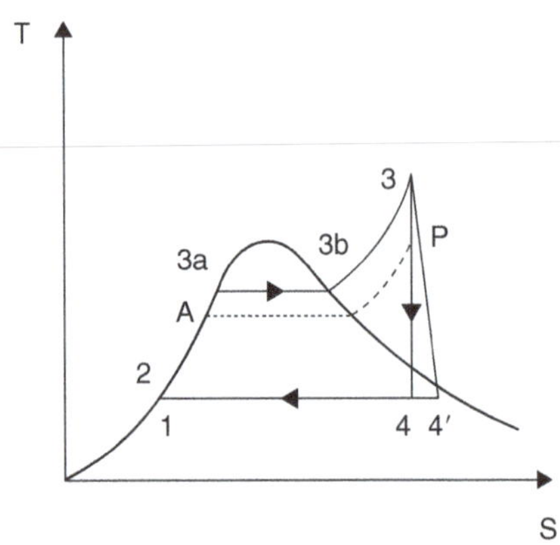

FIGURE 4.1.11 Extraction Hirn cycle.

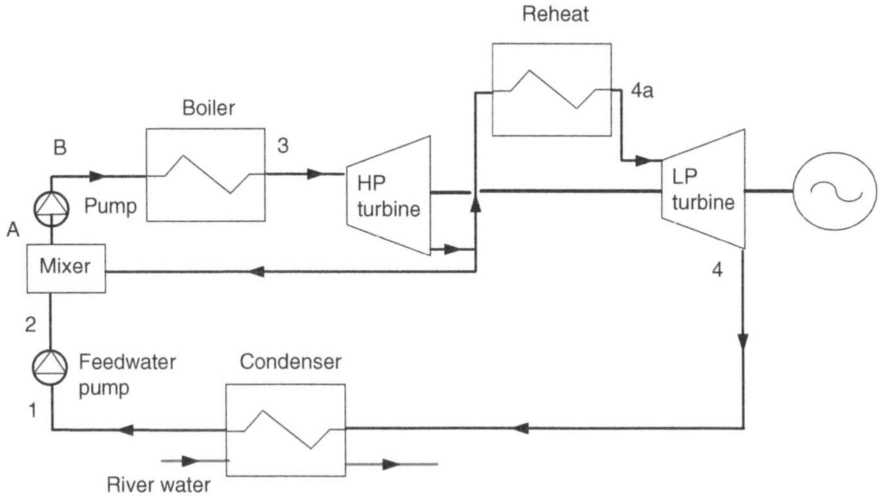

FIGURE 4.1.12 Extraction and reheat Hirn cycle.

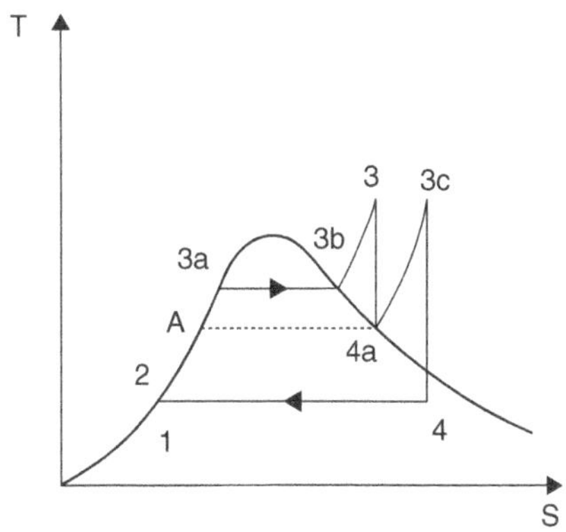

FIGURE 4.1.13 Extraction and reheat Hirn cycle in the entropy diagram.

between points 4a, 3c, 4, 1, and A is (1−ε) kg/s, and the flow rate between points A, 3a, 3b, 3, and 4a is equal to 1 kg/s.

Technologically, feedwater heaters can be of various types.

A first approach is to mix the two fluids, since they differ only in their state and not in their nature. The best heat exchange is thus obtained. However, it is necessary to equalize pressures in each mixer, which requires cascading compression of liquid. This solution, although attractive in thermal terms, was abandoned in favor of conventional shell-and-tube heat exchangers, with steam condensing outside the tubes.

There are therefore two types of feedwater heaters:

- Open feedwater heaters in which the hot fluid (the extracted steam) and the cold fluid (the condensed water) are mixed;

- Closed feedwater heaters in which the heat transfer between the two fluids takes place in a conventional exchanger. The condensed steam can then either be recompressed and mixed with the feedwater, or utilized in another feedwater heater.

In the first case, it is necessary to provide two pumps in the water circuit, while in the second, one is sufficient but heat exchangers must be installed.

In practice, in high-capacity power plants used for electricity generation, several feedwater heaters of both types are used (from 6 to 9), operating at temperature differences ranging from 30°C to 50°C.

At the condenser exit, an extraction pump compresses the liquid to a pressure at least equal to the saturation pressure at point A, to avoid partial evaporation in tubes. Various arrangements of heat exchangers and recovery pumps can then be used as we will see below. After being heated, the liquid is compressed to the desired pressure by a feedwater pump.

Because of extractions, the mass flow passing through the turbine stages is gradually reduced by about one-third in conventional thermal power plants, and 45% in pressurized water nuclear power plants. The result is a reduction of the previously mentioned strong constraint of having to convey increasingly large flow volumes during expansion. Extractions thus have an additional advantage.

Overall, extractions may contribute to an improvement of almost five percentage points in the internal efficiency of the Hirn cycle. Combined with reheat, the gain is about 7 percentage points, representing 20% higher efficiency than the basic cycle.

However, the cycle architecture in Figure 4.1.12 does not generally correspond to the layout used when there is only one extraction. To limit the irreversibility due to mixing, it is often preferable to perform bleeding at a lower pressure, and therefore downstream of the reheat, as we will see in the example provided below.

4.1.5.1 Hartlepool Power Plant Regenerative and Reheat Cycle

The synoptic view of a reheat and extraction cycle using an open feedwater heater modeled in Thermoptim is given in Figure 4.1.14, and its plot in the (h, P) diagram in Figure 4.1.15.

To develop this model, several modifications were implemented. The IP turbine is divided into two stages, and an extraction pump raises the water discharged from the condenser to IP. A splitter is positioned at the outlet of the first IP turbine stage, directing the stream into two paths—one toward an "exchange" process, representing the extraction, and the other toward the second IP turbine stage. Downstream of the extraction pump, an "exchange" process simulates feedwater heating by cooling the extraction stream in a heat exchanger. A mixer is used to model the feedwater tank, where the heated feedwater and the cooled extraction merge, and a feedwater pump then restores the pressure of the combined flow.

It is worth noting that within this model, an alternative approach could have been taken: the feedwater discharged from the extraction pump might have been mixed directly with the extraction stream in the feedwater tank, effectively creating an open feedwater heater.

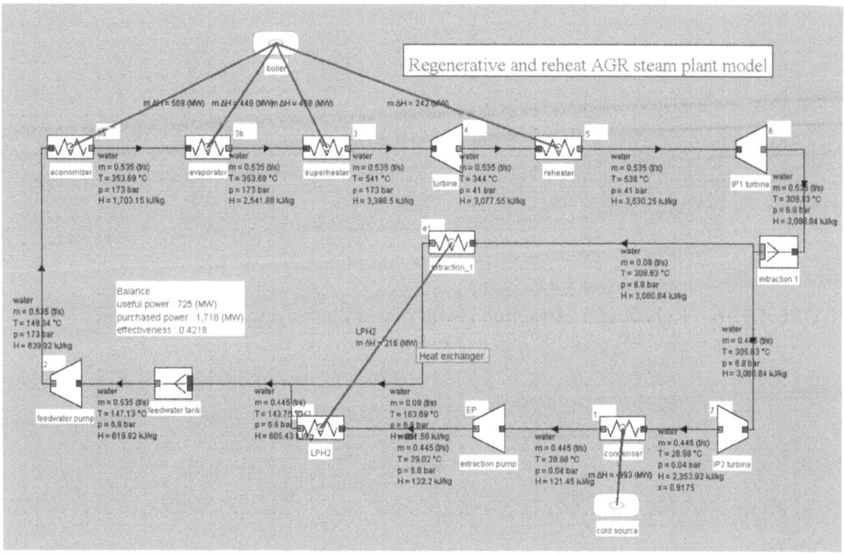

FIGURE 4.1.14 Synoptic view of the reheat and extraction cycle.

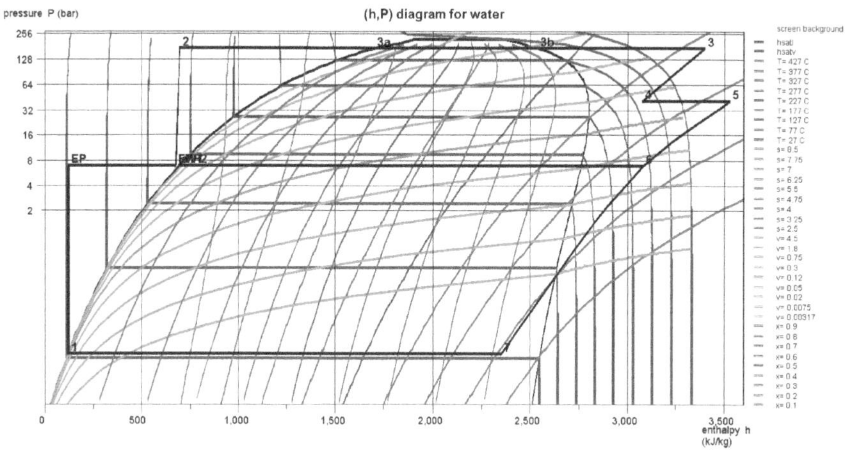

FIGURE 4.1.15 Cycle with reheat and extraction in the enthalpy diagram.

We note the slight decrease in power: 725 MW instead of 791 MW for the reheat cycle, due to the reduction of flow in the LP turbine. The efficiency is slightly higher: 42.2% instead of 39.85%.

Figure 4.1.16 gives the exergy balance of the new cycle. Its overall efficiency has increased compared with the reference one (61.3% instead of 53.6%), and above all, the irreversibilities in the economizer have been greatly reduced (21.3% instead of 49.2%).

component	Resource	Product	Exergy efficiency	Irreversibilities	% total	settings
feedwater pump	10.699	9.981	93.2929%	0.7176	0.1454%	
turbine	193.919	171.709	88.5468%	22.21	4.5008%	
IP1 turbine	264.926	240.434	90.7549%	24.493	4.9635%	
IP2 turbine	381.715	323.479	84.7437%	58.236	11.8014%	
extraction pump	0.3355	0.3041	90.6417%	0.0314	0.0064%	
LPH2	0	44.337	0	(-44.337)	(-8.9848%)	LPH2
extraction_1	78.188	0	0	(78.188)	(15.8447%)	LPH2
condenser	62.425	0	0	62.425	12.6503%	Tk = 10.00 °C
economizer	403.323	264.189	65.5032%	139.134	28.1953%	Tk = 700.00 °C
superheater	324.944	271.126	83.4379%	53.817	10.9059%	Tk = 700.00 °C
evaporator	318.161	245.994	77.3175%	72.167	14.6245%	Tk = 700.00 °C
reheater	171.727	145.343	84.6361%	26.384	5.3467%	Tk = 700.00 °C
global	1,280.58	787.114	61.4654%	493.466	1	

FIGURE 4.1.16 Reheat and extraction steam power plant cycle exergy balance.

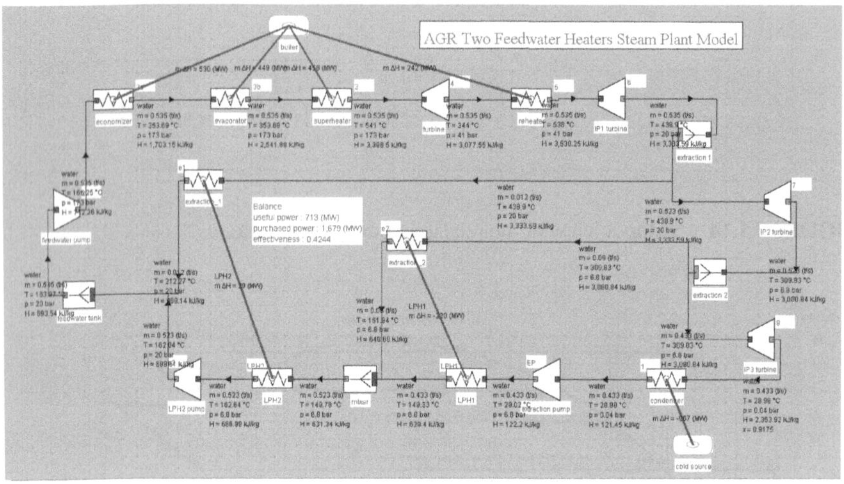

FIGURE 4.1.17 Synoptic view of the cycle with two extractions.

4.1.5.2 Hartlepool Power Plant Cycle with Two Extractions

In this new version of the Hartlepool cycle model, two extractions take place with closed feedwater heaters.

The same principle applies to the second extraction: the IP turbine section is divided into two to accommodate a new bleed point, so that the feedwater is preheated by the low-pressure extraction, then pressurized to the intermediate level, and finally reheated by the new extraction before reaching the feedwater tank.

The synoptic view and exergy balance are given in Figures 4.1.17 and 4.1.18.

By adjusting the extraction pressures and flow rates to match the available data as closely as possible, the parameters indicated in this synoptic view are obtained.

We note a small decrease in power: 713 MW instead of 725 MW. However, the efficiency is slightly higher: 42.44% instead of 42.18%.

Figure 4.1.18 gives the exergy balance of the new cycle. Its overall efficiency has also slightly increased compared with the previous one (61.7% instead of 61.3%).

A special feature of the automatically generated exergy balance is that the internal exchangers are represented by two lines (here “LPH2” and “extraction_1”) and that

component	Resource	Product	Exergy efficiency	Irreversibilities	% total	settings
feedwater pump	10.02	9.349	93.3001%	0.6713	0.1457%	
turbine	194.703	171.71	88.1907%	22.993	4.9911%	
IP1 turbine	115.347	105.22	91.2205%	10.127	2.1983%	
IP2 turbine	147.07	132.18	89.8755%	14.89	3.2322%	
IP3 turbine	373.423	314.76	84.2905%	58.663	12.7339%	
extraction pump	0.326	0.2948	90.4310%	0.0312	0.0068%	
LPH2 pump	0.848	0.7907	93.2449%	0.05728	0.0124%	
LPH1	0	40.404	0	(-40.404)	(-8.7705%)	LPH1
extraction_2	74.82	0	0	(74.82)	(16.2411%)	LPH1
LPH2	0	9.232	0	(-9.232)	(-2.0040%)	LPH2
extraction_1	12.198	0	0	(12.198)	(2.6478%)	LPH2
condenser	28.747	0	0	28.747	6.2401%	Tk = 20.00 °C
economizer	370.428	241.058	65.0755%	129.37	28.0822%	Tk = 700.00 °C
superheater	320.236	264.516	82.6005%	55.719	12.0949%	Tk = 700.00 °C
evaporator	313.548	238.834	76.1714%	74.714	16.2181%	Tk = 700.00 °C
reheater	169.24	141.922	83.8586%	27.318	5.9299%	Tk = 700.00 °C
global	1,202.199	741.516	61.6800%	460.683	1	

FIGURE 4.1.18 Two extraction cycle exergy balance.

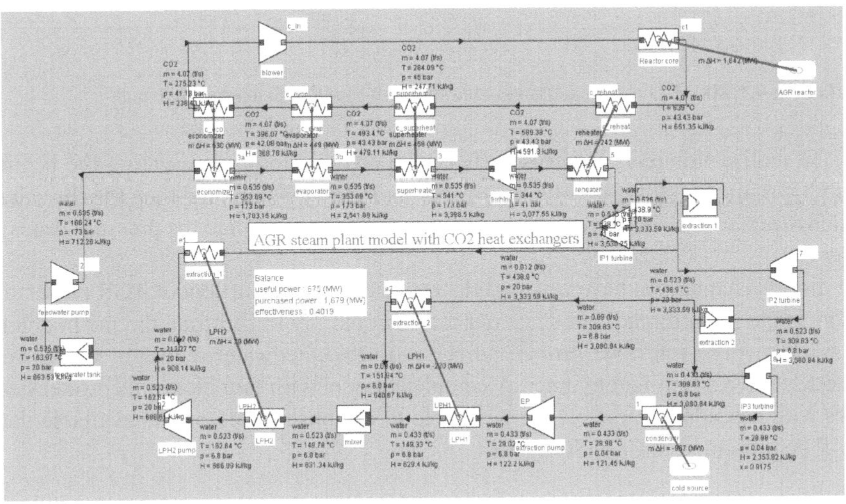

FIGURE 4.1.19 Synoptic view of the reheat and extraction cycle with the CO_2 circuit.

the contributions to the irreversibilities of the two fluids are represented algebraically, positively for the hot fluid and negatively for the cold fluid, with the irreversibility of the entire exchanger being obtained by summing these two values. That is why they are indicated in brackets in the latter exergy balance.

4.1.5.3 Hartlepool Power Plant Cycle with CO_2 Heat Exchangers

To date, we have not integrated the heat exchangers between the primary circuit of the CO_2-cooled reactor core and the secondary water circuit. We are now going to do so, especially since the primary circuit requires blowers to circulate the CO_2, and these blowers consume significant mechanical power (about 42 MW).

The new synoptic view and exergy balance are given in Figures 4.1.19 and 4.1.20.

Taking the CO_2 circuit and the corresponding exchangers into account has the effect of reducing the power output, which is now 675 MW, and efficiency, which is 40.2%. These figures are consistent with the data published on the Hartlepool plant.

component	Resource	Product	Exergy efficiency	Irreversibilities	% total	settings
feedwater pump	10.02	9.349	93.2993%	0.6714	0.1422%	
turbine	194.703	171.71	88.1907%	22.993	4.8708%	
IP1 turbine	115.347	105.22	91.2205%	10.127	2.1453%	
IP2 turbine	147.07	132.18	89.8755%	14.89	3.1543%	
IP3 turbine	373.423	314.76	84.2905%	58.663	12.4272%	
extraction pump	0.326	0.2948	90.4310%	0.0312	0.0066%	
blower	37.71	37.721	100.0000%	0	0	
mixer	49.715	49.713	99.9964%	0.001797	0.0004%	
LPH2 pump	0.847	0.7905	93.3288%	0.05651	0.0120%	
economizer	0	241.059	0	(-241.059)	(-51.0659%)	economizer
c_eco	279.152	0	0	(279.152)	(59.1355%)	economizer
evaporator	0	238.834	0	(-238.834)	(-50.5945%)	evaporator
c_evap	272.398	0	0	(272.398)	(57.7047%)	evaporator
LPH1	0	40.404	0	(-40.404)	(-8.5592%)	LPH1
extraction_2	74.82	0	0	(74.82)	(15.8499%)	LPH1
LPH2	0	9.232	0	(-9.232)	(-1.9557%)	LPH2
extraction_1	12.198	0	0	(12.198)	(2.5840%)	LPH2
reheater	0	141.922	0	(-141.922)	(-30.0647%)	reheater
c_reheat	162.167	0	0	(162.167)	(34.3534%)	reheater
superheater	0	264.516	0	(-264.516)	(-56.0350%)	superheater
c_superheat	293.211	0	0	(293.211)	(62.1137%)	superheater
condenser	28.747	0	0	28.747	6.0898%	Tk = 20.00 °C
Reactor core	1,147.101	969.207	84.4918%	177.895	37.6852%	Tk = 700.00 °C
global	1,175.848	703.793	59.8541%	472.055	1	

FIGURE 4.1.20 CO_2 heat exchanger steam power plant cycle exergy balance.

To realize this model, it was sufficient to create a loop representing the heating of CO_2 in the reactor, followed by its cooling in the reheater and then the three heat exchangers of the steam generator. The blowers help overcome the pressure drops of the CO_2.

In this chapter, we have retained the modeling of the Hartlepool AGR power station as the main thread, but since our aim was mainly to illustrate the methodology for improving cycles, we limited ourselves to two feedwater reheaters which were modeled in a simplified manner. A complete model with four closed feedwater heaters with cascaded drains, as we were able to establish based on the available data, will be presented in Case Study 8.1.

4.1.5.4 Feedwater Train

A feedwater train is the set of elements that increase the temperature of the feedwater leaving the condenser before it enters the economizer.

As we have seen, this arrangement plays a very important role in improving cycle performance. In Figure 4.1.21, the water supply system of the Hartlepool power plant is shown surrounded by a black rectangle. In this case, it is relatively simple, with only two heat exchangers, but we will see later that much more complex configurations exist.

As we explained above, if ε is the steam extracted, the fluid flow between points 4a, 3c, 4, 1, and A is $(1-\varepsilon)$ kg/s, and the flow rate between points A, 3a, 3b, 3, and 4a is equal to 1 kg/s.

The balance of the feedwater heater can be established as follows: it receives $(1-\varepsilon)$ kg/s of liquid in state 2 and ε kg/s of steam in state 4a. 1 kg/s of liquid at state A exits from it.

We therefore have:

$$(1-\varepsilon)h_2 + \varepsilon h_{4a} = h_A$$

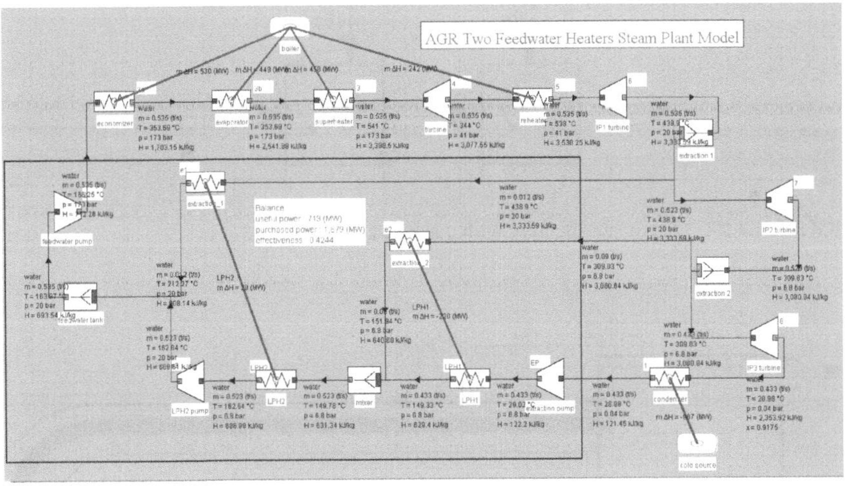

FIGURE 4.1.21 Water supply system of the cycle with two extractions.

We can thus determine ε.

In practice, in high-capacity power plants used for electricity generation, several extractions are used (from 6 to 9), with the various feedwater heaters operating at temperature differences ranging from 30°C to 50°C. The calculation of extractions is then done step by step, using the following recurrence formula:

$$\varepsilon_i = \left(1 - \sum_{1}^{i-1} \varepsilon_k\right) \frac{h_{Ai} - h_{Ai+1}}{h_{4ai} - h_{Ai+1}}$$

This relationship is valid in conventional high-temperature power plants, but we will see that it does not apply to water-cooled nuclear power plants due to different specificities.

4.1.5.5 Closed Feedwater Heaters with Cascaded Drains

The role of feedwater heaters is to preheat the feedwater to reduce the temperature difference within the economizer. In high-capacity power plants, there are a large number of feedwater heaters, from which condensates are extracted at increasingly higher temperatures and pressures as the feedwater is heated.

Rather than directly returning these condensates to the feedwater tank or the condenser, it is more thermodynamically efficient to use the hotter water exiting a downstream feedwater heater to preheat the water entering an upstream feedwater heater, thus achieving a counterflow of condensates relative to the feedwater (Figure 4.1.22).

The heat exchangers that enable this are of a particular type, as they receive two liquid flows and one steam flow. This configuration, called cascaded drain reheaters, is used in a large number of power plants, both nuclear and fossil-fueled.

The diagram in Figure 4.1.23 shows the internal operation of such a feedwater heater. The cold feedwater enters at the bottom left and flows inside the tube shown in

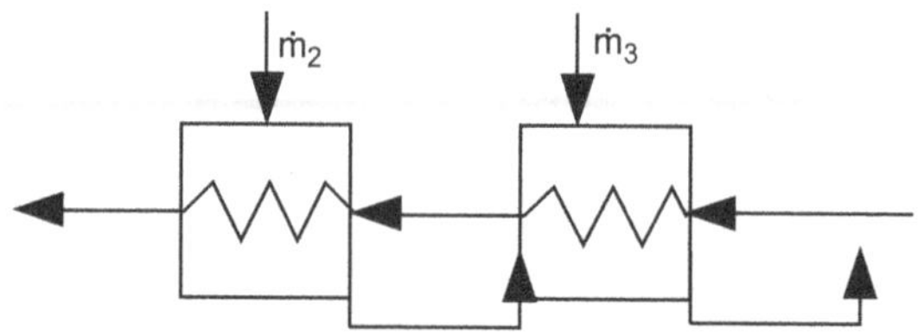

FIGURE 4.1.22 Feedwater heaters with cascaded drains.

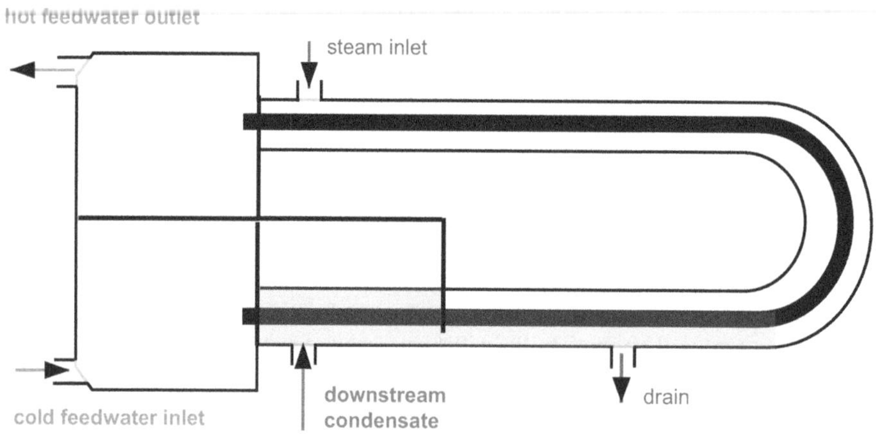

FIGURE 4.1.23 Feedwater heater with cascaded drain.

black. It is first preheated by the condensate from the downstream feedwater heater, then by heat exchange with the steam, and exits at the top left of the figure. The steam that condenses mixes with the downstream condensate, with the total flow exiting at the bottom right (drain).

From a functional perspective (Figure 4.1.24), the steam may be considered to fully condense in one heat exchanger before mixing with the condensates to preheat the main flow in another.

A Thermoptim model of such a component has been developed in the form of two external classes (FeedWaterReheaterInlet.java and FeedWaterReheater.java) and used in the detailed cycle models presented in the case studies of Chapter 8. The complete description of this model is available in the Thermoptim model library.

Let us now give an example of the use of these classes, for the cycle of the 77 MW module of NuScale.

As presented in the document submitted to the NRC for certification, the NuScale 77 model is particularly simple and serves as an easy-to-understand example (Figure 4.1.25). It includes four Turbine Stage Groups, three extractions, and three feedwater heaters. We will now explain how these are connected to the rest of the schematic and parameterized, with detailed explanations of the cycle provided in Case Study 8.3.

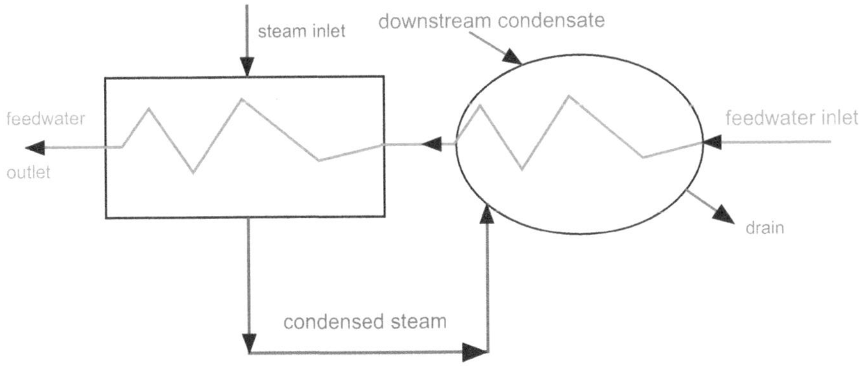

FIGURE 4.1.24 Closed feedwater heater with cascaded drain.

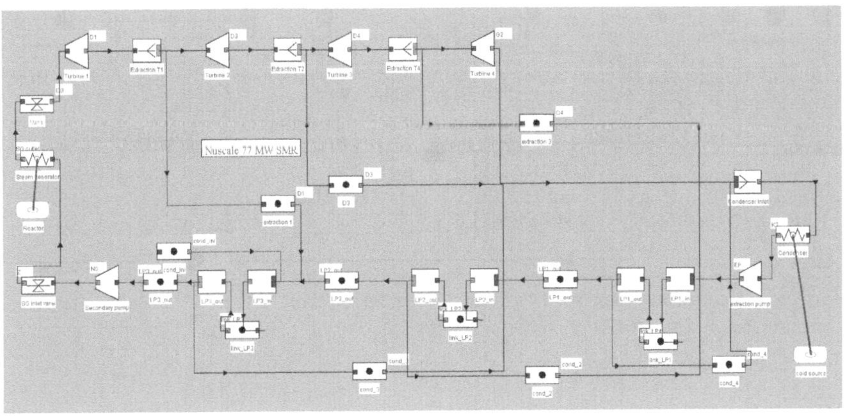

FIGURE 4.1.25 NuScale 77 model.

As we have just seen, the feedwater heaters are represented by two external nodes: an external mixer receiving three streams and an external divider outputting two streams—the heated feedwater and the mixed condensates. Among the incoming streams, two must be in the liquid state, representing the feedwater and the downstream condensates, while the third is steam. For the class to be calculated, this structure must be adhered to. This is why, in the proposed model, a zero-flow process-point "cond_ini" is connected to the third feedwater heater (Figure 4.1.26).

A detailed view of feedwater heater No. 2 is shown in Figure 4.1.26. The upper part clearly illustrates the two process-points representing the steam from the second extraction (D3) and the condensates from the downstream feedwater heater (cond_3), connected with the feedwater on the right side of the external mixer LP2_in, as well as the two process-points exiting the external divider: the heated feedwater (LP2_out) and the condensates (cond_2). The process-point connecting the two external nodes plays a purely passive role. It is only necessary for structural reasons.

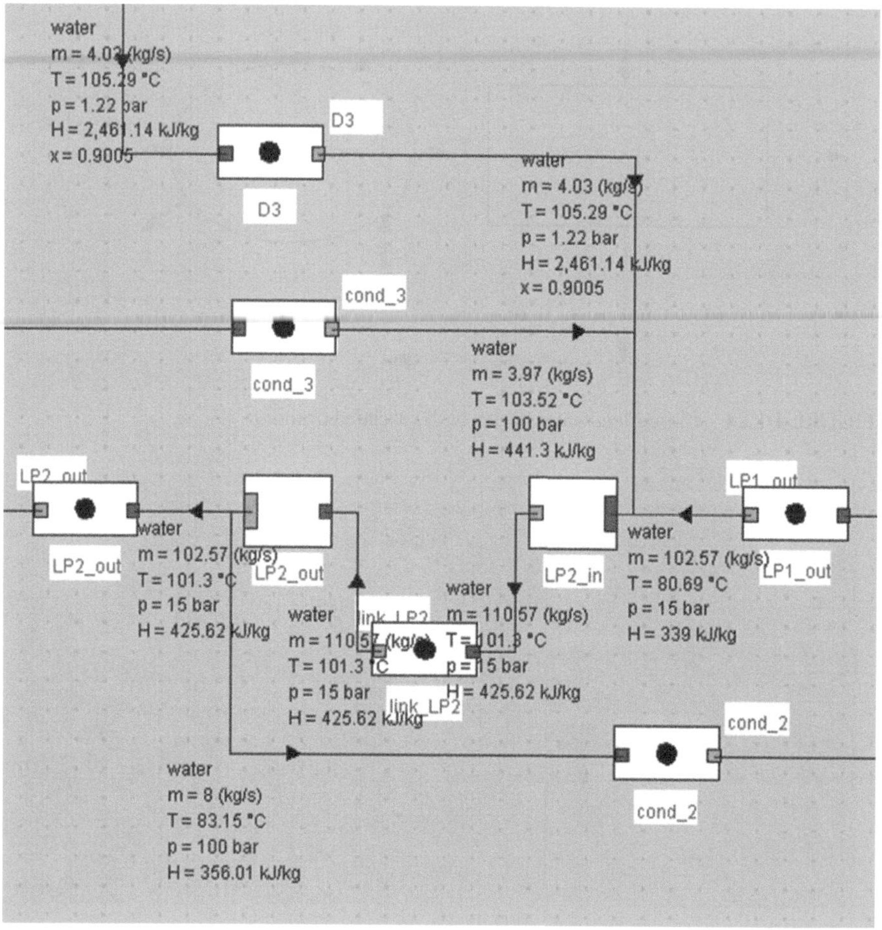

FIGURE 4.1.26 Zoom on the NuScale 77 model.

The screens for the two nodes are shown in Figure 4.1.27: on the left, the external mixer with its three streams, and on the right, the external divider, which includes the various parameterization fields and displays the calculation results.

It should be noted that when multiple feedwater heaters of this type are connected in series, as is the case in this model, solving the set of equations can be challenging because the calculations are coupled, with the condensates from downstream feedwater heaters interacting with the balancing of upstream feedwater heaters.

4.1.5.6 Feedwater Heaters Pumped Forward

A second configuration is sometimes used, called feedwater heaters pumped forward (Figure 4.1.28). In this case, the condensates exiting a heater, which are at a pressure lower than that of the feedwater, are repressurized by a pump and mixed with the feedwater, and the heater can be modeled as a simple heat exchanger.

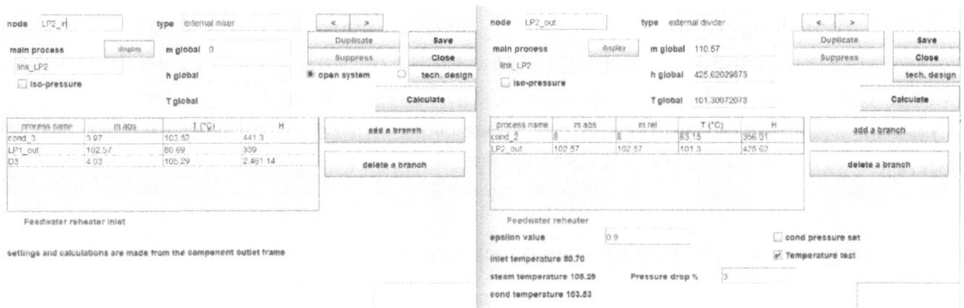

FIGURE 4.1.27 NuScale 77 model LP2 feedwater.

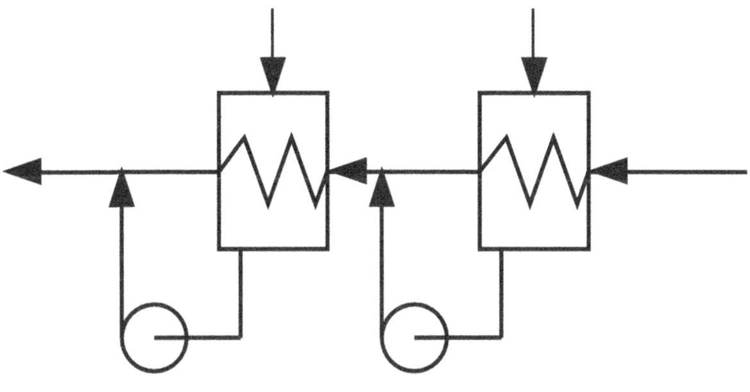

FIGURE 4.1.28 Feedwater heaters pumped forward.

4.1.6 Technical Constraints in AGRs

In the beginning of this section, we analyzed the variants of the basic steam power plant cycle, arriving at a more or less complete modeling of the cycle of the Hartlepool AGR power plant. Other types of nuclear reactors than AGRs can implement cycles close to this one, but not all.

This relatively high-temperature cycle is characterized by an initial superheat of nearly 200°C.

Such superheat can be achieved without any particular problem in a gas-vapor exchanger such as that of this plant, but can hardly be carried out in PWR or BWR type plants, for reasons which will be explained in the next chapter.

In our analysis of AGR plants, we have so far focused on optimizing the secondary cycle, without considering the other technological constraints of these plants, particularly the coolant, CO_2.

AGRs use carbon dioxide as a coolant to extract heat from the core, graphite as a moderator, and enriched uranium as fuel. These reactors operate at higher temperatures than WCRs (around 650°C in the CO_2 coolant), which requires the use of materials resistant to intense thermal conditions in the steam generators and turbine components.

These high temperatures and the CO_2 environment can cause material corrosion issues, especially for alloys in direct contact with the gas. This necessitates the use of specific materials, such as heat-resistant stainless steels and nickel alloys.

Despite the high performance of these cycles, all of these constraints resulted in high costs and operational limitations that led to the abandonment of this technology.

4.1.7 SFR, FNR, HTR-PM, MSR, FMSR Cycles

Other types of nuclear reactors than AGRs can use high-temperature steam cycles, such as SFR, FNR, HTR-PM, MSR, and FMSR.

For these types of reactors, it is not the steam thermodynamic cycle that poses a problem. These cycles are now very well-known and mastered. What is much less mastered is the technology of the reactors themselves, which are the subject of extensive research around the world, whether for large power plants or for small modular reactors, many of which are currently under development. However, since this book focuses mainly on secondary cycles, we will limit ourselves to general remarks on the technology readiness level of these reactors. We will discuss this in Chapter 6.

Among the high-temperature nuclear cycles commissioned in the past other than AGRs, two types of cycles using fast neutron reactors should be mentioned: the Russian BN-400 and BN-600 and the French SuperPhénix.

4.1.7.1 Sodium-Cooled Reactors

Sodium-cooled reactors are fast neutron reactors that have the advantage of behaving as breeders, which offers many benefits in terms of nuclear fuel use. They operate at relatively high temperatures, around 500°C–550°C, allowing for the use of a Hirn cycle with performance similar to that of fossil fuel plants or AGRs.

They use liquid sodium as a coolant to remove heat from the core. This metal has interesting thermal properties but poses particular challenges in the design of the steam cycle.

One of the major constraints is the risk of a violent chemical interaction between sodium and air or water. If sodium comes into contact with water in the steam generator, it can cause very rapid exothermic reactions, producing hydrogen and a risk of explosion. To prevent this, the primary (sodium) and secondary (water/steam) cycles are strictly isolated, and an intermediate sodium loop is often used to transfer heat from the primary circuit to the steam generator. This adds complexity to the system design and slightly reduces efficiency.

Sodium is corrosive at high temperatures, but it does not cause the same kind of corrosion as water or steam. The materials used must resist both the thermal effects and the specific chemical environment of sodium. It is also necessary to prevent the formation of deposits in the steam generators, which can affect cycle performance and increase maintenance costs.

SFR safety systems must be highly sophisticated to prevent sodium leaks. This imposes significant constraints on the design of heat exchangers and the management of potential leaks. Leak detection and sodium fire protection systems must be integrated into the steam cycle design.

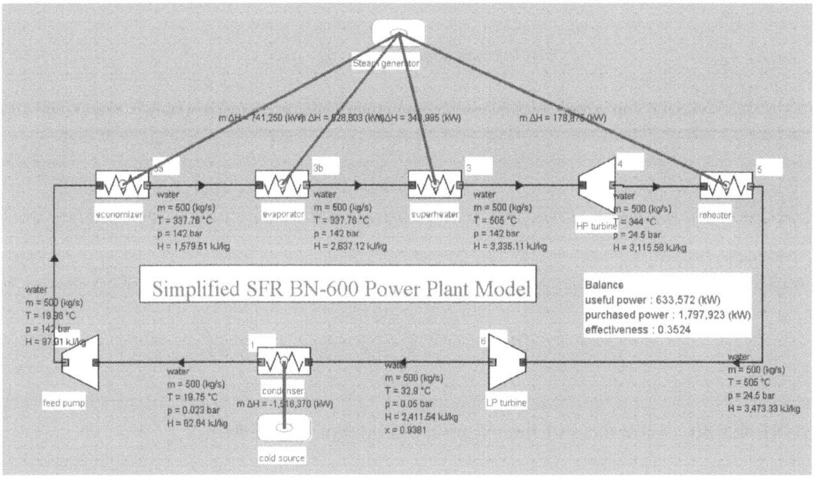

FIGURE 4.1.29 Synoptic view of the simplified model of SFR BN-600.

Steam generators in SFRs must be robust and capable of operating for long periods without failure, as maintenance operations can be complex and dangerous due to the reactivity of sodium.

Two countries have built operational sodium-cooled power plants: Russia and France. Figure 4.1.29 shows the synoptic diagram of a simplified model of the thermodynamic cycle of a Russian sodium-cooled fast neutron reactor, type BN-600. The temperature of the sodium at the core outlet reaches 550°C, but the presence of a double sodium cooling circuit for safety reasons means that the steam temperature is only 505°C.

The SuperPhénix plant will be the subject of a detailed model (Case Study 8.8). Its maximum temperature is equal to 407°C, and it has the particularity of using a moisture separator reheater like nuclear steam cycles at lower temperatures. The SCWR cycle that we will show in the next section does the same.

Among recent developments, let us mention China's CFR-600 which began construction in 2017 with operation planned for 2024–2025, while Russia's BN-800 has operated since 2016, successfully demonstrating MOX fuel capability under commercial conditions. India's PFBR achieved criticality in 2023, marking another milestone for this technology pathway.

4.1.7.2 A High-Temperature Small Modular Reactor: The HTR-PM

Another high-temperature steam cycle has been recently developed in China. This cycle is powered by a high-temperature nuclear reactor cooled by helium. It is called High-Temperature Reactor Pebble-bed Module (HTR-PM).

The HTR-PM power plant in Shandong's Shidao Bay uses a high-temperature gas-cooled reactor coupled with a Hirn steam cycle for electricity generation.

It is an intrinsically safe Small Modular Reactor with a unit thermal output of 250 MWth, but can be combined with up to six reactors of the same type to achieve high capacities. The concept provides for two reactors to be coupled to a single 210 MWe turbine.

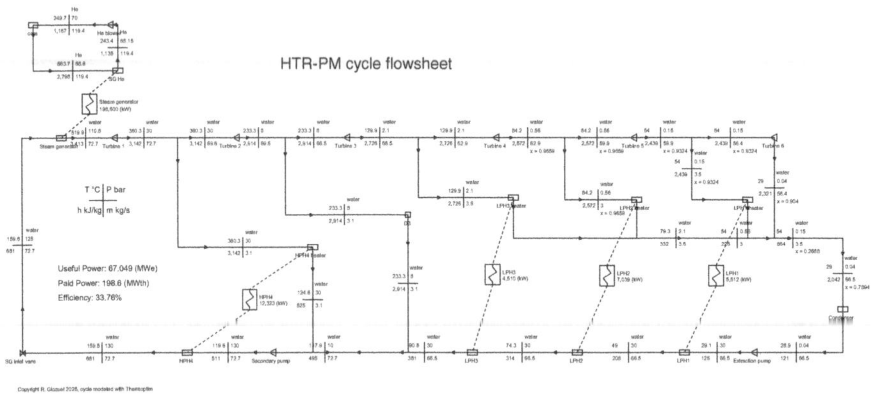

FIGURE 4.1.30 Flowsheet of the simplified model of HTR-PM.

HTR-PMs are high-temperature thermal neutron reactors using pebbles of 8.5% enriched uranium fuel embedded in graphite (which plays the role of a moderator) and using helium as a coolant. Fuel pebbles comprise a 5 cm diameter inner zone and a 5 mm outer graphite shell. In the inner zone, TRISO[1] type coated fuel particles are dispersed into the graphite matrix.

The safety limit of the fuel temperature is 1,620°C, but the normal operation value is 1,200°C. Primary helium pressure is 70 bar, and helium inlet and outlet temperatures are set to 250°C and 750°C, respectively.

This makes it the first Generation IV reactor in service in the world. The very high temperature level that it should reach makes it possible to envisage very high conversion efficiencies. However, as it is a small modular reactor, the thermodynamic cycle chosen is relatively simple and does not involve steam reheats.

The HTR-PM uses a once-through type steam generator. The feedwater line comprises three LP reheaters, a deaerator, and an HP reheater.

The design inlet conditions of the HP turbine are 132.4 bar and 535°C–570°C, leading to thermal efficiencies up to 44%–47.6% in power generation mode and 32%–54.9% in cogeneration configurations.

However, very little information is available about its thermodynamic cycle. The only data available lead to much lower performance for this prototype plant: 67 MW for the first of the two modules powering a single turbine, and an efficiency of about 33.76%.

This information was deduced from the data provided in the article by Dong et al. (2023), as explained in Case Study 8.9.

Based on this data, we were able to build a model that is perfectly consistent with the results published about tests conducted in 2023.

The flowsheet in Figure 4.1.30 corresponds to our model.

4.2 SUPERCRITICAL CYCLES

The optimum technical and economic capacity of flame Hirn cycles until recent years matched boiler conditions of about 560°C and 165 bar, leading, with reheat and without extraction, to a thermodynamic efficiency close to 40%. To significantly increase efficiency, it is possible to use so-called supercritical cycles in which water pressure exceeds the critical pressure of 221.2 bar.

The result is obviously much higher stress for boiler tubes. Progress in creep tube resistance can provide technological solutions unthinkable a short time ago. For example, the pipe thickness needed to withstand a pressure of 225 bar and a temperature of 600°C changed from 250 mm with steel P22 at 2.25% Cr, to 93 mm with P91 steel with 9% Cr, and 68 mm with steel HCM12A at 12% Cr (Jayet-Gendrot et al., 1999).

Another constraint faced by supercritical boilers is the following: due to the absence of an evaporator, you cannot cool the furnace by pipe screens traversed by boiling water with very high heat transfer coefficients. This is why a different technology is used, without a drum separator, in so-called mono-tubular boilers (improperly named because in reality the layers of tubes are arranged in parallel), or in English "once-through" to indicate the absence of recirculation. These tubes are fluted with internal and external fins, mounted in spiral bundles.

Supercritical cycles are not new (40% of former Soviet Union plants were supercritical, as well as more than 150 plants in the US). The recent trend is to seek increasingly high boiler conditions and two reheats. To provide context, we modeled in Thermoptim such an advanced supercritical steam cycle. The efficiency reaches 44% (against 40% for a conventional subcritical cycle). Figures 4.2.1–4.2.3 show the cycle synoptic view and its plot in Mollier and (h, P) diagrams.

In this example, we have simplified things by keeping an isentropic expansion efficiency of 0.85 for the different turbines. As the expansion takes place entirely in the dry zone, higher values could have legitimately been taken.

Thus, this technology provides efficiency gains between 6% and 10% depending on the steam pressure and temperature conditions, for a cost increase of 3%–5%.

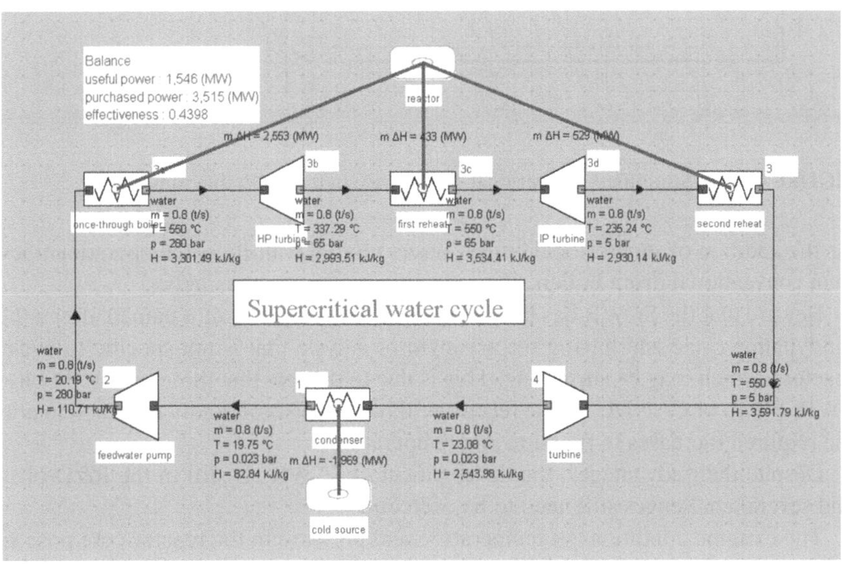

FIGURE 4.2.1 Synoptic view of a simplified supercritical steam cycle.

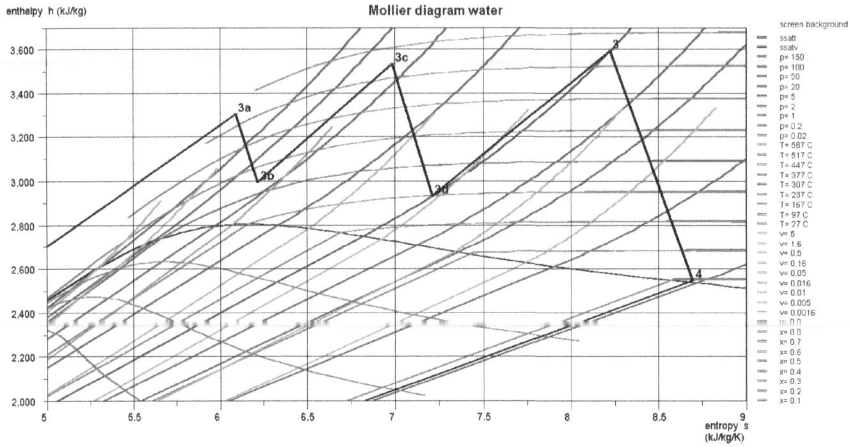

FIGURE 4.2.2 Simplified supercritical steam cycle in the Mollier diagram.

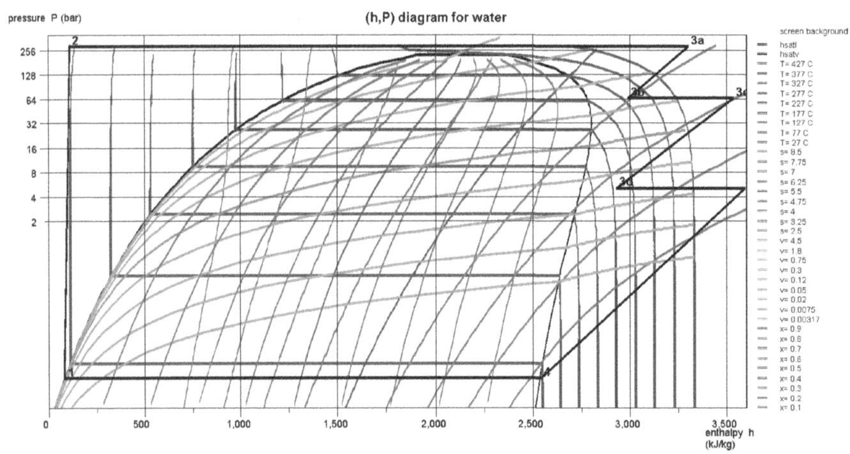

FIGURE 4.2.3 Simplified supercritical steam cycle in the (h, P) diagram.

As the addition of supercritical boilers offers more flexibility at the operational level than conventional drum boilers, their use is spreading increasingly.

Beware that the SCWR has the particularity of being a reactor named after a thermodynamic cycle and having the acronym of a cycle that is not specific to nuclear reactors, which may be surprising. This is due to the fact that these are evolutions of PWR, BWR, or CANDU whose reactor will have to be profoundly modified to allow the required increases in pressure and temperature.

Despite their advantages, the development of SCWRs is still in the R&D phase, and several challenges still need to be overcome.

The extreme conditions of temperature and pressure in the reactor core pose significant constraints for materials. Corrosion and the resistance of materials to these conditions and to irradiation are major areas of research.

Although supercritical water is used in conventional thermal power plants (coal or gas), its behavior in the presence of nuclear radiation is still poorly understood and is the subject of in-depth research. Heat transfer issues and flow stability are particularly critical.

The density of supercritical water changes considerably with temperature, which affects neutron moderation. Reactor control, especially under transient or accident conditions, is complex and requires in-depth research in thermohydraulics and reactor physics.

The technology readiness level of SCWRs in 2025 is generally situated between TRL 3 and TRL 5 according to sources.

To provide a concrete example of a supercritical cycle, we have modeled the so-called Canadian SCWR cycle, a 1,200 MWe pressure-tube type reactor with an efficiency of 47.1%. It is presented in Case Study 8.10.

It should be noted, however, that unlike the detailed WCR cycles that we will examine later, particularly in the case studies, this is a theoretical cycle that has not yet been realized. Moreover, it seems to us that the assumptions regarding the isentropic efficiencies of the turbines are somewhat optimistic.

The cycle involves eight feedwater heaters, four at low pressure and four at high pressure. The model we have established based on the data provided in the schematic of the cycle provided on p. 264 of the G4NR Handbook gives very similar results. Some differences arise from the values of the water properties, which are slightly different. Our model leads to a delivered power and efficiency that are slightly higher, equal to 1,243 MWe and 48.7%, respectively.

The steam expansion path is plotted in the Mollier diagram shown in Figure 4.2.4.

The detailed flowsheet of our model, divided into two parts for better readability, is shown in Figure 4.2.5.

The complete diagram in SVG vector format is also available at the following address.[2]

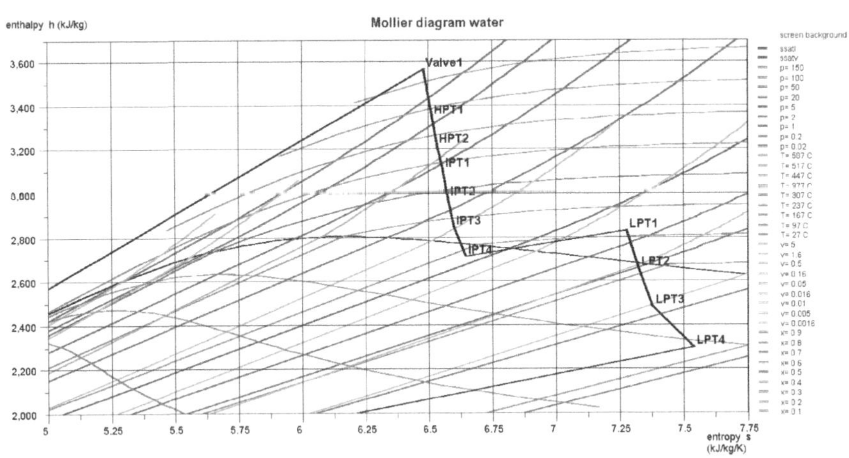

FIGURE 4.2.4 Canadian SCWR cycle in the Mollier (h, s) diagram.

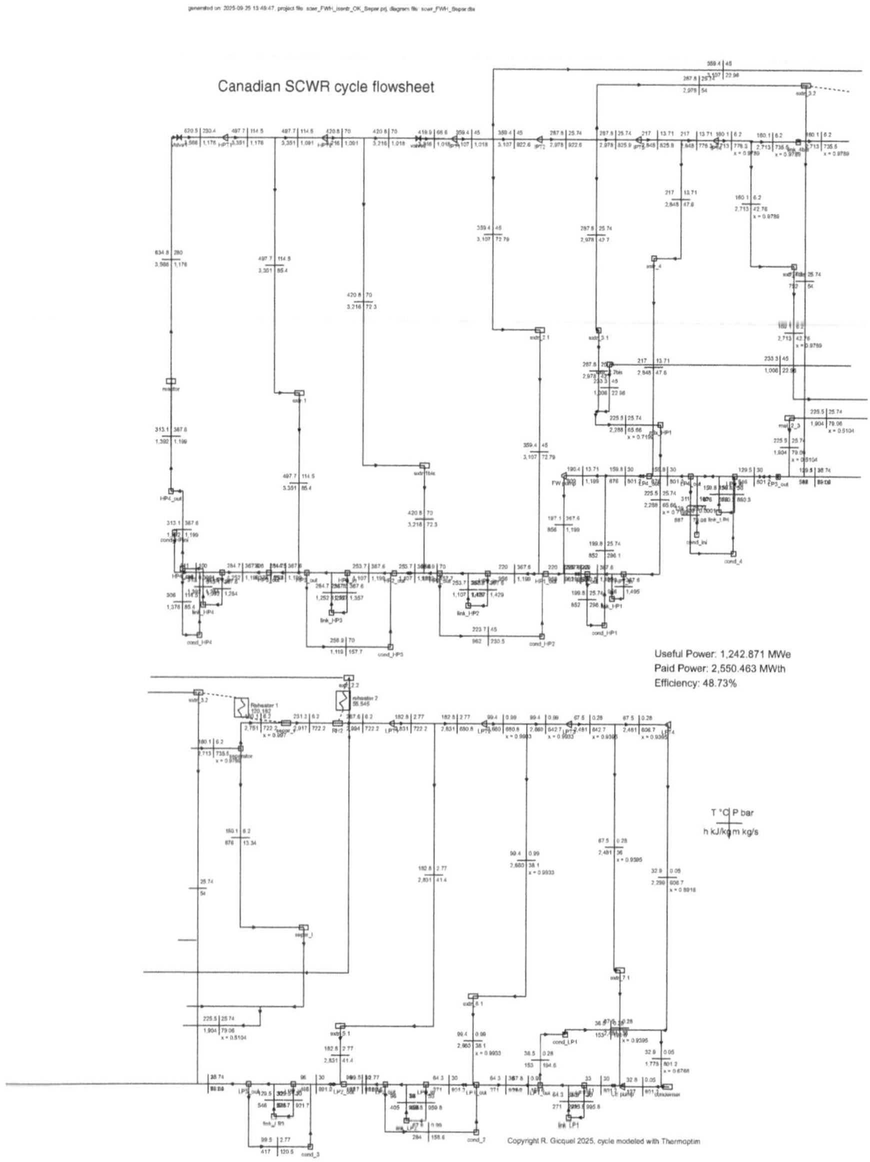

FIGURE 4.2.5 Flowsheet of the Canadian SCWR cycle.

4.3 CONCLUSION

This chapter presented methodologies for improving the performance of steam cycles in nuclear power plants, focusing on high-temperature variants of the Hirn cycle. Exergy analysis of the base cycle revealed that nearly half of the irreversibilities occur in the economizer. Two main modifications were examined: reheat cycles, which improve efficiency by several points, and regenerative cycles with feedwater heaters and steam extractions, which significantly reduce irreversibilities. The Hartlepool AGR power plant served as a case study, with progressive modeling that included cascade configurations and heat exchangers with a CO_2 coolant. Various advanced reactor technologies and supercritical cycles achieving 47%–48% efficiency were also introduced, accompanied by Thermoptim models and exergy analyses demonstrating the quantitative impact of the various improvements.

Further resources on the models presented in this chapter are available on this book's website.[3]

NOTES

1 TRISO (Tri-Structural Isotropic) fuel is an advanced nuclear fuel primarily used in high-temperature reactors (HTRs), such as Pebble Bed Reactors and Prismatic Block Reactors. It consists of tiny uranium particles encased in three protective layers: pyrolytic carbon (PyC), silicon carbide (SiC), and another PyC layer. These layers provide mechanical strength, corrosion resistance, and containment of fission products, ensuring high safety and performance standards.

The technology was developed in the 1960s in the USA and Germany and has since been tested globally. TRISO particles are spherical, about 1 mm in diameter, and compacted into graphite matrices to form fuel compacts. The manufacturing process involves precise deposition of layers through thermal cracking of gases, ensuring uniformity and durability.

TRISO's design minimizes the risk of radioactive leakage and allows for operation at higher temperatures compared to conventional fuels, making it a promising option for Generation IV nuclear reactors.

2 https://server.s4e2.com/crc/tcnpp/resources/svg/index.html.

3 Available at: https://server.s4e2.com/crc/tcnpp/chapters/chapter4/_resources/index.html.

BIBLIOGRAPHY

Dong, Z., Zhang, Z., Dong, Y. et al. (2025). "Testing the feasibility of multi-modular design in an HTR-PM nuclear plant." *Nature Communications*, 16, 2778. DOI: 10.1038/s41467-025-58194-7.

Jayet-Gendrot, S., Arnoldi, F., Billard, P., Dorier, C., Dutheillet, Y., Lelait, L. & Renaud, D. (1999). "Des matériaux innovants pour un sujet brûlant." *Revue Épures,* DRD EDF, Paris.

Pioro, I. (2011). "Lecture SC06 – Introduction to thermodynamics." In *Science and Technology of SCWRs: International School,* International Atomic Energy Agency (IAEA), Trieste, Italy. Available online: https://indico.ictp.it/event/a10196/session/12/contribution/9/material/0/1.pdf.

Zhang, Z., Dong, Y., Li, F., Zheng, Y. et al. (2021). "The Shandong Shidao Bay 200 MWe high-temperature gas-cooled reactor pebble-bed module (HTR-PM) demonstration power plant: An engineering and technological innovation." *Nuclear Engineering and Technology,* 53(5), 1292–1306. DOI: 10.1016/j.net.2021.02.026.

5 Medium-Temperature Steam Power Cycles

INTRODUCTION

In Chapter 3, we presented the basic steam power plant cycle and demonstrated its limitations. In Chapter 4, we applied the methodology and tools we introduced to improve the performance of the Hartlepool Advanced Gas-cooled Reactor cycle.

We will now study other steam cycles associated with nuclear reactors.

As previously mentioned, high-temperature steam cycles coupled with nuclear reactors are very similar to those using conventional flame boilers and are therefore well understood.

The same cannot be said for medium-temperature cycles, which have particular characteristics that are important to consider and which we will develop here.

The thermodynamic cycles coupled with medium-temperature nuclear reactors are essentially cycles where the reactor is cooled by water, hence their designation as WCRs.

5.1 MAIN CATEGORIES OF WCRs

Water-cooled nuclear power plants using steam cycles are divided into four main categories:

- PWRs, including Russian-designed VVERs
- BWRs
- RBMK (graphite-moderated pressure tube reactors)
- CANDU

Refer to Sections 2.2.4–2.2.7 and 2.2.10 for an overview of these types of reactors.

In BWRs and RBMKs, steam is produced directly in the reactor core and feeds the turbines via a direct cycle. In PWRs and CANDUs, a primary coolant transfers heat to a secondary circuit through a steam generator (SG), thus producing steam for the turbines. The primary coolant is light water (H_2O) in PWRs and heavy water (D_2O, deuterium oxide) in CANDUs. A key distinction lies in the moderator, which slows down neutrons to sustain the chain reaction: BWRs and PWRs use light water as both a moderator and a coolant, and CANDUs use heavy water, while RBMKs combine a graphite moderator with a light water coolant—a design unique to the former USSR.

Another difference between these technologies is the void coefficient, which measures the effect of steam bubble formation in the coolant on the reactor's reactivity. This coefficient is negative in BWRs and PWRs, but strongly positive in RBMKs and moderately positive in CANDUs. This means that partial vaporization of the coolant reduces

DOI: 10.1201/9781003606024-6

the thermal power released by the reactor in the first two cases, whereas it increases power in RBMKs and, to a lesser extent, in CANDUs. These phenomena, complex due to neutron interactions and local void variations, increase the risk of runaway reactions, particularly for RBMKs, as illustrated by the Chernobyl accident (1986), and highlight the importance of precise reactivity control and feedback mechanisms specific to each technology. In CANDUs, this risk is low thanks to a moderate coefficient and features such as control rods and the large thermal mass of the moderator.

In BWRs and PWRs, the fuel is arranged in a single vessel containing both the cooling and moderating water. In contrast, in RBMKs and CANDUs, fuel assemblies are placed in pressure tubes through which the coolant circulates. This configuration allows certain tubes to be individually isolated for fuel assembly extraction or insertion without shutting down the reactor, an advantage for operational availability.

- BWRs and PWRs/VVERs require uranium enrichment of 3.5%–5%, with averages around 4.5% for 12–24 month cycles.
- RBMKs use lower enrichment (1.8%–2.8%, average 2.4%) thanks to the graphite moderator and online refueling capability.
- CANDUs, as noted, use natural uranium (0.72%).

Although these plants belong to the second generation, most have evolved to meet third-generation requirements: ABWR and ESBWR for BWRs, EPR and AP1000 for PWRs, VVER-1200 for VVERs, and EC6 or CANFLEX for CANDUs, improving efficiency and safety.

The following summary shows the global distribution of installed capacity for the four main types of WCRs used for electricity production, according to data from the International Atomic Energy Agency and the World Nuclear Association.

- PWRs/VVERs overwhelmingly dominate the global nuclear fleet, representing over three-quarters of installed capacity.
- BWRs are the second most common type, though their share is much lower than that of PWRs.
- Pressurized heavy water reactors (PHWR)/CANDU and RBMK represent smaller shares, with specific geographic distributions.
- PWRs/VVERs are found in many countries, notably the United States, France, China, Russia, and South Korea.
- BWRs are mainly used in the United States, Japan, and Sweden.
- PHWR/CANDU are present in Canada, India, Romania, Argentina, China, and South Korea.
- RBMK exists exclusively in Russia.

Steam cycle conditions vary slightly by category: BWRs and RBMKs produce saturated steam at approximately 280°C and 70 bar, while PWRs and CANDUs generate steam via SGs, often at optimized temperatures and pressures (up to 300°C and 60–70 bar for the former, somewhat less for the latter).

We will now provide details on the technologies corresponding to these four types of plants, starting with PWRs, the most widespread worldwide as previously noted.

5.2 ARCHITECTURE OF PWRs

In second-Generation PWR plants, the pressure in the SG is approximately 60 bar, and the steam temperature barely exceeds 275°C. The sketch of a PWR is shown in Figure 5.2.1.

5.2.1 Primary Circuit

On the right side of the diagram is the containment vessel, including three main components:

- the reactor with its control system (control rod drive mechanism);
- the SG;
- the pressurizer.

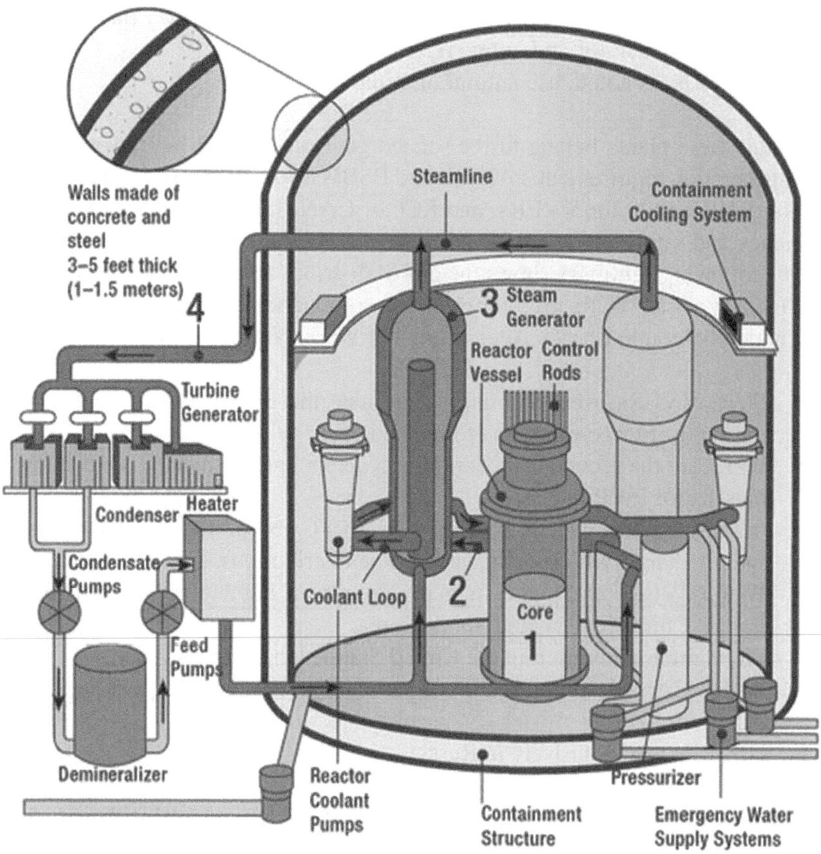

FIGURE 5.2.1 Sketch of a PWR, nuclear regulatory commission from the US, public domain, via wikimedia commons.

These three devices are connected by the primary circuit, including connecting pipes and reactor coolant pumps, which circulate the coolant in the direction indicated by the arrows.

The SG is connected to the secondary circuit located outside the containment structure, on the left of the figure, which corresponds to the thermodynamic cycle followed by steam, symbolized in the diagram by a turbine, condenser, feedwater pump, and heater.

In a PWR, heat extraction from the nuclear core involves two circuits for safety reasons. The cooling fluid (water under pressure) in contact with the reactor core becomes radioactive due to fission products migrating through the cladding and products dissolved in water (mainly due to corrosion) that capture neutrons. To prevent contaminated water from contacting the external environment and passing through electricity production cycle components, the working fluid is separate from the coolant.

The choice of this fluid is based on water's qualities as a coolant (high heat capacity), environmental compatibility, and ease of use (stability, safety, availability). However, it imposes a strong constraint: the need to ensure that water remains in the liquid state in the reactor vessel to avoid local superheating of the nuclear fuel due to the presence of steam, which would induce low heat exchange coefficients.

For this reason, water must be maintained at a pressure higher than the saturation pressure at its maximum temperature in the reactor core. The entire primary circuit must withstand this maximum pressure, resulting in severe mechanical stresses. The circuit is sized accordingly, and the system pressure is regulated with great precision to avoid either overpressure that could cause circuit leakage or rupture, or pressure decrease that risks boiling in the core and fuel melting. The pressurizer's role is precisely to ensure this function.

A compromise must be found between security constraints, cycle thermodynamic efficiency, and installation costs. In existing plants, the maximum temperature of the thermodynamic cycle is approximately set at 280°C and the primary circuit at 330°C. Specifically, the temperature of the primary circuit changes from 290°C (zero power) to 325°C (maximum power).

To ensure non-boiling primary water, the coolant system pressure is set at 155 bar, corresponding to a saturation temperature of 345°C, which provides a small safety margin. Such pressure is already high and imposes severe technological constraints at all levels.

To accurately control the primary circuit pressure, the pressurizer is a reservoir containing water in the two-phase state, with pressure and temperature linked by the saturated pressure law.

To control the pressure, one simply regulates the temperature, which is accomplished by heating or cooling (by spray) water in the pressurizer. It communicates with the entire primary system and sets its pressure.

The remainder of the primary circuit consists of pump unit/SG sets. Pumps serve only to ensure that primary water circulates throughout the reactor and simply offset pressure drops (approximately 8 bar).

5.2.2 Steam Generator

The SG transfers the total power of the reactor to the secondary circuit with a very low temperature difference, as the thermodynamic cycle performance improves with higher temperatures.

We have seen that primary water temperature varies between 290°C and 325°C. The presence of feedwater heaters in the secondary circuit (see below) means that secondary water enters the generator at approximately 220°C.

The current performance of SGs used in PWRs (Figure 5.2.1) leads to a maximum outlet temperature of approximately 275°C.

Given the low temperature differences between primary and secondary circuits, the need to transfer significant power prevents achieving any superheat in the SG, because the exchange coefficients between the primary liquid and superheated steam would be too small.

Therefore, the type of SG used by virtually all manufacturers of nuclear boilers corresponds to Figure 5.2.1.

Primary water enters the SG in the lower part of the unit and passes inside perforated plates designed to distribute the flow between inverted U-shaped tubes.

Water from the secondary circuit enters in the liquid state at approximately 220°C and 70 bar pressure. It is distributed around the SG periphery by a torus provided with tubes that allow it to flow through a bundle acting as an economizer. It then follows an upward flow around the bundle and partially vaporizes with a steam quality between 0.2 and 0.4, in a nucleate boiling regime leading to very high heat exchange coefficients.

The two-phase emulsion then passes through various separators. The liquid fraction falls to the bottom of the SG and is recirculated with the feedwater (the recirculation rate is between 2 and 4.5). The vapor fraction reaches a quality above 0.997 and exits at the top, directed toward the turbine high-pressure (HP) section.

5.3 ARCHITECTURE OF BWRs

One of the main differences between BWRs and PWRs is the absence of a pressurizer. Feedwater flows vertically from bottom to top through the reactor, topped by separators whose role is to ensure that only steam is directed to the turbines, both to limit blade erosion by possible droplets and to minimize the carryover of radioactive isotopes contained in the core cooling water, such as nitrogen-16.

The containment vessel is located on the left of Figure 5.3.1, housing the core cooled by pressurized water. On the right, we see the thermodynamic cycle, with the turbine (8-9), below which are the condenser (12) and the pump (15), and the alternator (10).

In BWRs, water boils directly in the core, producing steam at approximately 70 bar, regulated by the steam control system and separators. The pressure in the reactor vessel is thus slightly less than half that of PWRs, reducing mechanical constraints. However, since the steam produced in the core directly feeds the turbines, a major technical challenge is managing radioactivity in the steam circuit, requiring shielding and specific maintenance procedures.

Additionally, reactivity control is more complex than in PWRs due to thermohydraulic transients, such as density oscillations caused by boiling, which can induce local power variations. BWRs also face internal component corrosion and, in older designs, less robust containments, as seen at Fukushima.

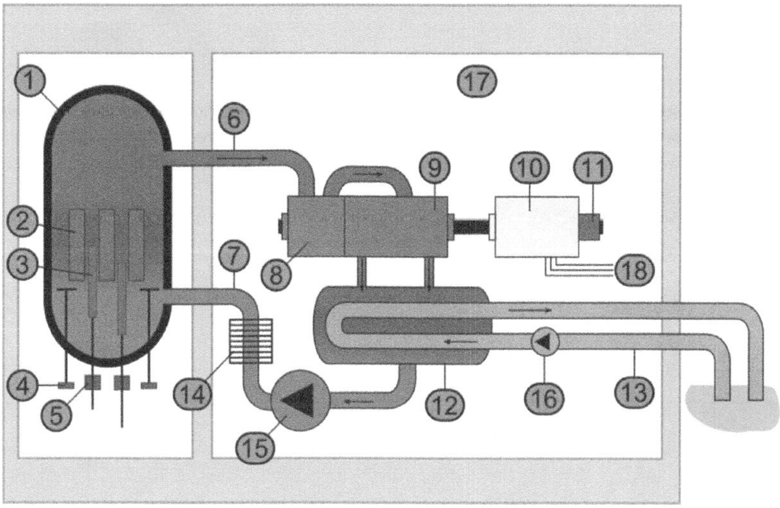

FIGURE 5.3.1 BWR reactor schematic, Robert Steffens (alias RobbyBer 8 November 2004), SVG: Marlus_Gancher, Antonsusi using a file from Marlus Gancher. See File talk: Schema Siedewasserreaktor.svg#License history, GFDL <http://www.gnu.org/copyleft/fdl.html>, via wikimedia commons.

To prevent radioactive steam from escaping at turbine and pump seals, an auxiliary exchanger (often an evaporator) produces pure steam for these seals.

5.4 ARCHITECTURE OF RBMK REACTORS

As previously mentioned, RBMKs and CANDUs are pressure tube reactors, where each fuel assembly is inserted into an independent metal tube. In RBMKs, these zircaloy tubes are approximately 8 cm in internal diameter with 4 mm thick walls, numbering over 1,650. They contain the fuel assemblies and allow coolant water to circulate at 69 bar, with an outlet temperature of approximately 284°C. As in BWRs, water boils directly in these channels, producing steam for the turbines. Separators play a similar role as in BWRs (Figure 5.4.1). The moderator is graphite, which can reach a maximum temperature of 730°C. It surrounds the pressure tubes to slow neutrons, facilitating the chain reaction.

As already noted, RBMKs have a strongly positive void coefficient, which has led to the abandonment of this technology.

5.5 ARCHITECTURE OF CANDU REACTORS

CANDU reactors, developed in Canada, are PHWR that use heavy water as both a moderator and a coolant, and natural uranium as fuel. The choice to develop this nuclear technology in Canada was made to avoid dependence on foreign countries for large metal vessels like those used in BWRs and PWRs, or for enriched uranium production.

FIGURE 5.4.1 The RBMK water and steam networks illustrated in 3D, Tadpolefarm, CC BY-SA 4.0 <https://creativecommons.org/licenses/by-sa/4.0>, via wikimedia commons.

These reactors differ from previous types by their use of heavy water, which is an excellent moderator that slows neutrons sufficiently to allow absorption by natural uranium, making enrichment unnecessary. This results in 25%–30% less natural uranium consumption compared to light water reactors. However, producing heavy water is extremely expensive, requiring complex industrial processes such as distillation or isotopic exchange (e.g., the Girdler sulfide process). The initial cost and replacement of losses (approximately 1%–2% per year due to leaks or degradation) represent a significant expense.

As shown in Figure 5.5.1, a CANDU reactor includes:

1. Fuel assemblies
2. A calandria containing heavy water used only as a moderator
3. Control rods for regulating reactor power
4. A pressurizer similar in function to that in PWRs
5. An SG where the coolant (heavy water) transfers heat to the thermodynamic fluid (light water)
6. A feedwater pump to pressurize the light water
7. A primary pump for circulating heavy water through the pressure tubes
8. A system for extracting fuel assemblies
9. The heavy water in the calandria is maintained at 70°C–80°C
10. The calandria tubes contain pressure tubes where the fuel assemblies are placed; the annular space between the two types of tubes is filled with CO_2, serving as thermal insulation. The pressure tubes are traversed by heavy water used as both a moderator and a coolant
11. Pipes supply the turbine with HP, high-temperature steam
12. Feedwater pipes supply the light water

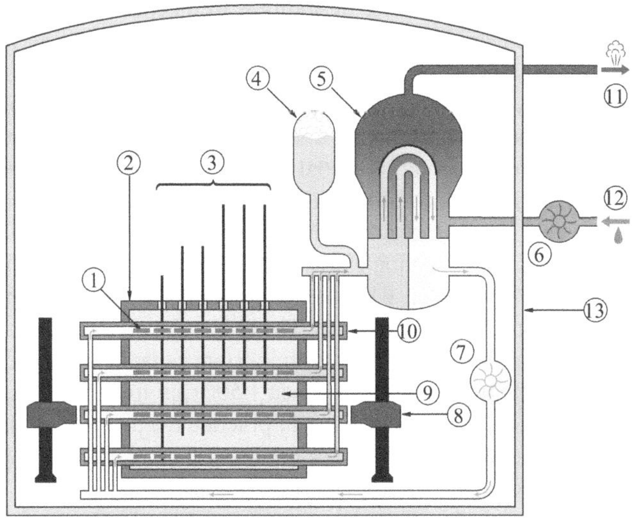

FIGURE 5.5.1 PHWR/CANDU reactor schematic, Inductiveload, CC BY-SA 2.5 <https://creativecommons.org/licenses/by-sa/2.5>, via wikimedia commons.

As previously noted, CANDUs have a slightly positive void coefficient: if the heavy water in the primary circuit vaporizes due to a loss of moderator/coolant, core reactivity increases, which can accelerate the chain reaction. This phenomenon imposes high safety requirements and accident management standards.

5.6 IMPLICATIONS FOR THE SECONDARY CIRCUIT

This review of WCR technologies has allowed us to understand the main differences between them. Nevertheless, the main characteristics of their secondary cycles are very similar: the steam exiting the reactor or SG is always nearly saturated with very low moisture content, around 0.5%–1%, and its pressure ranges between 50 and 80 bar. The feedwater temperature is generally between 150°C and 230°C.

Let us now examine the implications of this observation for the thermodynamic cycles used.

The main distinction between high-temperature steam cycles and medium-temperature cycles coupled with nuclear reactors is that it is impossible to achieve significant superheating in the latter, mainly for safety reasons. It is considered that the heat exchange coefficients on the steam side would be too low to ensure effective cooling of the nuclear reactor or the heat transfer fluid if there is one, which excludes this solution.

In addition, the temperatures themselves are limited in many nuclear reactors for reasons also related to reactor safety. In the most common reactors worldwide, PWRs and VVERs, the mechanical stresses supported by the containment vessel limit the pressure of the primary circuit, and therefore its temperature, because the aim is to prevent any risk of boiling. In BWRs, the constraint is the same.

In CANDU and RBMK reactors, in addition to mechanical strength concerns, there is the risk of runaway neutron reactions in the event of boiling of the reactor cooling water.

Among other nuclear plant cycle features, we have already mentioned the large flow rate values. In fact, on one hand the useful work per unit of flow rate is approximately 60% lower than in conventional coal or oil plants, and on the other hand their optimum technical and economic unit capacities are much larger (900 or 1,300 MWe against approximately 600 in the others). Mass and volume flow rates are very high. At turbine low-pressure (LP) section exit, velocities must be limited to increase efficiency and protect the condenser, which leads to designing giant turbines (diameter up to 7 m) with very large passage sections. The only solution to meet the mechanical constraints at the end of the blades for these important capacities is therefore to halve the rotation speed of the turbine, which is then 1,500 rpm for a 50 Hz network.

It follows from these considerations that in practice the pressure of the secondary cycle is between 50 and 80 bar for all these types of reactors and that superheating is almost impossible to achieve.

As we will see, this translates into many constraints in the thermodynamic cycle, first of all the fact that the expansion in the turbine takes place mainly in the wet zone and no longer in the dry steam zone.

5.7 MOISTURE SEPARATOR REHEATER

Generally, the state of the steam entering the HP turbine is two-phase, with a quality close to 0.995, or a humidity of 0.5%.

Complete expansion of steam from this state would lead to too low a steam quality, which would be both disadvantageous in terms of performance and fatal to the mechanical strength of turbine blades. The solution that is almost always implemented to improve this situation is to reheat the partially expanded steam by taking live steam at the outlet of the SG.

However, before this heating is carried out, the partially expanded vapor is stripped of its liquid phase, which would be counterproductive to revaporate.

This drying of the expanded steam followed by superheating is carried out in a device called an MSR, which is a special feature of all these thermodynamic cycles.

An MSR (Figure 5.7.1) receives on the left steam partially expanded with a quality close to 0.87, the liquid phase of which is separated by chevrons and directed to feedwater heaters, while the steam passes through a heat exchanger traversed internally by a low flow of saturated steam at HP (and therefore higher temperature), which condenses. For a 1,500 MW unit, two MSRs of 370 tons each are needed, 24.8 m long, 6 m high, and 4.3 m wide.

The shape of the chevrons creates a tortuous flow path for the steam. Steam is a gas and flows easily through the chevrons, despite being forced to change direction multiple times. Moisture (water) entrained in the steam is a liquid, which is denser than steam, and consequently is not able to change direction as easily as steam. As moisture flows through the chevrons, it impinges upon the chevron plates and drips down to the base of the MSR due to gravity. The water accumulated at the base of the MSR is referred to as the condensate.

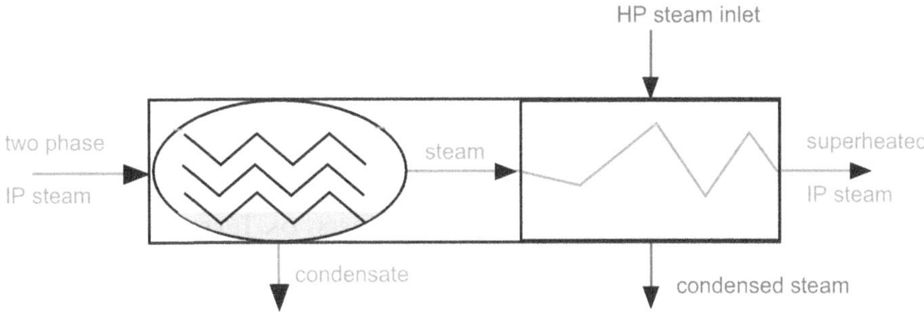

FIGURE 5.7.1 Sketch of an MSR.

This sketch suggests how an MSR can be modeled in Thermoptim: a phase separator followed by a heat exchanger intended to superheat the steam, while the liquid phase is used to preheat the feedwater, as we will detail later.

5.8 EXPANSION IN THE WET STEAM REGION

As previously discussed, a distinctive feature of the steam generated in WCR nuclear power plants is the absence of initial superheating.

This results in significantly higher moisture content at the end of the expansion process compared to conventional thermal power plants, increasing the risk of turbine blade erosion and reducing isentropic efficiency.

These characteristics have multiple implications throughout the cycle.

First, a large portion of the expansion occurs in the wet steam region, where the thermodynamic properties of steam deviate substantially from those of an ideal gas.

Although condensation phenomena in turbines are not yet fully understood, it is generally accepted that condensation begins once the so-called Wilson zone is crossed, corresponding to a moisture content of approximately 3%–5%.

In the HP and LP turbines of nuclear power plants, however, steam moisture typically far exceeds this threshold.

Below this limit, wet steam can often be treated as if it behaves similarly to dry steam, which itself can be approximated as an ideal gas. Beyond this point, such assumptions no longer hold.

It should also be noted in passing that this implies that the usual expression of Stodola's law is no longer valid and must be modified. This point is mentioned in Annex 2 and fully developed in the first volume of this set.

Several drawbacks can be identified:

- Reduced polytropic efficiency due to elevated moisture content
- Increased mechanical and chemical stress on turbine blades
- Increased parasitic droplet condensation, requiring moisture extraction
- Two feedwater heating extractions are in the liquid phase rather than vapor, resulting in suboptimal flow rates for feedwater heating

The latter two issues are closely interconnected.

5.8.1 Reduced Polytropic Efficiency

The reduction in turbine polytropic efficiency is mainly due to the condensation of steam droplets on the turbine blades. This phenomenon can be accounted for using Baumann's rule, presented in Appendix 2.

5.8.2 Increased Mechanical and Chemical Stress on Turbine Blades

Droplets present in the steam flow cause accelerated erosion of turbine blades, with severity increasing as moisture content rises. In addition to mechanical wear, the presence of condensate leads to a significant increase in the concentration of corrosive chemical species, which further degrade blade surfaces.

5.8.3 Condensate Removal in Turbines

As in the MSR previously discussed, droplets are expelled by centrifugal force during expansion in the wet steam region. These droplets accumulate on the turbine casing and must be removed for safety and performance reasons. Their behavior can be modeled using a moisture separator with low droplet removal efficiency.

This process leads to an apparent increase in steam enthalpy between the outlet of the upstream stage and the inlet of the downstream stage, due to the removal of the low-enthalpy condensate.

One possible model is to represent it with a phase separator (Figures 5.8.1 and 5.8.2).

The separator efficiency must be calculated such that the liquid extraction flow rate equals the condensate value. It may range from 0.05 to 0.2, while that of an MSR is close to 0.95.

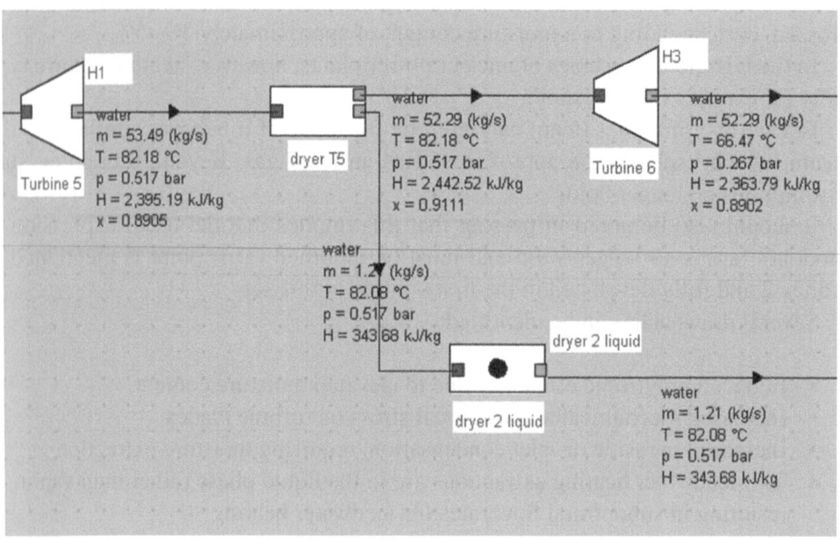

FIGURE 5.8.1 Schematic of a condensate removal system.

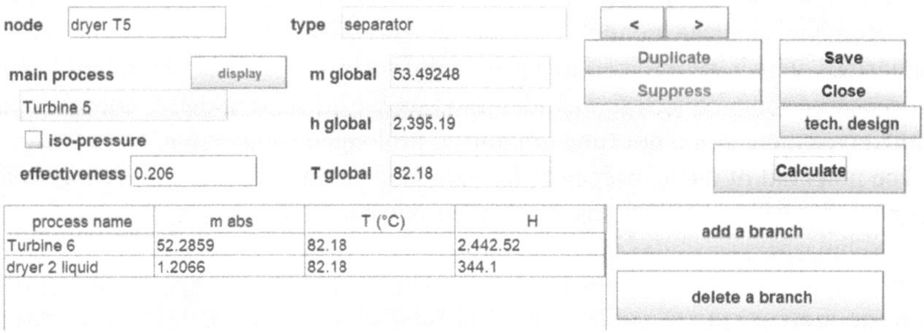

FIGURE 5.8.2 Condensate removal screen.

Note that the enthalpy at the turbine outlet differs from that of the downstream turbine inlet.

If an extraction takes place between the two turbines, it should be placed downstream of the separator, as the condensation takes place inside the turbine and not at its outlet.

5.8.4 Suboptimal Flow Rates for Feedwater Heating

As seen in Section 4.1.5, the calculation of extraction flows for feedwater heating in conventional steam power plants can be based on relatively simple rules, typically resulting in a temperature rise of several tens of degrees in each reheater.

In water-cooled nuclear power plant cycles, however, the situation is much more constrained. At least two large extraction flows (the liquid exiting the MSR separator and the condensed steam at the outlet of the reheater) are liquid instead of steam.

Under these conditions, optimizing the feedwater heating system becomes significantly more complex. Guided Exploration TCNPP-3 presents the complete feedwater system of a PWR NPP.

5.9 SECONDARY CIRCUITS OF WCRs

We will now provide examples of secondary cycles for WCRs, starting with simple examples and progressively increasing their complexity.

Although we have indicated that most of these cycles use an MSR, there are some exceptions, justified by economic or other considerations.

5.9.1 Naval Propulsion

What we have outlined for the four main technologies of civilian WCRs is not an absolute rule, and some cycles deviate slightly from this pattern. This is particularly the case for reactors used in naval propulsion, where optimization criteria differ significantly from those of power plants for electricity production.

Most nuclear reactors installed on ships provide not only propulsion but also heat supply for onboard needs, particularly to power seawater desalination boilers. Thus, they can be considered true cogeneration systems.

The development of nuclear propulsion for maritime applications dates back to the very origins of nuclear technology. During World War II, the limited autonomy of submarines, which were forced to periodically surface to operate their diesel engines and recharge their batteries, was identified as a critical weakness. Diesel engines, requiring oxygen, could not function during prolonged submersion.

The potential of nuclear reactors for extended submerged navigation was quickly recognized, serving as a key driver for their development.

In submarines, specific constraints related to space, maneuverability, acoustic stealth, reliability, and maintenance at sea have led to design specifications different from those of land-based reactors, justifying simpler thermodynamic cycles than those of civilian reactors.

The United States, France, and the United Kingdom, after some prototypes using Advanced Gas-cooled Reactors, have adopted very compact PWR cycles without separators or superheating to minimize auxiliary equipment and mechanical constraints. These use highly enriched uranium (20%–93% U-235), enabling operation without refueling for cycles of 10–40 years.

As it is practically impossible to obtain detailed schematics of military vessel cycles due to their confidential nature, we will limit ourselves here to some general indications.

For surface ships (such as aircraft carriers), PWR cycles with a separator but without superheating are generally used, except in some cases. The pressure in the primary circuit is typically in the range of 30–60 bar, corresponding to a turbine inlet temperature of 235°C–270°C, with thermodynamic efficiency below or slightly above 20%. These reactors must also produce the steam needed for aircraft catapulting or for operating arrestor cables during landings.

To optimize efficiency and power output density, the Soviets initially developed reactors cooled by a lead-bismuth eutectic for their Alpha-class submarines, achieving an efficiency of 40% due to higher temperatures (400°C–500°C). However, this technology was abandoned due to operational constraints: corrosion of steels, solidification of the lead-bismuth (melting point ~125°C), complexity of startups, and maintenance challenges related to radioactivity (production of polonium-210). It was replaced by PWRs.

Nowadays, there is a renewed interest in nuclear naval propulsion. The country with the most experience in this field is Russia, particularly for icebreakers. The type of reactor used corresponds to Small Modular Reactors (SMRs) of the PWR category.

5.9.2 Thermodynamic Cycles of Electricity Generation WCRs

In light of the objectives of this book and the prominence of water-cooled nuclear power plants, which represent over 95% of the world's civil nuclear fleet, it seems appropriate to offer the reader a broad set of examples of secondary cycles representative of these systems.

In this section, we present a summary of key results for these systems, with more detailed discussions provided in the case studies where we will give detailed flowsheets of such cycles for a number of plants.

We will progressively build the secondary circuit of a WCR with an MSR, starting with a highly simplified process flow diagram and then expanding it to achieve a model structure that corresponds to the four categories of WCRs identified at the beginning of this chapter. The power plant we will study is the EPR Flamanville 3, commissioned in late 2024 in France. This is a Generation III PWR of 1,680 MW, designed with significantly improved safety standards.

5.9.2.1 Russian Novovoronezh NPP

We will start with one of the very first civil nuclear cycles commissioned in the world, that of the Soviet VVER 210 reactor, with a separator but without a reheater, which equipped the Novovoronezh power plant in Russia. Since it is undoubtedly the oldest for which the details of the thermodynamic cycle are publicly available, it holds significant historical interest. The original schematic and the plot in the Mollier diagram of this cycle are given on page 221 of the third edition of (Margulova, 1978), Figures XII.2 and XII.3.

The first saturated steam turbine for nuclear power plants in the USSR was the K-70-29. Such turbines were installed at the first and second units of the Novovoronezh Nuclear Power Plant and at the first NPP in the GDR. The K-70-29 turbine had an initial steam pressure of 29 bar and generated 70 MWe. The Novovoronezh plant of 210 MW was equipped with three such turbines.

The plant consisted of six reactor loops and, consequently, six SGs, from which the steam was sent to three turbines, each with a power of 70 MW.

The turbine condensate was pumped by the extraction pump through the LP feedwater heaters installed in series to the deaerator, and then taken up by the feed pump to the HP feedwater heaters in the SG. All the heaters and the deaerator were supplied with steam extracted from the turbines. The steam taken from the turbines was also used for cogeneration, heating water for an urban network. Two heaters were installed in series (one main, one peak), activated only during peak heating periods. The thermal power of the reactor was 760 MW, which corresponds to an efficiency of $(210 \div 760) \times 100 = 27.6\%$.

This turbine had one medium-pressure section and one two-flow LP section.

A separator was installed between the turbine sections. The liquid content after the separator was assumed to be around 1% (which means that the drying was not very thorough), considering the very high steam flow rates through the turbine separator and the need for its compactness. There was no steam reheater.

The relatively small size of the outlet blade allowed for a liquid content of 12.5% after the LP stage, despite the high speed of the machine. With a liquid content of 1% before the LP turbine, the steam pressure was 1.85 bar downstream of the separator. Before the HP turbine, the initial pressure was 29 bar, and accounting for pressure loss in the control devices, it was 27.5 bar at the turbine inlet. As a result of steam expansion in the HP turbine to a pressure of 2 bar, the steam wetness reached 12.3%, approximately the same value as after the LP turbine.

It should be noted that at the end of expansion, the curve in the Mollier diagram shows a slight increase in quality at constant pressure. This corresponds to various phenomena that are difficult to model (leaving losses due to steam exit velocity, delay in condensation, etc.), which exist in most existing WCR cycles and are mentioned in this set's first volume.

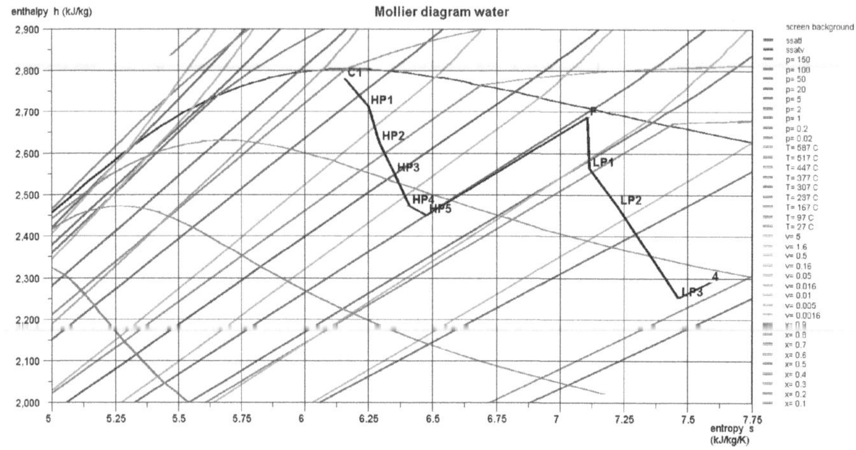

FIGURE 5.9.1 K70-29 turbine expansion plot in the Mollier diagram.

At that time, the state of development, research, and effectiveness of internal turbine moisture-collecting devices did not allow for accounting for moisture removal when plotting the turbine working process in the (h, s) diagram, so the actual liquid content after the turbine was less than the calculated one.

Given that the exact architecture of the water station and the extraction rates are unknown, it is not possible to perform a complete modeling of this cycle. However, its trace in the Mollier diagram allowed us to estimate the polytropic efficiencies of its various Turbine Stage Groups (TSGs), and we prepared a possible model based on our assumptions (Figure 5.9.1).

5.9.2.2 WCR Model with MSRs and Two Extractions

One of the simplest process flow diagrams of the secondary circuit of a WCR for electricity production is shown in Figure 5.9.2. On the left are the economizer and the evaporator, heated by the nuclear reactor, followed by a divider separating the HP turbine and the live steam flow necessary to superheat the intermediate-pressure (IP) steam, with two pressure regulating valves. The HP turbine is located upstream of the separator, with a second turbine section placed between the outlet of the reheater and the condenser.

An extraction pump raises the pressure of the condensed water before it enters the feedwater tank along with the two flows exiting the MSR.

The feedwater is then repressurized before entering the nuclear reactor.

As explained in Section 5.7, an MSR performs two complementary functions:

- Separating the liquid and vapor phases of the incoming IP wet steam
- Superheating the IP vapor phase using a bleed of HP steam from the nuclear reactor

The model of the MSR is given in the upper right of the figure, surrounded by a rectangle. It includes a separator and a heat exchanger between IP steam and the extraction of saturated HP steam (called "reheat" in the process flow diagram).

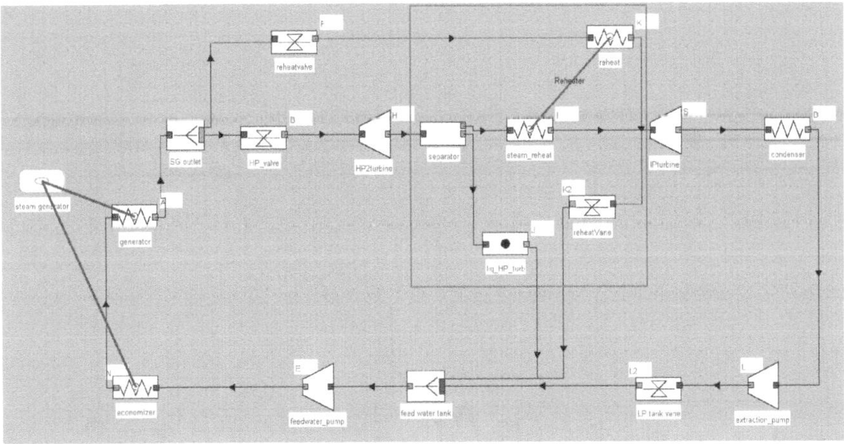

FIGURE 5.9.2 Diagram of the WCR secondary circuit.

At the reactor outlet, a fraction of the flow is diverted to the superheater, while the main stream is expanded at IP. A separator recovers the vapor which is superheated by exchange with the fraction of flow which was extracted before. The superheated steam is expanded and then condensed, before being directed toward the feedwater tank where the other streams flow. Water is then delivered under pressure to feed the SG.

To model an MSR in Thermoptim, we need a phase separator, which is a special kind of node, and a heat exchanger.

The three components of the MSR (Figure 5.9.3) are:

- the separator, which separates the liquid and vapor phases
- the "reheat" process, which represents the fraction of the flow of HP saturated steam leaving the reactor, which condenses
- the "steam_reheat" process, which represents the IP steam which is superheated

The link between these two processes corresponds to the heat exchanger called reheater.

The separator screen is given in Figure 5.9.4. It receives the flow leaving the HP turbine at medium pressure and at the corresponding saturation temperature, with a quality equal to 0.857 here. A parameter quantifies the effectiveness of the phase separation. Indeed, a small part of the liquid is not condensed and remains in the outgoing steam.

The result of the calculation is that 13.6% of the flow rate entering the separator is found in liquid state and is directed toward the feedwater tank, while 86.4% corresponds to saturated steam at IP and is directed toward the reheater.

The screen of the reheater heat exchanger is given in Figure 5.9.5.

In the left part appear the characteristics of the hot fluid, which is the steam coming out of the reactor, and in the right part are displayed those of the cold fluid, leaving the separator, whose temperature goes from 184.45°C to 257.7°C, which corresponds to a superheat of approximately 73.2°C.

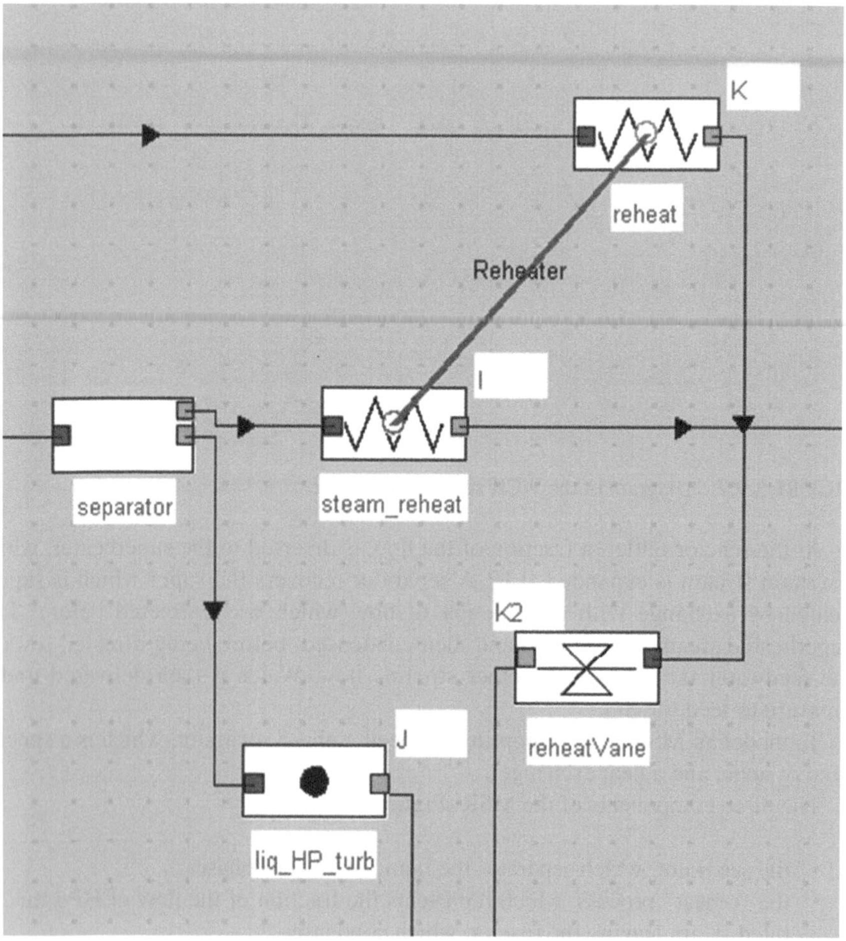

FIGURE 5.9.3 Layout of the MSR model.

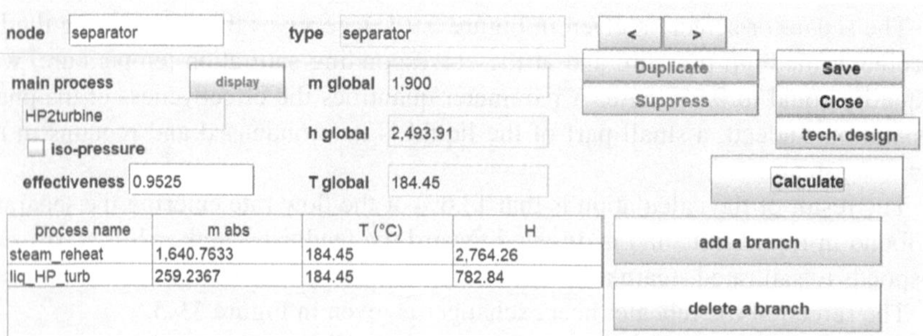

FIGURE 5.9.4 Separator screen.

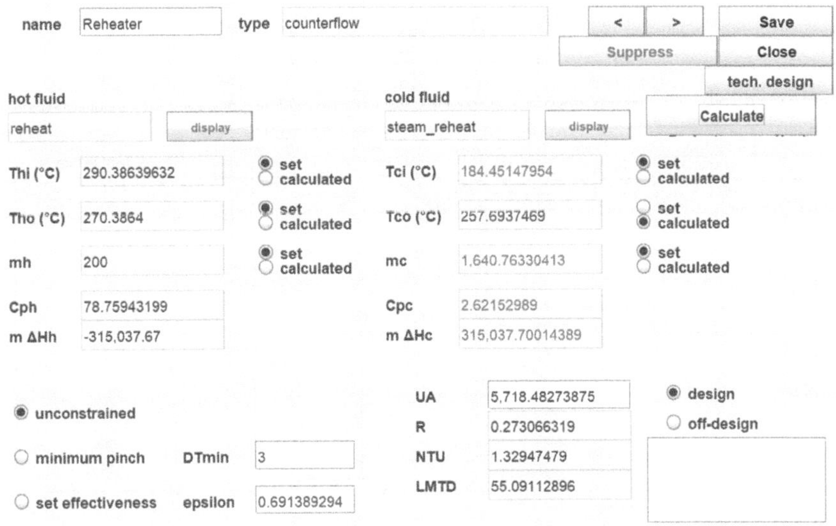

FIGURE 5.9.5 Reheater screen.

To set the superheater, we set all the known elements, namely the flow rates of the two fluids, the state of the water entering and leaving the hot process (the HP steam which condenses), and the state of the water entering the cold process (the medium-pressure steam that we want to superheat).

We have thus set five constraints, so that the exchanger can be calculated.

The superheated IP steam is directed to the LP turbine, where it produces mechanical power before being condensed, while the two other flows leaving the MSR are mixed in the feedwater tank with the partially compressed condensed water, before being compressed to HP by the feedwater pump (Figure 5.9.2, where the MSR is shown surrounded by a rectangle).

The efficiency of the cycle is relatively low (32.3%), but this is explained by the low temperature of the feedwater due to the absence of feedwater heaters. Instead of being close to 227°C as in the actual cycle, it is only 74.4°C.

The flowsheet of this WCR secondary circuit is given in Figure 5.9.6.

5.9.2.3 WCR Model with MSRs and Four Extractions

It is possible to improve this model by adding extractions and feedwater heaters. In the model we are presenting now, both HP and LP turbines are decomposed into two TSGs, with two complementary extractions, one at HP and the other at the outlet of the IP turbine.

In addition to the feedwater tank, the model also includes two closed feedwater heaters, a LP one and a HP one. The new flowsheet is shown in Figure 5.9.7.

As you can see, its efficiency is significantly improved compared to the previous model, reaching 34.1%, due to the fact that the feedwater now enters the SG at approximately 210°C instead of 74°C.

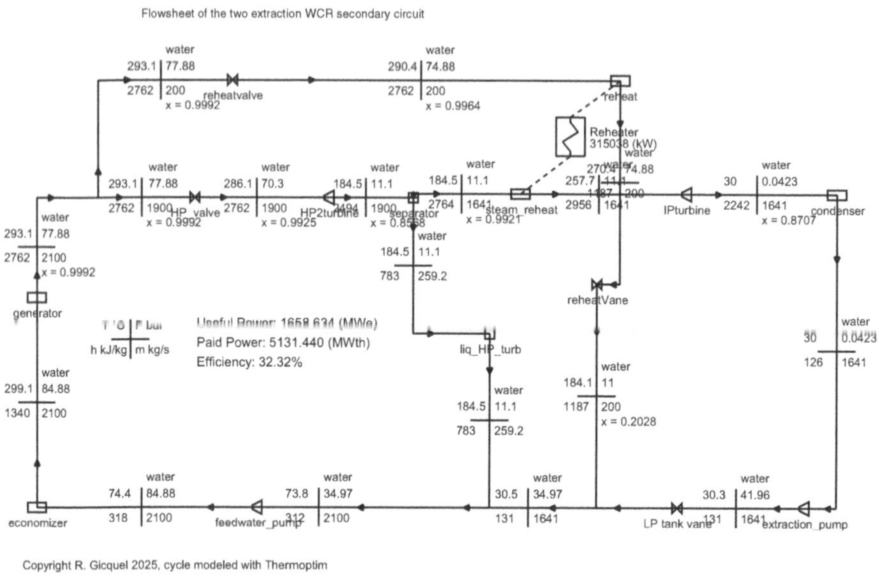

FIGURE 5.9.6 Flowsheet of the two extraction WCR secondary circuit.

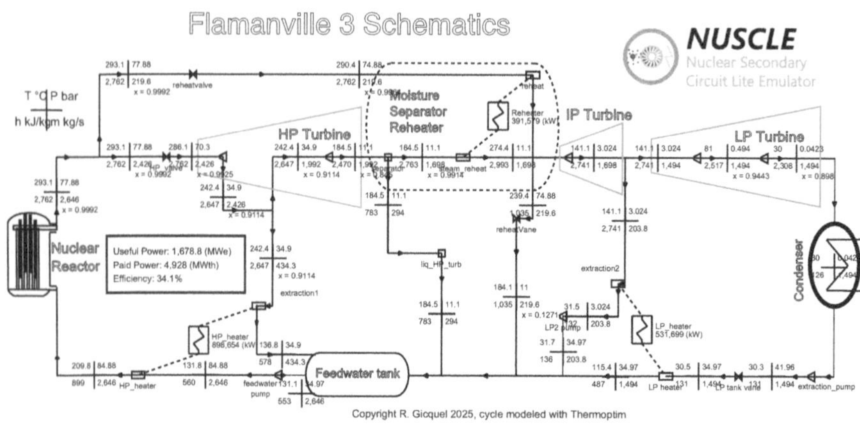

FIGURE 5.9.7 Flowsheet of the WCR secondary circuit.

This model is much more precise than the previous one, but it is still far from representing all the complexity of the real cycle, as we will see later when analyzing the detailed model of the EPR Flamanville 3.

5.9.2.4 NUSCLE: A Pedagogical Modeling Tool

The four-extraction model presented above achieves an efficiency of 34.1% and represents a good pedagogical compromise between simplicity and realism. Although Thermoptim does support off-design analysis, this capability was not included in the previous model.

Rather than extending the Thermoptim model, we chose to develop **NUSCLE** as a **standalone, fully open-source tool dedicated specifically to WCR analysis**. NUSCLE implements the same four-extraction, five TSG model structure while adding comprehensive off-design calculation capabilities. This independent tool comes with complete documentation: the full equation set, solution methodology, and operational data for about ten reference plants.

5.9.2.4.1 Capabilities and Educational Value

NUSCLE (Nuclear Secondary Circuit Lite Emulator) enables users to:

- Explore how key parameters affect the performance of various WCR types (PWR, BWR, VVER, RBMK, CANDU)
- **Analyze part-load behavior** using Stodola law for TSG modeling and NTU-effectiveness models for heat exchangers
- Compare different reactor secondary cycles with reasonable accuracy
- Access and modify the complete set of equations for deeper understanding (a key advantage of the open-source approach)

The tool is coupled with the IF97 water property library for high accuracy. Its graphical interface (Figure 5.9.8) displays the cycle diagram with interactive value readouts

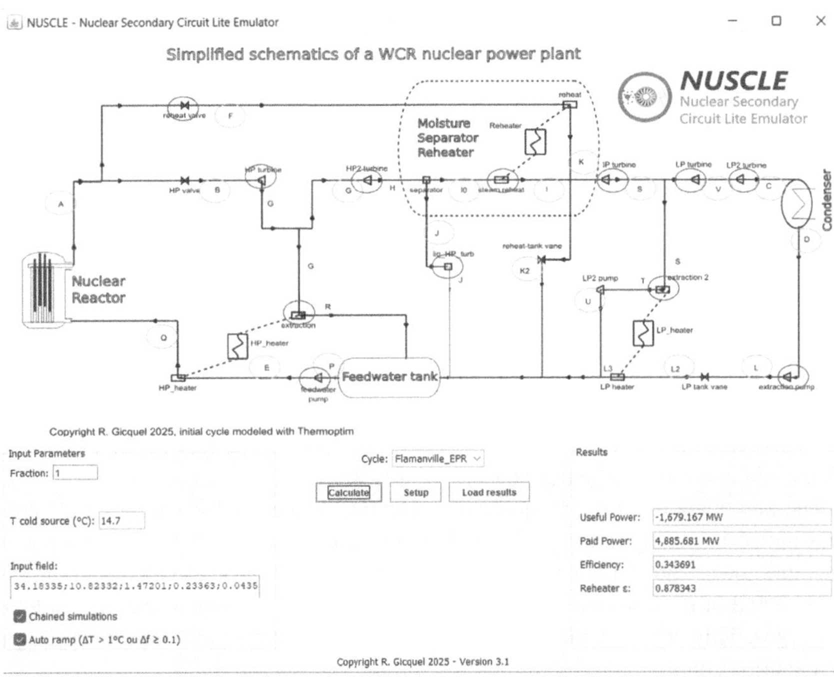

FIGURE 5.9.8 NUSCLE interface.

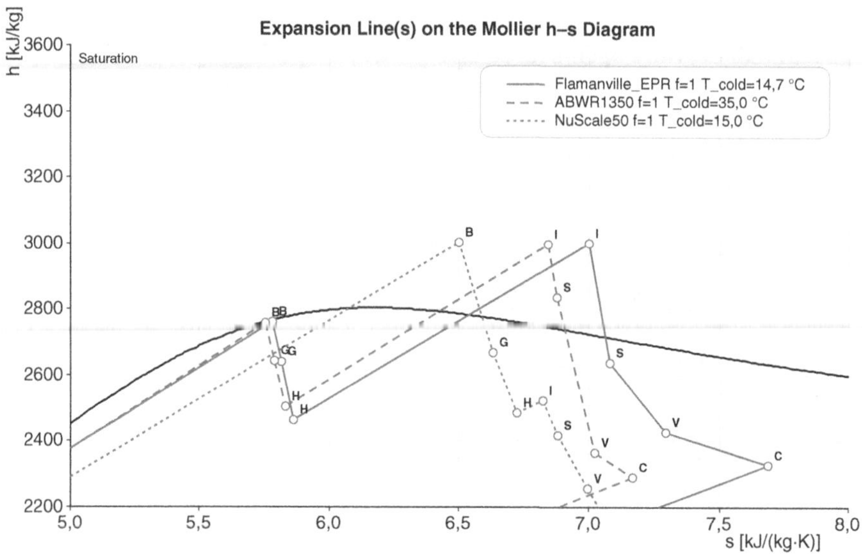

FIGURE 5.9.9 Expansion line(s) on the Mollier h–s diagram.

and enables sensitivity studies as well as off-design condition analysis—capabilities that few educational software programs achieve with such accessibility.

- The simulation results can be exported in several formats:
- a detailed text file containing almost all calculated quantities;
- a summary window displaying the thermodynamic states of the main points, the mass flow rates and power of the turbines, the overall performance indicators, and the global convergence criterion of the model (Figures 5.9.14 and 5.9.15);
- a graphical window showing the expansion line of the turbines on the Mollier h–s diagram (Figure 5.9.9).

5.9.2.4.2 Technical Foundation

The model requires solving several hundred highly nonlinear equations. Given the load factor and condenser inlet cooling water temperature as inputs, seven principal variables must be determined: the condensing pressure, five TSG inlet pressures, and the high-temperature reheater outlet temperature. These principal variables allow all other cycle quantities to be derived.

The key challenge lies in equation hierarchy. Generic solvers treating all equations indiscriminately fail beyond one hundred equations. NUSCLE's solution separates unknowns into three categories: pre-calculable parameters, principal variables recalculated at each iteration, and dependent variables computed once the solution converges. This organization ensures convergence when properly initialized despite the equation count.

The model originated from Thermoptim, which generated 256 equations (about ten redundant). These were then supplemented to implement off-design calculations and the solution algorithm.

5.9.2.4.3 Application to Flamanville 3 EPR

Configured to represent the Flamanville 3 EPR, NUSCLE's thermodynamic diagram is shown in Figure 5.9.10. Although simplified, the model successfully captures the main cycle characteristics: maximum pressure, electrical power output, separator pressure, superheating temperature, and feedwater temperatures.

Although the detailed Flamanville cycle model is only studied later in Section 5.9.2.7, it is already interesting to compare the results it provides with those of NUSCLE.

Figure 5.9.11 illustrates simulated cycles plotted in the Mollier chart for power levels of 30%, 50%, 70%, and 100%. The simulation curves closely match actual data for HP and IP turbines. Discrepancies in the LP expansion line arise because the simplified model includes only two stage groups instead of the real system's six.

The global performance comparison (Figure 5.9.12) shows that generated power is accurately reproduced, but cycle efficiency is lower in the simplified model—an expected outcome. With only four versus eleven steam extractions, achieving feedwater temperatures close to actual values requires significantly increasing the HP

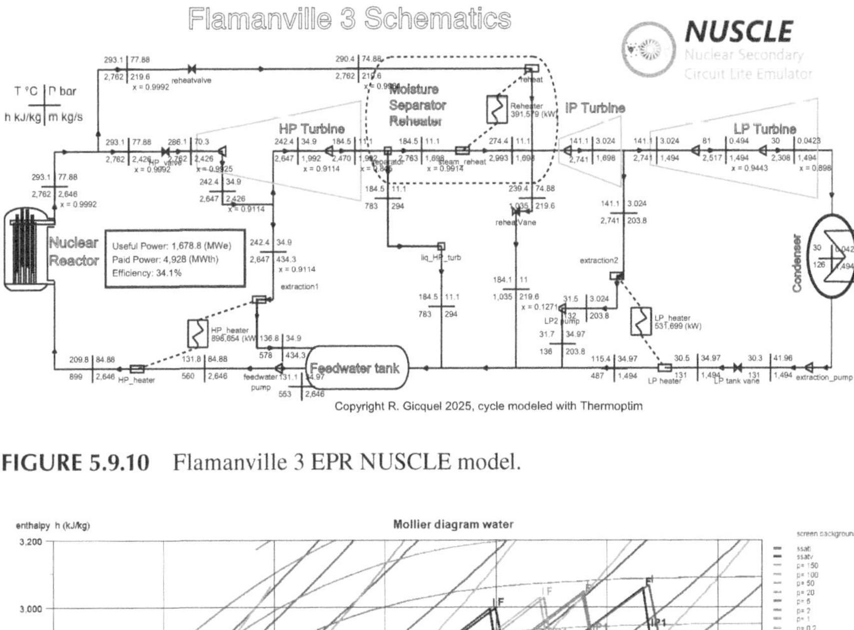

FIGURE 5.9.10 Flamanville 3 EPR NUSCLE model.

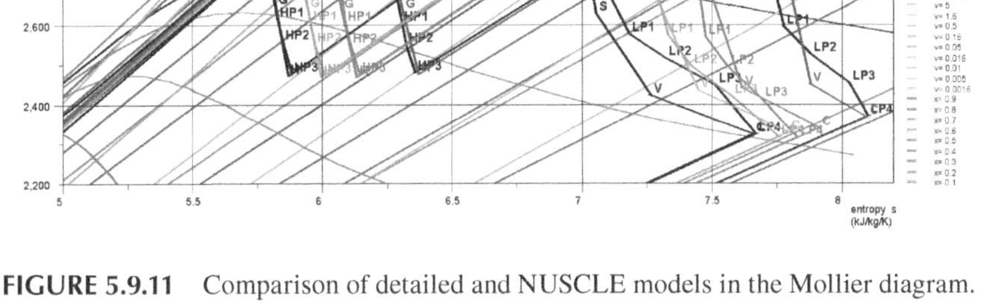

FIGURE 5.9.11 Comparison of detailed and NUSCLE models in the Mollier diagram.

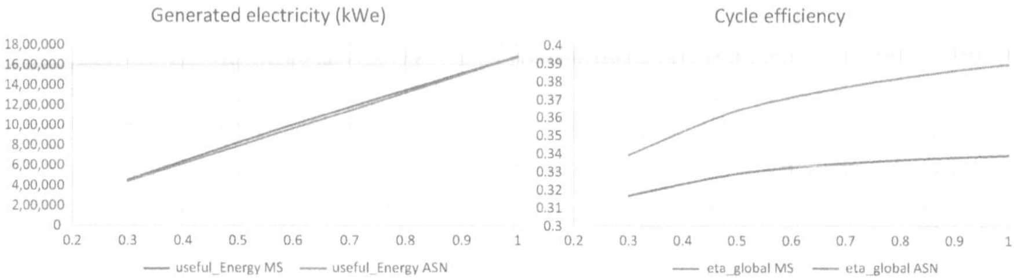

FIGURE 5.9.12 Comparison of detailed and NUSCLE model global performances.

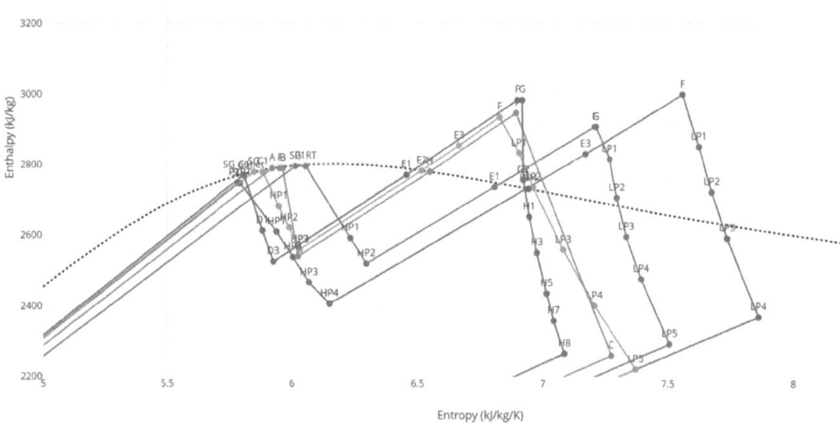

FIGURE 5.9.13 Comparison of WCR cycle plots.

extraction and thus the flow passing through the nuclear reactor to compensate for it and maintain the steam flow through remaining turbine stages. The reactor thermal input is increased, thus explaining the lower efficiency.

Despite these limitations, NUSCLE successfully demonstrates the fundamental thermodynamic behavior of WCR secondary circuits while remaining computationally accessible for educational purposes.[1]

5.9.2.5 Comparative Results for Various WCR Types

Examples of BWR, VVER, RBMK, and CANDU cycles are provided in the case studies, presenting both simplified and detailed flowsheets for each technology.

In this section, we give just a few comparative results obtained with these models.

Figure 5.9.13 shows the comparison of the plots of PWR, BWR, VVER, RBMK, and CANDU models in the Mollier diagram.

The online interactive version allows you to display extensive information and to select only the cycles that interest you. It is available at this address: https://www.s4e2.com/diag/fla3/HS-cycles-plotter.html.

Simulation results provided by NUSCLE for four WCRs are given in Figures 5.9.13 and 5.9.14. These screens summarize the simulation results. The upper part shows the values of the main points of the model, and the lower part shows the flow rates and

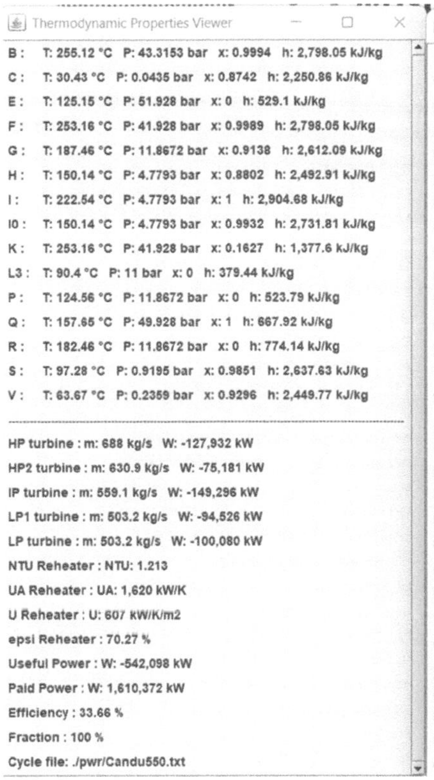

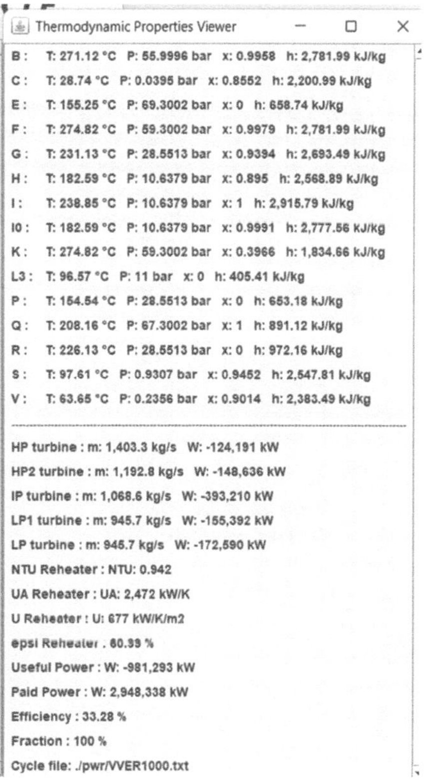

FIGURE 5.9.14 CANDU and VVER NUSCLE model results

power output of the turbines, followed by the characteristics of the steam reheater, and finally the elements of the performance summary.

5.9.2.6 Example of an SMR: The NuScale 77 MW Module

Not only do naval propulsion nuclear reactors deviate from the rules we have established for WCRs. Certain SMRs, in particular, adopt cycles with slight superheat or without an MSR. Two examples of such cycles without a separator and with superheating are those of the NuScale 50 and 77 MW SMRs.

The NuScale concept is an integral PWR, where all major components (core, SG, pressurizer) are contained within a single vessel measuring 23 m in height and 4.5 m in diameter. The reactor core is cooled by natural convection (thermosiphon), eliminating the need for circulation pumps, which enhances its safety.

The SG in the NuScale design consists of helical heat exchangers integrated within this vessel. It produces steam at 34 bar, superheated to 307°C, with the highest temperature in the reactor being 310°C. The initial superheat is thus approximately 66°C for the NuScale 50 MW and 45°C for the 77 MW version.

In the 50 MW NuScale, the feedwater temperature is raised to 149°C using three feedwater heaters.

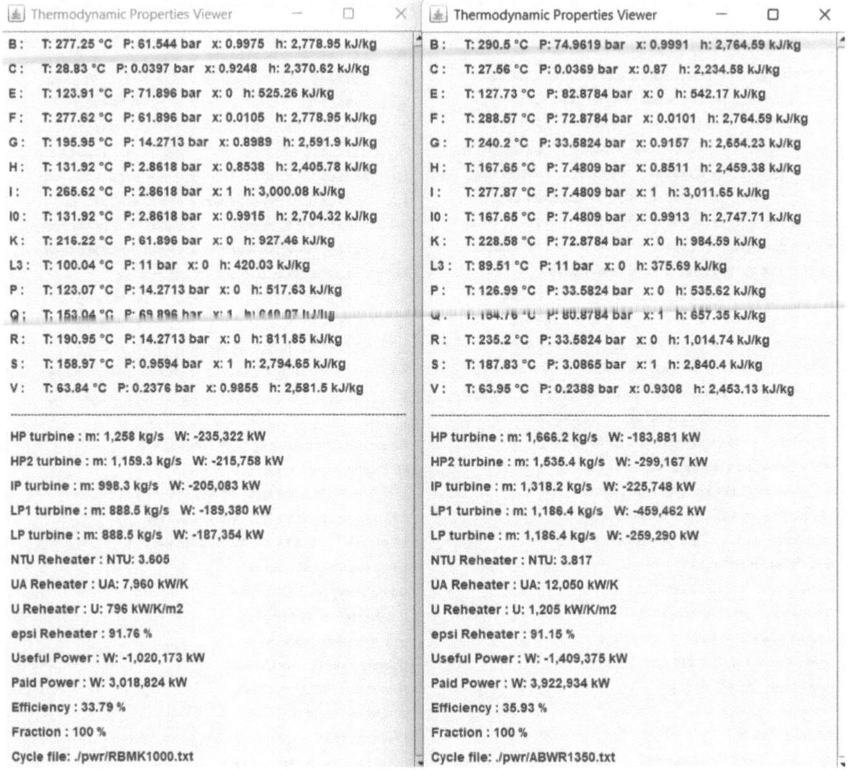

FIGURE 5.9.15 RBMK and ABWR NUSCLE model results

There is no separator in the strict sense, but condensates are removed at two stages.

Detailed models of the NuScale 50 and 77 are presented in Case Studies 8.2 and 8.3, respectively.

5.9.2.7 Detailed Model of Flamanville 3 EPR

We now have all the elements necessary to understand the detailed model of Flamanville 3. You will see that this model is much more complex than those we have discussed so far.

The thermodynamic cycle provided is based on detailed official data submitted by EDF to the French Nuclear Safety Authority (ASN) for licensing approval.[2]

Its detailed flowsheet, divided into two parts for better readability, is shown in Figure 5.9.16.

We recommend downloading it and working from this version, which is much easier to read.

This cycle involves nine TSGs, 11 extractions, two reheaters in the MSR, and six feedwater heaters: four LP feedwater heaters of the simple heat exchanger type (noted as LP1 to LP4 in the figure), and two HP closed feedwater heaters with cascaded drains (HP1 and HP2 reheater).

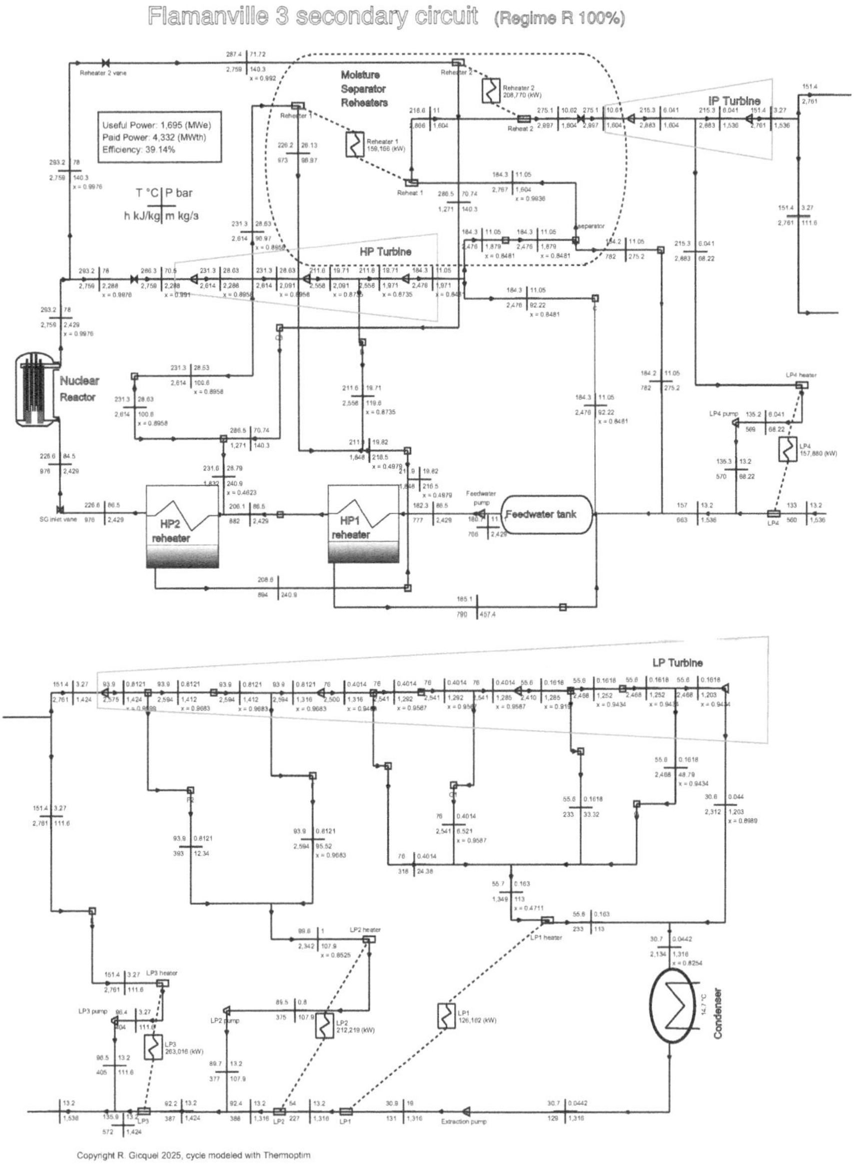

FIGURE 5.9.16 Detailed flowsheet of Flamanville 3 EPR.

5.9.2.7.1 Steam Circuit

The nuclear reactor is shown on the left of the figure. It receives, in its lower part, 2,429 kg/s of feedwater at 84.5 bar and 226.6°C. It exits at 78 bar and 293.2°C, with a quality of 0.9976.

The steam is then divided into two flows: the main flow of 2,288 kg/s, directed to the HP turbine, and a secondary flow of 140.3 kg/s, used to heat the second reheater.

The HP turbine consists of three TSGs. At the outlet of the first TSG, at 28.63 bar, 231.3°C, and a quality of 0.8958, two extractions are performed: one of 96.97 kg/s to supply the first reheater, and another of 100.6 kg/s to supply steam to the second HP feedwater heater.

At the outlet of the second HP TSG, at a pressure of 19.71 bar and a quality of 0.8735, a new extraction of 119.6 kg/s is used to supply steam to the first HP feedwater heater.

At the outlet of the HP turbine, at a pressure of 11.05 bar and a quality of 0.8481, an extraction of 92.22 kg/s is used to heat the feedwater tank.

The main flow then enters the separator of the MSR. The liquid phase (275.2 kg/s) also joins the feedwater tank, while the vapor phase enters the first reheater at 184.3°C, exiting at 218.6°C before entering the second reheater, which raises it to 275.1°C at a pressure of 10.62 bar.

It then enters the IP turbine, consisting of two TSGs. At the outlet of the first, at a temperature of 215.3°C and a pressure of 6.041 bar, an extraction of 68.22 kg/s is used to heat the LP4 feedwater heater.

The main flow enters the second IP TSG, exiting at 3.27 bar and 151.4°C.

At the outlet of the IP turbine, an extraction of 111.6 kg/s is used to heat the LP3 feedwater heater.

The remaining flow enters the LP turbine, consisting of six TSGs. At the outlet of the first TSG, at 0.8121 bar and a quality of 0.9683, a condensate evacuation of 12.34 kg/s takes place, followed by an extraction of 95.52 kg/s of steam, which, mixed with the condensates, is used to heat the LP2 feedwater heater.

The main flow enters the second TSG. At the outlet, at 0.4014 bar and a quality of 0.9587, a condensate evacuation of 24.38 kg/s takes place, followed by an extraction of 6.52 kg/s of steam, which, mixed with the condensates, is used in part to heat the LP1 feedwater heater.

The main flow enters the third TSG. At the outlet, at 0.1618 bar, a condensate evacuation of 33.32 kg/s takes place, followed by an extraction of 48.79 kg/s of steam, which, mixed with the condensates, is also used to heat the LP1 feedwater heater.

The remainder passes through the fourth TSG and exits the LP turbine at a pressure of 0.0442 bar, with a quality of 0.8989.

The condensates exiting the LP1 feedwater heater are mixed with the expanded steam before entering the condenser, from which they exit in the liquid state at 30.7°C.

5.9.2.7.2 Feedwater Circuit

At the outlet of the condenser, an extraction pump compresses the 1,316 kg/s of feedwater to a pressure of 19 bar, then it is preheated in the LP1 feedwater heater to 54°C, and then in the LP2 feedwater heater to 92.2°C. Once condensed, the steam entering this second heat exchanger is repressurized and mixed with the feedwater, the combined flow being preheated in the LP3 feedwater heater to 133°C, and here too mixed with the repressurized condensed steam. The same process occurs at the LP4 feedwater heater; then, the total flow of 1,536 kg/s at 157°C enters the feedwater tank. These feedwater heaters are thus of the pumped forward type.

The feedwater tank receives four fluids: the three mentioned previously, as well as condensates from the HP1 feedwater heater. It outputs 2,429 kg/s of water at 11.1 bar and 180.7°C, which is brought to a pressure of 86.5 bar by the feedwater pump and then directed successively to the two HP feedwater heaters, exiting at 206.1°C and then 226.6°C, before entering the SG. These two feedwater heaters are of the cascaded drains type.

The cycle is thus completed.

In this model, we have not accounted for pressure losses in the feedwater piping, which remain low.

The overall balance is summarized in the legend located in the upper left part of the process diagram.

5.9.2.7.3 Regulation of the Primary Circuit Flow Rate

Generally, in a PWR reactor, power control according to electrical demand is achieved through four complementary means:

- Primary pump flow modulation that influences heat transfer and thus moderator temperature. A reduction in the primary flow decreases the removal of heat from the core, which tends to increase the temperature of the moderator and therefore reduce reactivity (negative feedback)
- Control rods that insert more or less into the core to directly adjust reactivity through neutron absorption, and thus the thermal power generated by the reactor core
- Boric acid concentration in the primary circuit, diluted or concentrated for fine and continuous reactivity control
- Secondary pressure control that modifies heat exchange conditions in the SG and indirectly affects primary temperature. Indeed, as the heat exchange occurs in a two-phase regime in a large part of the SG, varying the pressure of the secondary circuit changes the saturation temperature, which affects the average temperature difference between the two fluids and thus the thermal power exchanged

These systems work in a coordinated manner through automated control. This control exploits thermohydraulic coupling between primary and secondary circuits, enabling flexible and efficient load following according to grid demand.

A purely thermal regulation of the SG would operate as follows during a reduction in the demanded power output:

- To maintain the correct rotational frequency of the turbine, its inlet valves would reduce the steam flow rate. The thermal power produced on the primary side would initially be in excess, and the SGs would produce too much steam. The only way to reduce this excess is to decrease the temperature difference within the SG, by increasing the saturation temperature of the steam and thus the pressure of the secondary side, with equilibrium being gradually restored.

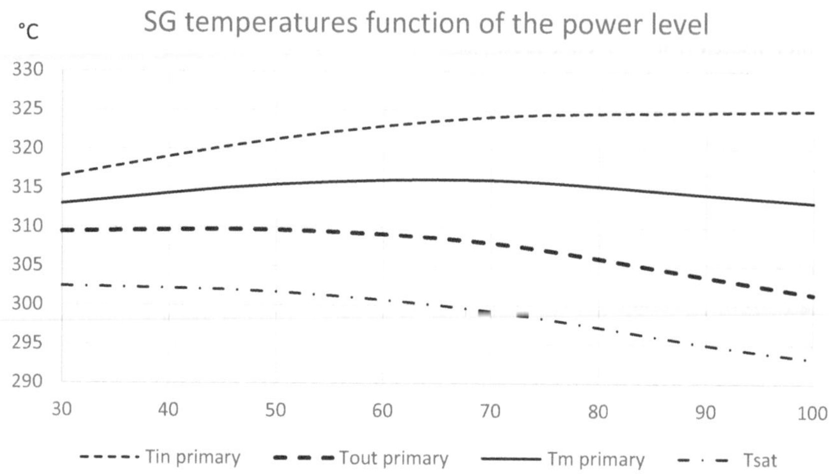

FIGURE 5.9.17 SG temperatures.

- This would result in an increase in the return temperature on the primary side.
- The arrival of hotter water would lead to a drop in reactivity, and consequently a reduction in the neutron flux and the thermal power generated by the reactor.

The system would therefore be globally stable.

However, this regulation mode, where the return temperature of the primary circuit varies with the load, has several drawbacks, including the modification of the water volume in the primary circuit, which can notably cause malfunctions in the pressurizer.

The regulation used therefore aims to limit variations in the average temperature of the primary circuit by modulating the temperature difference within the heat exchanger.

In most second-generation PWRs, the flow rate of the primary circuit is kept constant, and regulation is achieved by adjusting the control rods, boron injection, and the pressure of the secondary side.

In order to study the interdependence between the parameters which govern the equilibrium of the SG, we created a model of its off-design behavior, and we used it to analyze the detailed data related to the Flamanville 3 EPR.

This work is the subject of Case Study No. 8.10. All calculations done, we obtain the curves of Figure 5.9.17, which show the evolution of the temperatures in the SG. It clearly shows that the average temperature of the primary circuit varies much less than the other ones.

5.9.2.7.4 Flowsheets of the Part-Load Flamanville Cycles

The available data corresponds to six complementary operating flowsheets for two turbines, labeled A and B. We used those for turbine B, in regime R, at five load rates: 100%, 90%, 70%, 50%, and 30%.

This page of this book's website gives you access to the flowsheets of the five cycles, created in SVG format.[3]

5.10 ORC POWER PLANTS

5.10.1 LIMITS OF STEAM CYCLES

Steam power plants use water as a working fluid, which turns out to be an excellent thermodynamic fluid, with its critical temperature of 374°C and its high latent heat of vaporization at ambient pressure and temperature. Its low viscosity makes it possible to limit the consumption of auxiliarics, and its reduced price and non-toxicity finish placing it in a very good position compared to all its competitors. However, for certain applications, other fluids may sometimes prove more suitable.

The main field of application for steam cycles is that of producing electricity from either fossil fuels, for capacities of around 300–800 MW, or from nuclear reactors, for even higher capacities.

5.10.2 ADVANTAGES OF ORCs

Organic Rankine Cycles (ORCs) are variants of steam cycles, which are used when the hot source from which one wishes to produce mechanical power is at low or medium temperature, or when the installed capacity is low and steam installations are no longer economical.

As their name suggests, ORCs are steam cycles, which is why we discuss them in this chapter.

When the temperature of the hot source drops, or the installed capacity decreases, typically below ten MW, the performance of steam cycles deteriorates, and it becomes preferable to resort to other thermodynamic fluids.

As many of these are organic in nature, it is customary to label these cycles organic, but other types of fluids, such as ammonia or carbon dioxide, can be used.

In large installations, the expansion machine is generally a turbine. In the middle range, volumetric devices are preferably used, such as screw machines, and, for very small capacities, scrolls or piston machines.

ORCs have aroused growing interest worldwide since the end of the 1980s, particularly in Anglo-Saxon countries, for four main classes of applications, which account for most of the achievements, in terms of installed capacity and number of installations:

- geothermal power plants;
- biomass combustion plants;
- heat recovery from effluents;
- thermodynamic solar power plants.

Their use in the nuclear field is now being considered, particularly for SMRs.

Unlike water, the dew curve of some organic fluids has a slope such that the expansion takes place entirely in the vapor phase, and even deviates from the saturation curve.

Under these conditions, it may be advantageous to use a regenerative cycle.

In this cycle (Figure 5.10.1), the steam leaves the turbine in state 4, in the superheated state. The energy corresponding to the desuperheating (4 – 4') is recovered to preheat the condensed liquid before entering the SG (2 – 2').

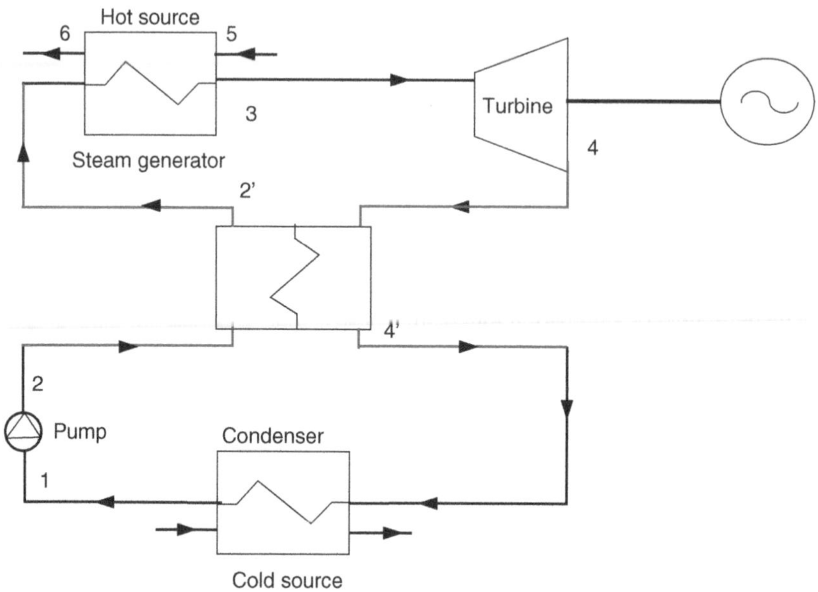

FIGURE 5.10.1 Regenerative ORC.

5.11 BINARY CYCLES

We pointed out in Section 5.2.3 that one of the drawbacks of steam is the great value of its specific volume at low temperature, which results in:

- first, LP stages have large exhaust sections and huge fins that work at the strength limit of the material;
- second, leaving losses by residual velocity may exist when the condensing temperature is very low (see Section 9.4.5 of this set's first volume).

To overcome these drawbacks, we consider the use of binary water/ammonia cycles in which steam is replaced, in the terminal portion of the LP expansion (when its specific volume is very high), by ammonia which, under the same conditions of temperature, would be about 120 times denser. A Thermoptim model of such a cycle is given in Figure 5.11.1.

In this cycle, ammonia plays a dual role:

- it transfers heat between the LP steam turbine outlet and the cold source, i.e., ambient air;
- it serves as working fluid expanding in a second turbine that replaces the LP steam turbine stages.

Steam is condensed in a condenser/boiler consisting of bundles of tubes on which occur steam condensation and ammonia vaporization, the temperature difference

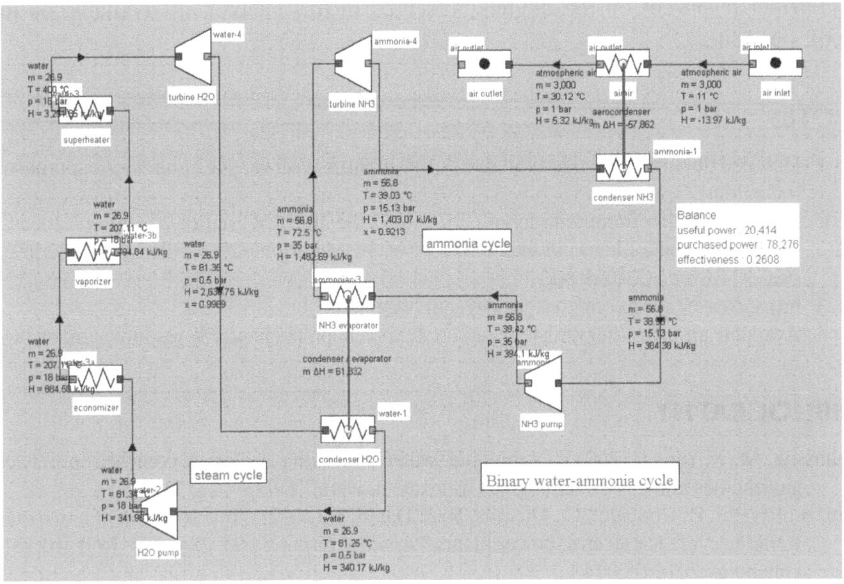

FIGURE 5.11.1 Flowsheet diagram of a binary water-ammonia cycle.

between the two fluids being close to 7°C. Ammonia vapor is then expanded to produce work in an ammonia turbine driving an alternator.

After expansion, the vapor condenses in contact with the cold source. Heat transfer between the condensing ammonia vapor and air takes place in an air condenser made up of finned tubes. After condensation, ammonia is pressurized, the bottom cycle being thus closed. Guided Exploration TCNPP-4 Exploring a Water-Ammonia Binary Cycle presents such a cycle.

If the use of wet cooling towers is called into question because of the growing environmental constraints (scarcity of water resources and global warming), the use of binary cycles of this type could become widespread. Recent summer heat waves have shown that the solutions being implemented will not continue to be sufficient in the future if temperature continues to rise. Other fluids than ammonia may also be considered.

5.12 CONCLUSION

This chapter examined the medium-temperature steam cycles associated with WCRs, which constitute over 95% of the global civilian nuclear fleet. Unlike high-temperature cycles, these cycles feature unique thermodynamic characteristics linked to nuclear constraints: the impossibility of significant superheating for safety reasons, expansion into the wet steam region, and the necessity of MSRs. The four main reactor categories were analyzed (PWR, BWR, RBMK, CANDU) through increasingly detailed thermodynamic modeling. Although these cycles achieve lower thermodynamic efficiencies than conventional power plants (32%–35% compared to 40%–45%), they ensure safe and reliable operation given the specific constraints of nuclear technology.

Further resources on the models presented in this chapter are available on this book's website.[4]

NOTES

1 Detailed documentation is available online at: https://server.s4e2.com/crc/tcnpp/nuscle/index.html.

2 https://www.asn.fr/content/download/194137/file/10.%20Circuit%20vapeur%20et%20transformation%20de%20l%E2%80%99%C3%A9nergie%20-%20Rapport%20de%20s%C3%BBret%C3%A9.pdf.

3 https://server.s4e2.com/crc/tcnpp/resources/epr/

4 Available at: https://server.s4e2.com/crc/tcnpp/chapters/chapter5/_resources/index.html.

BIBLIOGRAPHY

Johnsson, M. & Yan, J. (2001). "Ammonia–water bottoming cycles: A comparison between gas engines and gas diesel engines as prime movers." *Energy*, 26, 31–44.

Liu, B., Rivière, P., Coquelet, C., Gicquel, R. & David, F. (2012). "Investigation of a two-stage Rankine cycle for electric power plants." *Applied Energy*, 100, 285–294. DOI: 10.1016/j.apenergy.2012.05.044.

Margulova, T. Kh. (1978). *Atomic Power Stations: Textbook for Universities*, 3rd ed. Vysshaya Shkola, Moscow. 360 p. (in Russian).

Margulova, T. Kh. & Podushko, L.A. (1982). *Nuclear Power Plants M 25: A Textbook for Technical Schools.* Energoizdat, Moscow. 264 p., ill. (in Russian).

Troïanovski, B.M., Filippov, G.A. & Boulkine, A.E. (1985). *Steam and Gas Turbines of Nuclear Power Plants.* Energoatomizdat Publishing, Moscow. 256 p.

6 Improvement of Closed-Cycle Gas Turbines

INTRODUCTION

This chapter addresses thermodynamic cycles where the working fluid remains in a gaseous state, unlike steam cycles where water transitions between vapor and liquid phases. It comprises two main sections:

- Improvement of Closed-Cycle Gas Turbines, where variations of the simple cycle presented in Section 3.3 are studied
- Supercritical CO_2 cycles, where cycles involving this fluid with particular properties are examined

In Section 3.3, we presented the basic closed gas turbine cycle and demonstrated its limitations. We will now apply the methodology and tools we have introduced to improve the performance of this cycle.

General technological explanations on turbines are available in Appendix 2.

In Chapter 3, we studied a simple closed gas turbine cycle corresponding to a high-temperature reactor (HTR) nuclear power plant. We also compared the simple cycle with the Carnot cycle and concluded that improvements can be obtained through staged compression with intercooling, staged expansion with reheating, regeneration, and combined cycle arrangements. We will now elaborate on these approaches.

6.1.1 Energy and Exergy Balances of the Simple Cycle

Modeling the simple helium gas turbine cycle in Thermoptim presents no difficulty and leads to the synoptic view presented in Figure 6.1.1.

In the exergy balance of Figure 6.1.2, losses are given as a percentage of total irreversibilities. It shows that they are mainly concentrated in the cooler and reactor core, and are substantially equal in the compressor and turbine (their polytropic efficiencies being equal).

Ways to improve gas turbine performance can be deduced from the exergy balance: recover the exergy available in the cooler. We will now discuss possible solutions.

DOI: 10.1201/9781003606024-7

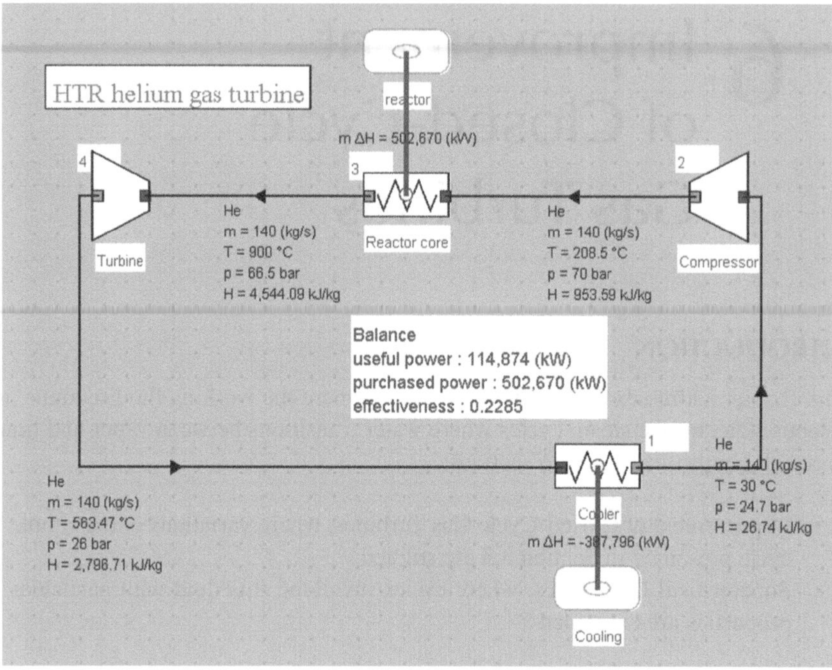

FIGURE 6.1.1 Synoptic view of the closed-cycle gas turbine.

component	Resource	Product	Exergy efficiency	Irreversibilities	% total	settings
Compressor	129,759.03	119,888.462	92.3932%	9,870.568	3.6924%	
Turbine	252,638.126	244,632.89	96.8313%	8,005.236	2.9946%	
Cooler	175,841.714	0	0	175,841.714	65.7790%	Tk = 20.00 °C
Reactor core	382,195.975	308,591.379	80.7417%	73,604.596	27.5340%	Tk = 950.00 °C
global	382,195.975	114,873.861	30.0563%	267,322.114	1	

FIGURE 6.1.2 Exergy balance of the closed-cycle gas turbine (exerBalanceGTHesimple-en.png).

6.1.2 Cycle Improvements

In the reference cycle, the pressure losses in the two heat exchange processes were considered non-negligible and were assumed to equal approximately 5% of the inlet pressure.

The overall efficiency is, as can be seen, quite low: 22.85%.

The plant capacity is approximately 115 MW.

6.1.2.1 Regenerative Gas Turbine

We will now present a cycle modification that greatly increases its performance.

Note that the helium exits the turbine at 563.5°C, while its temperature at the compressor outlet is 208.5°C.

It is therefore possible to partially preheat the helium before it enters the reactor core, which reduces nuclear fuel consumption. This requires only the insertion of a heat exchanger between the two flows.

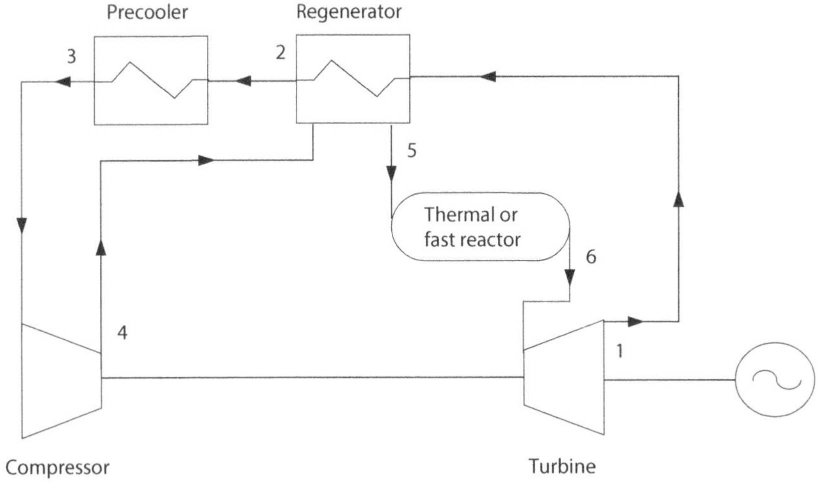

FIGURE 6.1.3 Regenerative gas turbine cycle.

This device is called a regenerator, and this variant of the simple cycle is called the regenerative cycle (Figure 6.1.3).

As explained in Section 3.1.2.2, the main advantage of regenerative cycles lies in their ability to more closely approximate the Carnot cycle. This operating principle is similar to that employed in steam cycles, where steam extractions are used to improve efficiency.

6.1.2.1.1 Regenerator Effectiveness

The effectiveness of the regenerator is the ratio $\varepsilon = \frac{(h_5 - h_4)}{(h_1 - h_4)}$, which characterizes the fraction of the available enthalpy that can be effectively recovered.

By introducing the ratio $\theta = \frac{T_6}{T_3}$ of the absolute turbine inlet temperature to the compressor inlet temperature, it is possible to plot the efficiency of the gas turbine with a regenerator as a function of the compression ratio r.

The heat to be supplied to the cycle is reduced in this case to $h_6 - h_5$. With the previous assumptions and notations, we obtain:

$$h_6 - h_5 = c_p(T_6 - T_5)$$

$$T_5 - T_4 = \varepsilon(T_1 - T_4)$$

$$T_4 = T_3 r^{\beta c} \quad T_6 = \theta T_3$$

$$T_1 = T_3 \theta r^{-\beta t}$$

$$\eta = \frac{c_p(T_6 - T_1) - c_p(T_4 - T_3)}{c_p(T_6 - T_5)}$$

Numerator:

$$T_6 - T_1 + T_3 - T_4 = \theta T_3 - T_3\theta r^{-\beta t} + T_3 - T_3 r^{\beta c}$$

$$T_6 - T_1 + T_3 - T_4 = [\theta(1 - r^{-\beta t}) + 1 - r^{\beta c}]T_3$$

Denominator:

$$T_5 = T_4 + \varepsilon(T_1 - T_4) = [r^{\beta c} + \varepsilon(\theta r^{-\beta t} - r^{\beta c})]T_3$$

$$T_6 - T_5 = \theta T_3 - [r^{\beta c} + \varepsilon(\theta r^{-\beta t} - r^{\beta c})]T_3$$

$$T_6 - T_5 = [\theta - r^{\beta c} - \varepsilon(\theta r^{-\beta t} - r^{\beta c})]T_3$$

$$T_6 - T_5 = [\theta - r^{\beta c} - \varepsilon\theta r^{-\beta t} + \varepsilon r^{\beta c})]T_3$$

$$\eta = \frac{\theta(1 - r^{-\beta t}) + 1 - r^{\beta c}}{\theta(1 - \varepsilon r^{-\beta t}) + (\varepsilon - 1)r^{\beta c}} \qquad (6.1.1)$$

Figure 6.1.4 shows the efficiency of the regenerative gas turbine given by equation (6.1.1) for two values of effectiveness: 0.9 and 0.75. The four curves in each family correspond to different values of θ: 3, 3.5, 4, and 4.5. In this figure, the working fluid is helium.

We note that the benefit of the regenerator is greater when the compression ratio is low. Indeed, beyond a certain limit, which itself is a function of θ, the fluid heating in the compressor is such that temperature $T_{2'}$ at the compressor outlet becomes higher than temperature $T_{4'}$ at the expansion outlet. The regenerator can no longer function and becomes useless.

This condition, independent of ε, is given by:

$$T_4 \leq T_1 \quad r^{\beta c} \leq \theta r^{-\beta t} \quad \text{or} \quad r \leq \theta^{1/(\beta_c + \beta_t)}$$

As shown in the graph of Figure 6.1.4, all curves corresponding to the same θ and different values of ε intersect at the point where $T_4 = T_1$.

The above analytical results are, of course, no longer valid when the working fluid is not comparable to a perfect gas, but the limit on the use of regeneration is not called into question.

6.1.2.1.2 Thermoptim Model

The synoptic view obtained when modeling this cycle with Thermoptim is given in Figure 6.1.5.

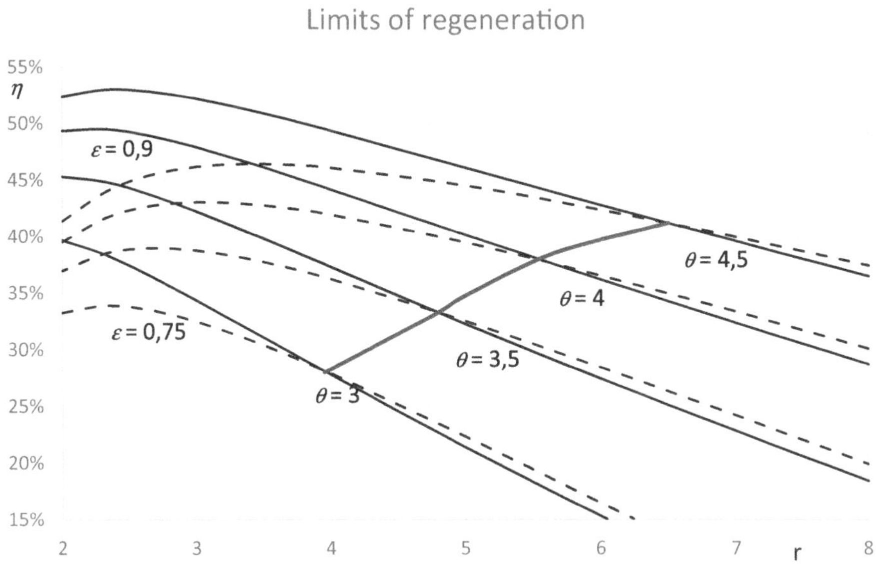

FIGURE 6.1.4 Limits of the regenerative Brayton cycle.

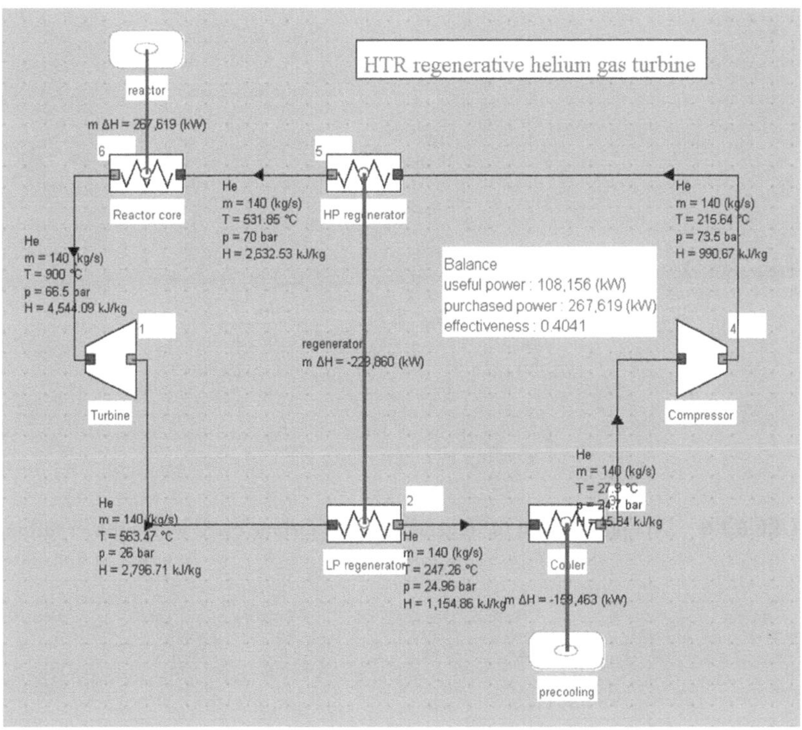

FIGURE 6.1.5 Synoptic view of the regenerative gas turbine cycle.

The efficiency has almost doubled: from 22.85% to 40.41%. The regenerator makes it possible to recover nearly the equivalent of the heat provided by the nuclear reactor.

A similar cycle is the subject of directed exploration C-M2-V2.

It should be noted that, for reasons which will be explained later, the turbine is often divided into two complementary units: the High Pressure (HP) section is directly coupled to the compressor, while the Low Pressure (LP) section, called a free turbine or power turbine, drives the alternator alone. One advantage of this configuration is that the rotational speeds of the high-pressure assembly and the power turbine can be independent, which is an advantage in terms of machine regulation (Figure 6.1.6).

This variant has no impact on the efficiency and net power of the cycle, which remain the same.

6.1.2.1.3 *Exergy Balance of the Regenerative Cycle*

Figures 6.1.7 and 6.1.8 allow for a comparison of the exergy balances of the cycles with and without regeneration.

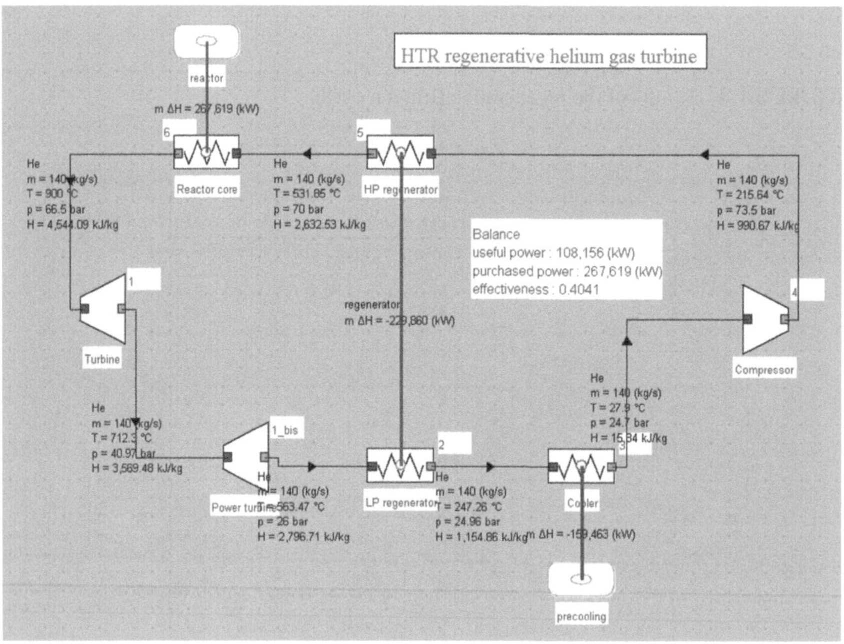

FIGURE 6.1.6 Synoptic view of the regenerative gas turbine cycle with a power turbine.

component	Resource	Product	Exergy efficiency	Irreversibilities	% total	settings
Compressor	129,759.03	119,888.462	92.3932%	9,870.568	3.6924%	
Turbine	252,638.126	244,632.89	96.8313%	8,005.236	2.9946%	
Cooler	175,841.714	0	0	175,841.714	65.7790%	Tk = 20.00 °C
Reactor core	382,195.975	308,591.379	80.7417%	73,604.596	27.5340%	Tk = 950.00 °C
global	382,195.975	114,873.861	30.0563%	267,322.114	1	

FIGURE 6.1.7 Exergy balance of the simple gas turbine.

component	Resource	Product	Exergy efficiency	Irreversibilities	% total	settings
Compressor	136,476.67	126,147.356	92.4314%	10,329.314	11.2954%	
Turbine	140,574.796	136,445.29	97.0624%	4,129.506	4.5157%	
LP regenerator	132,171.575	0	0	(132,171.575)	(144.5332%)	regenerator
HP regenerator	0	119,385.353	0	(-119,385.353)	(-130.5511%)	regenerator
Cooler	43,715.359	0	0	43,715.359	47.8039%	Tk = 20.00 °C
Reactor core	203,479.179	182,992.352	89.9317%	20,486.828	22.4029%	Tk = 950.00 °C
global	203,479.179	112,031.95	55.0582%	91,447.229	1	

FIGURE 6.1.8 Exergy balance of the regenerative gas turbine.

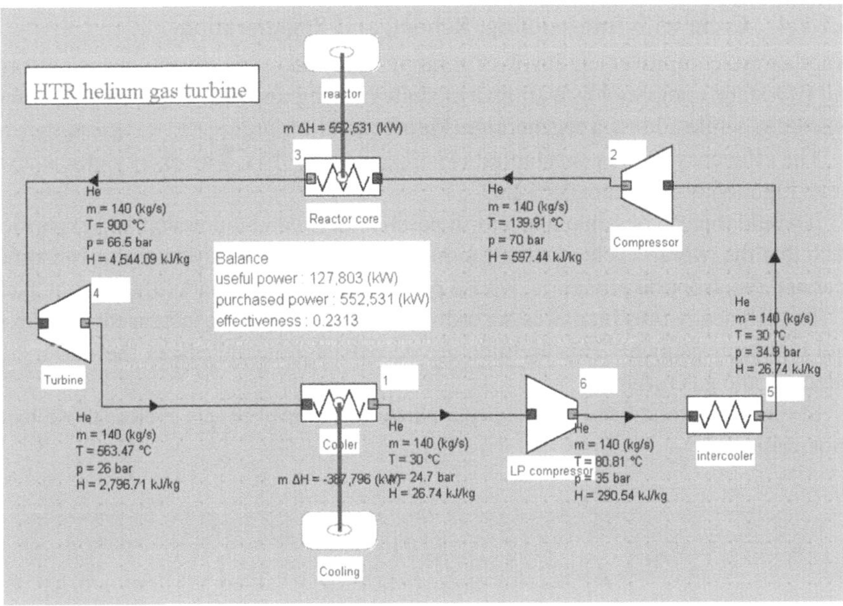

FIGURE 6.1.9 Synoptic view of a gas turbine with an intercooler.

The addition of the regenerator has greatly increased the exergy efficiency, which reaches 55%. The main gain is in the cooler, which nevertheless continues to account for approximately 47% of losses.

It should be recalled that a special feature of the automatically generated exergy balance is that internal exchangers are represented by two lines (here "LP regenerator" and "HP regenerator"), and the contributions to the irreversibilities of the two fluids are represented algebraically: positively for the hot fluid and negatively for the cold fluid. The irreversibility of the entire exchanger is obtained by summing these two values. This is why they are indicated in brackets in the exergy balance.

6.1.2.2 Cycle with an Intercooler

We know that compression work can be reduced by conducting staged compression with intermediate cooling.

Figure 6.1.9 shows the simulation results obtained for the simple gas turbine, with an intermediate pressure equal to 35 bar and compressed helium cooling to 30°C.

The efficiency gain is modest (23.13% instead of 22.85%), but the capacity is increased by approximately 17%. The efficiency increases slightly because the heat that the reactor must provide is greater due to helium exiting the compressor at a lower temperature.

6.1.2.3 Reheat Cycle

In addition to intercooled compression, it is also possible to use a reheat cycle (Figure 6.1.10). The efficiency is slightly increased (26.4%), and the capacity gain is approximately 15%.

6.1.2.4 Cycle with Intercooling, Reheat, and Regeneration

Another cycle improvement involves combining the above modifications by performing two-stage compression with intermediate cooling and two-stage expansion with reheating, while adding a regenerator (Figure 6.1.11).

The efficiency gain is substantial (47.4%), Figure 6.1.11. The exergy efficiency is quite high (65.6%) (Figure 6.1.12).

It would thus be possible to perform para-isothermal compression and expansion, such that the overall cycle would be similar to a regenerative cycle with two isothermal and two isobaric processes, whose performance equals that of the Carnot cycle.

The efficiency thus increases accordingly, but at the cost of increased complexity and higher investment. This architecture departs significantly from the initial simplicity of the gas turbine.

In the next section, we will study a particular family of gas cycles: those using supercritical CO_2.

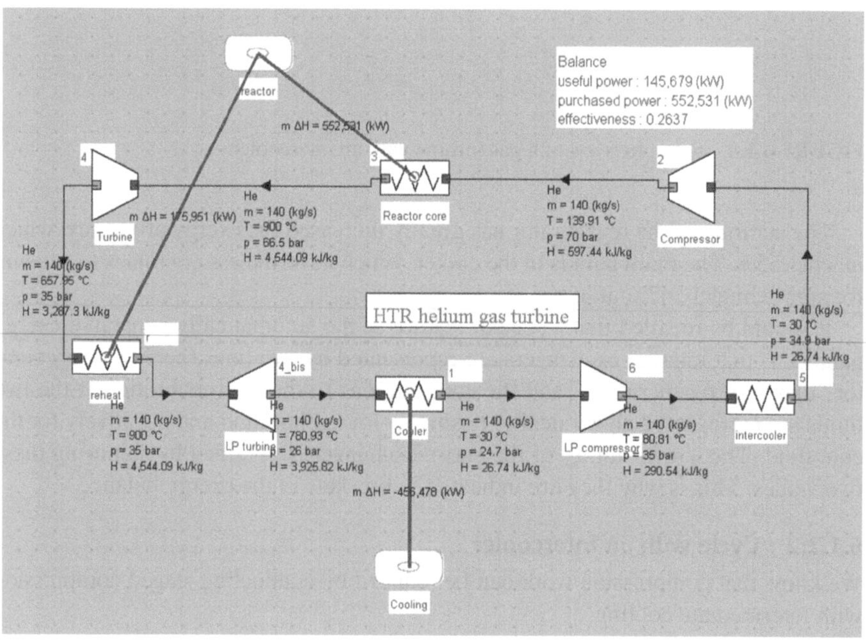

FIGURE 6.1.10 Efficiency and power of a gas turbine with intercooling and reheat.

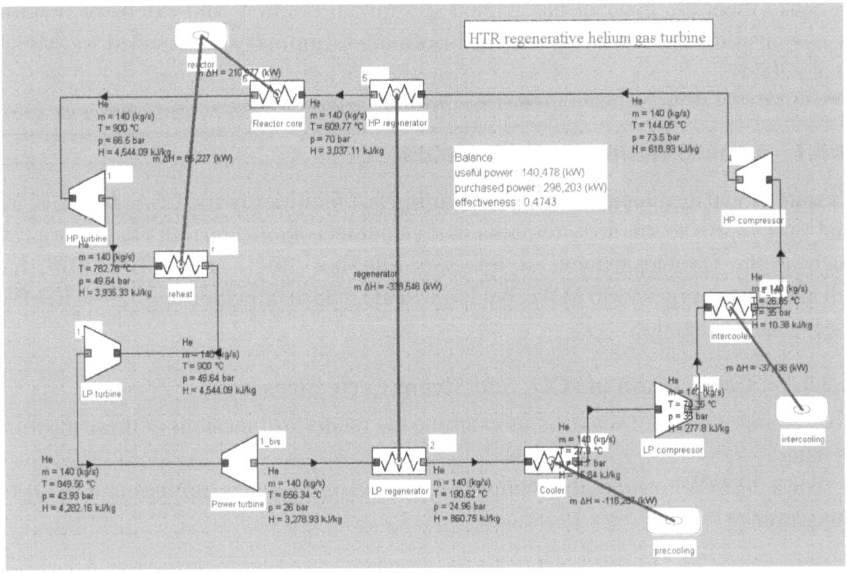

FIGURE 6.1.11 Synoptic view of a gas turbine with intercooling, reheat, and regeneration.

component	Resource	Product	Exergy efficiency	Irreversibilities	% total	settings
HP compressor	85,195.96	78,162.056	91.7439%	7,033.904	8.8368%	
HP turbine	87,714.568	85,226.56	97.1635%	2,488.008	3.1257%	
Power turbine	144,924.551	140,452.21	96.9140%	4,472.341	5.6187%	
LP turbine	37,709.672	36,669.92	97.2427%	1,039.752	1.3063%	
LP compressor	36,674.67	33,367.457	90.9823%	3,307.213	4.1549%	
LP regenerator	193,869.192	0	0	(193,869.192)	(243.5608%)	regenerator
HP regenerator	0	174,632.977	0	(-174,632.977)	(-219.3941%)	regenerator
Cooler	27,096.138	3,901.939	14.4004%	23,194.199	29.1392%	Tk = 30.00 °C
intercooler	3,676.301	1,234.965	33.5926%	2,441.337	3.0671%	Tk = 30.00 °C
Reactor core	160,412.344	146,038.871	91.0397%	14,373.473	18.0576%	Tk = 950.00 °C
reheat	64,800.475	62,789.06	96.8960%	2,011.415	2.5270%	Tk = 950.00 °C
global	225,212.819	145,614.962	64.6566%	79,597.857	1	

FIGURE 6.1.12 Exergy balance of a gas turbine with intercooling, reheat, and regeneration.

6.2 SUPERCRITICAL CO_2 CYCLES

Introduction

In the early 2000s, MIT conducted research on cycles using supercritical carbon dioxide, which they determined offered superior performance compared to other options for generating electricity from nuclear reactors at moderate temperatures between 650°C and 800°C (Dostal et al., 2003).

From a thermodynamic perspective, the primary advantage of such cycles is the significantly reduced compression work required when the working fluid is in the supercritical liquid state, compared to conventional Brayton cycles where the working fluid remains gaseous.

Proponents of these cycles argue that efficiencies in this temperature range exceed those of steam cycles and that the machinery is considerably more compact.

Since then, the concept has gained substantial traction, and extensive literature is now available on this subject. An informative summary is provided in (Molière et al., 2024).

6.2.1 Characteristics of sCO_2 Cycles

Despite global development efforts spanning test facilities in the USA, Japan, Korea, and Europe, the technology remains in the engineering and testing phase. Programs such as the US Department of Energy's SunShot (2011) and STEP (2016) have advanced prototypes to 10 MWe, but large-scale industrialization (e.g., 50–450 MWe) faces ongoing delays.

6.2.1.1 Comparison of sCO_2 and Steam Cycle Sizes

To establish a sense of scale, let us examine the relative dimensions of these machines in detail:

For a 20 MW turbine, the diameter ratios between steam turbines and sCO_2 turbines are:

- HP casing: ratio of 2.3 (steam versus sCO_2)
- LP casing: ratio of 11.7 (steam versus sCO_2)

This dramatic reduction is explained by the high density of supercritical CO_2 (57 kg/m^3 at the exhaust) compared to LP steam (0.077 kg/m^3).

A 10–15 MW sCO_2 turbine has a shaft diameter of only 76 mm, blade heights between 6 and 14 cm, and a 4-stage rotor length of approximately 250 mm. These dimensions contrast sharply with conventional steam or gas turbines, whose diameters increase significantly in the LP stages due to the low density and high compressibility of the fluids, as previously mentioned.

The expansion ratio in the turbine is much lower (approximately 2.6) than in steam cycles (approximately 7,300). However, the mass flow rate is considerably higher.

While the steam turbine is significantly larger than the supercritical CO_2 machine, the heat exchangers of the latter are of substantial size. Based on a heat transfer coefficient h of approximately 100 W/m^2/K for this gas, the high-efficiency recompression cycle modeled below in Thermoptim would require a 210,000 m^2 High Temperature (HT) regenerator, a 180,000 m^2 Low Temperature (LT) regenerator, and an 80,000 m^2 precooler, not to mention the Intermediate Heat Exchanger (IHX), whereas the steam cycle condenser would have an area of only tens of thousands of m^2.

Printed Circuit Heat Exchangers (PCHE) are considered the most promising solution for sCO_2 systems due to their compact structure, high thermal transfer efficiency, heat exchange surface density reaching up to 2,500 m^2/m^3, and their resistance to high pressures (up to 600 bar) and temperatures (up to 1,000°C).

6.2.1.2 Main Variants of sCO_2 Cycles

Several main types of supercritical CO_2 cycles are considered: the simplest is a Brayton cycle with a regenerator (Figure 6.2.1), with principal variants involving recompression, pre-compression, or partial cooling.

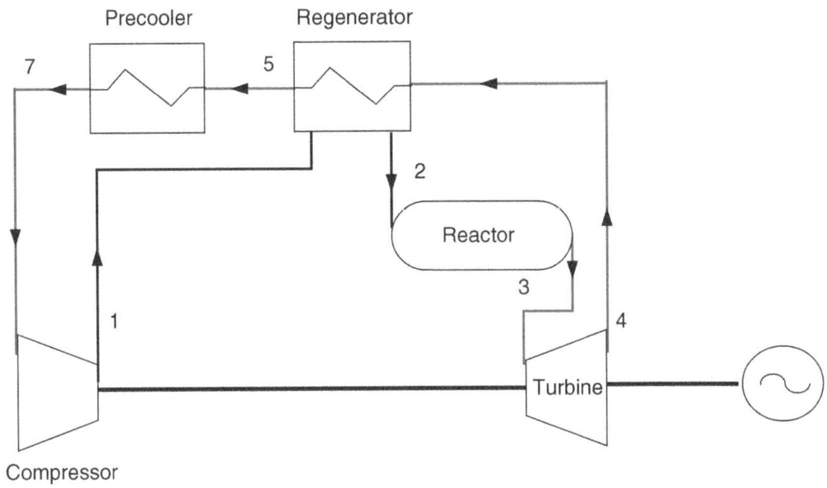

FIGURE 6.2.1 Supercritical CO_2 regenerative cycle.

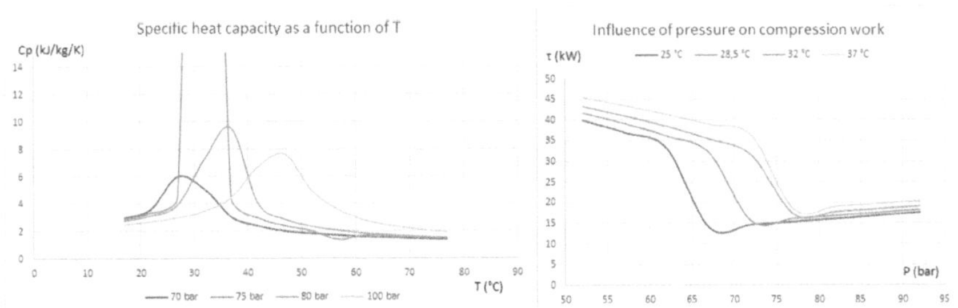

FIGURE 6.2.2 Significant variation of the specific heat capacity c_p and compression power.

In the following sections, we compare the performance of these four cycle variants using similarly configured examples.

6.2.1.3 sCO_2 Properties

A particular characteristic of supercritical CO_2 is that its specific heat capacity c_p varies significantly above the critical point, as shown in the left portion of Figure 6.2.2, where the evolution of this property is plotted as a function of temperature for different pressures. This results in challenges in sizing heat exchangers and the potential for temperature crossovers to occur within them. This represents an important design constraint.

The influence on compression work is indicated in the graph on the right, which shows the evolution as a function of pressure in bar of the isentropic compression work for 1 kg/s of CO_2 at the indicated temperature. As can be seen, the compression work is approximately halved.

Two disadvantages related to CO_2 properties stem from the values of its critical coordinates. For the fluid to be in a supercritical state, its pressure must exceed the critical pressure of approximately 78 bar, which limits the compression ratio if excessively high pressures and associated leakage risks are to be avoided. Furthermore, the cold source temperature of the cycle must be below the critical temperature of 31.1°C for the sCO_2 to be completely condensed in the cooler before compression: cold source temperatures above approximately 25°C cannot be used, which restricts applications in warm climates or during summer.

6.2.2 Simple Regeneration Cycle

6.2.2.1 Cycle Description

In a 350 MWe supercritical CO_2 simple regenerative cycle (Figure 6.2.3), a flow of 3 t/s of CO_2 at 200 bar enters the reactor at a temperature of 376°C and exits at 650°C. It is then expanded to a pressure of 77 bar in a turbine. Regeneration occurs between the expanded CO_2 and that entering the reactor. The CO_2 then enters the cooler, exiting at 32.5°C, before being drawn into the compressor, which raises it to 200 bar.

The efficiency of this cycle remains relatively low, close to 34% with a regenerator effectiveness of 0.9 and turbomachinery polytropic efficiency of 0.9 (Figures 6.2.3 and 6.2.4).

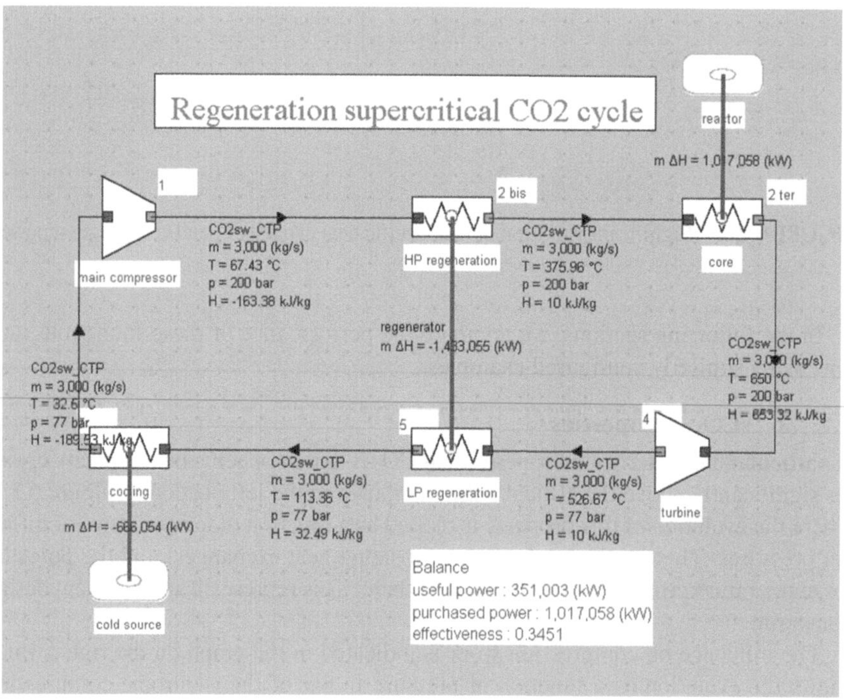

FIGURE 6.2.3 Synoptic view of the supercritical CO_2 regenerative cycle.

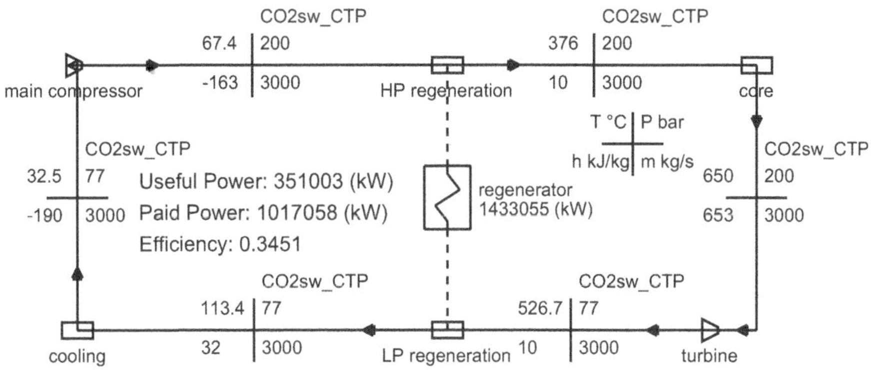

FIGURE 6.2.4 Flowsheet of the supercritical CO_2 regenerative cycle.

component	Resource	Product	Exergy efficiency	Irreversibilities	% total	settings
turbine	445,299.934	429,441.76	96.4388%	15,858.174	3.4202%	
main compressor	78,438.28	60,818.09	77.5362%	17,620.19	3.8002%	
HP regeneration	0	512,648.87	0	(-512,648.87)	(-110.5646%)	regenerator
LP regeneration	694,188.794	0	0	(694,188.794)	(149.7179%)	regenerator
core	814,667.939	634,208.235	77.8487%	180,459.704	38.9203%	Tk = 1,200.00 °C
cooling	68,186.467	0	0	68,186.467	14.7060%	Tk = 20.00 °C
global	814,667.939	351,003.48	43.0855%	463,664.459	1	

FIGURE 6.2.5 Exergy balance of the supercritical CO_2 regenerative cycle.

The fluid used for this model is "CO_2sw_CTP." It is an external mixture from the CTP Lib library. Its model corresponds to a highly accurate equation of state for CO_2 proposed by R. Span and W. Wagner, as the internal model of CO_2 as a condensable vapor in Thermoptim is insufficiently accurate in the supercritical region. We will analyze this cycle and its improvement prospects in the following sections.

6.2.2.2 Exergy Balance of the Regenerative Cycle

The exergy balance is given in Figure 6.2.5:

We note that a characteristic of the automatically generated exergy balance is that internal heat exchangers are represented by two lines (here "HP regeneration" and "LP regeneration" for the regenerator), and that the contributions to irreversibilities of the two fluids are represented algebraically, positively for the hot fluid and negatively for the cold fluid, with the irreversibility of the entire exchanger obtained by summing these two values. This is why they are indicated in parentheses. The temperatures of the hot and cold sources were taken as 1,200°C for the reactor core and 20°C for the cooling fluid, respectively. Note that the choice to consider the hot source at 1,200°C is arbitrary given the very high temperatures achievable with nuclear fuel.

An initial observation is that the overall exergy efficiency is approximately 43%, indicating that the cycle is already quite efficient.

The exergy balance shows that the main irreversibilities, other than those in the reactor core (38%), occur in the regenerator (39.7%), which operates with a temperature difference (LMTD) of 88°C as shown in this figure. Such a temperature difference severely penalizes the cycle.

We will now examine how this cycle can be improved.

6.2.3 Pre-Compression Cycle

A first improvement consists of splitting the compression with intermediate cooling, an operation known to be beneficial. This cycle is called pre-compression.

6.2.3.1 Cycle Description

In such a cycle (Figure 6.2.6), compression is two-stage with intercooling. After passing through the high-temperature regenerator (4–5), a first compressor compresses the CO_2 to approximately 100 bar (5–10). The pre-compressed CO_2 is cooled in the second low-temperature regenerator (10–9), followed by exchange with the cold source (9–7). It is then compressed (7–8) in the main compressor and then heated (8–2) by exchange with the flow exiting the pre-compressor (10–9). A second exchange (2–6) with the flow exiting the turbine allows the CO_2 heating to continue before entering the reactor.

As shown in Figure 6.2.6, with the same assumptions as before regarding heat exchanger effectiveness and turbomachinery polytropic efficiency, the power output of this cycle decreases to 307 MW, but its efficiency reaches 39.6%.

The advantage of this cycle is that when the temperature difference between the two fluids decreases in the regenerator, the insertion of pre-compression allows it to be increased, thus enhancing regeneration. However, a problem remains, as shown in Figure 6.2.7: as the CO_2 heat capacities of the two fluids become very different, the internal irreversibilities of the heat exchanger are large.

6.2.3.2 Exergy Balance of the Pre-Compression Cycle

Figure 6.2.8 shows the exergy balance of the new cycle. An initial observation is that the overall exergy efficiency is approximately 50%. The distribution of irreversibilities is as follows: 39.6% in the core, 25% in the high-temperature regenerator where

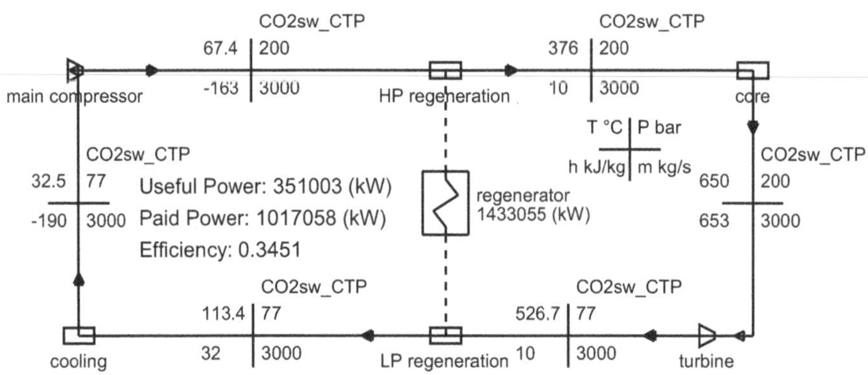

FIGURE 6.2.6 Synoptic view of the supercritical CO_2 pre-compression cycle.

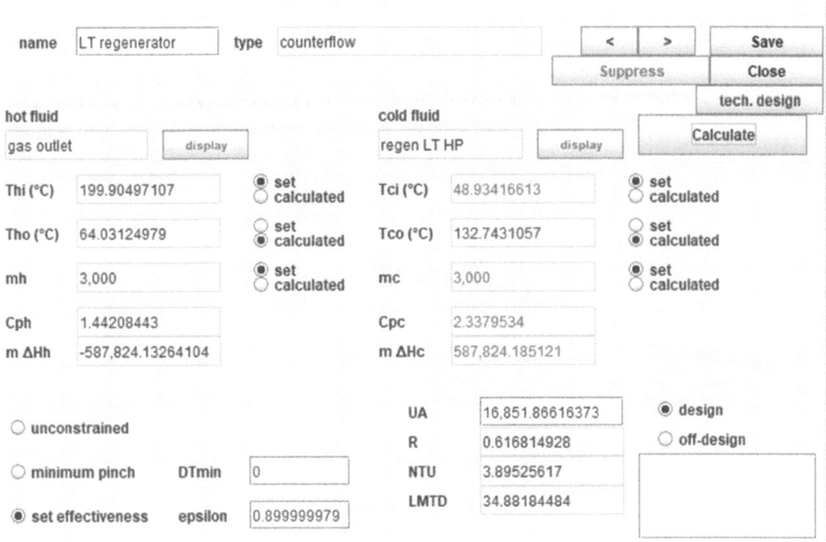

FIGURE 6.2.7 LT regenerator screen.

component	Resource	Product	Exergy efficiency	Irreversibilities	% total	settings
pre-compressor	69,686.9	64,862.129	93.0765%	4,824.771	1.5338%	
turbine	445,299.934	429,441.76	96.4388%	15,858.174	5.0412%	
main compressor	52,692.05	39,896.626	75.7166%	12,795.424	4.0676%	
regen gas HT	632,806.172	0	0	(632,806.172)	(201.1660%)	HT regenerator
regen gas	0	556,317.288	0	(-556,317.288)	(-176.8506%)	HT regenerator
regen LT HP	0	109,502.759	0	(-109,502.759)	(-34.8104%)	LT regenerator
gas outlet	150,176.548	0	0	(150,176.548)	(47.7404%)	LT regenerator
core	621,631.882	496,899.339	79.9347%	124,732.543	39.6519%	Tk = 1,200.00 °C
precooler	39,195.488	0	0	39,195.488	12.4601%	Tk = 20.00 °C
global	621,631.882	307,062.809	49.3962%	314,569.073	1	

FIGURE 6.2.8 Exergy balance of the supercritical CO_2 pre-compression cycle.

the mean temperature difference is 59.3°C, 13.3% in the low-temperature regenerator where the mean temperature difference is 35°C, and 12.5% in the precooler.

Figure 6.2.9 shows the plot of the pre-compression cycle in the CO_2 (T, s) and (h, ln(P)) diagrams. These are not the usual Thermoptim diagrams, as the CO_2 is modeled using the CTP Lib library. Explanations on how to open this type of diagram are given in Guided Exploration TCNPP-6: Exploring Supercritical CO2 Nuclear Cycles which complements the presentations made in this section. The backgrounds for them were generated using the CreateMixtureCharts external class.[1]

This cycle represents an improvement over the previous one. However, a problem remains, as shown in Figure 6.2.7 which displays the low-temperature regenerator screen. The specific heat capacities of CO_2 for the two fluids are very different; the cold fluid only warms by approximately 80°C, while the hot CO_2 cools by 135°C. The exchanger is unbalanced, and the only way to rebalance it (and reduce internal irreversibilities due to temperature heterogeneity) is to decrease the cold fluid flow rate. This is precisely what the recompression cycle accomplishes, which we will now examine.

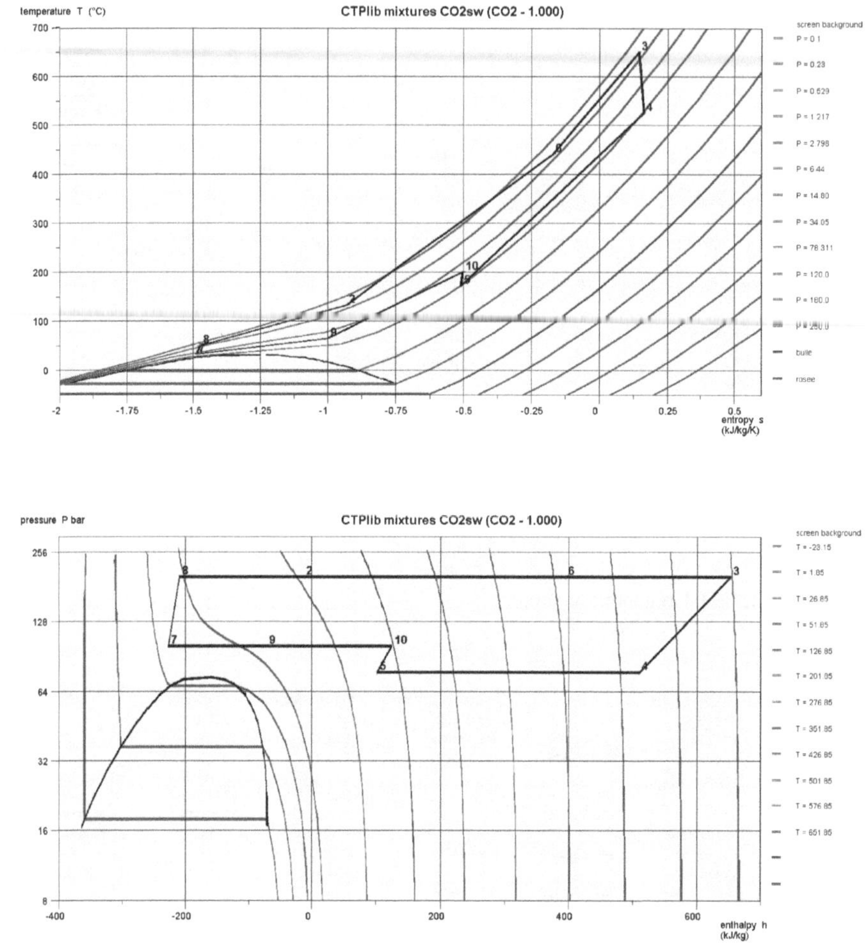

FIGURE 6.2.9 CO_2 pre-compression cycle in the (T, s) and (h, ln(P)) diagrams.

6.2.4 Recompression Cycle

6.2.4.1 Cycle Description

In a recompression cycle (Figure 6.2.10), compression is two-stage with intermediate cooling of only a portion of the fluid, which allows recycling of a larger amount of heat.

The main flow of CO_2 (3 t/s) exiting the turbine passes through a high-temperature regenerator (4–5) and then a low-temperature one (5–6), before being split in two.

60% of the flow is cooled in a precooler (6–7), then compressed in the main compressor from 77 to 200 bar (7–8), and heated in the LT regenerator (8–9).

The remaining CO_2 flow is compressed to 200 bar in the recompressor (6–8b) and then mixed with the other stream exiting the LT regenerator (1), and the total flow is heated in the HT regenerator (1–2) before entering the reactor.

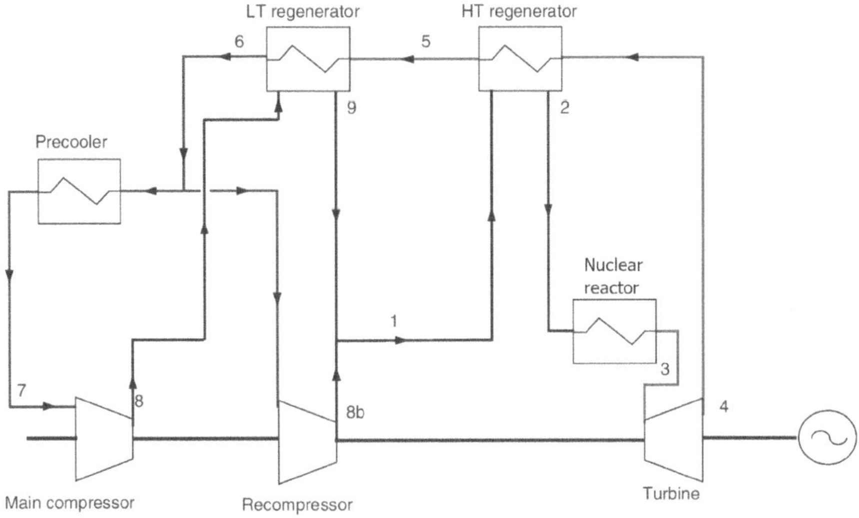

FIGURE 6.2.10 CO_2 supercritical recompression cycle.

The CO_2 exiting the reactor is then expanded in the turbine (3–4), and the cycle is thus closed.

In this case, the reactor inlet temperature is higher, and with the same assumptions as before regarding the polytropic efficiency of turbomachinery (0.9) and exchanger effectiveness (0.9), the power output remains close to 300 MWe, and the efficiency exceeds 45% (Figure 6.2.11). Even higher values are anticipated, on the order of 50%, with exchanger effectiveness around 95%.

6.2.4.2 Exergy Balance of the Recompression Cycle

Figure 6.2.12 shows the exergy balance of the new cycle. An initial observation is that the overall exergy efficiency is approximately 56.7%, which represents a significant improvement compared to previous cycles. The distribution of irreversibilities is as follows: 44.3% in the core, 15% in the high-temperature regenerator with a temperature difference of 42°C, 12.5% in the low-temperature regenerator with a temperature difference of 22.8°C, and 13% in the precooler.

The performance improvement results from the fact that the LT regenerator is much better balanced in terms of enthalpy than in the pre-compression cycle because only a fraction of the flow traverses the high-pressure section, thus compensating for the increased thermal capacity of CO_2 in the supercritical region at low temperatures (Figure 6.2.13).

Figure 6.2.14 shows the plot of the recompression cycle in the CO_2 (T, s) and (h, ln(P)) diagrams.

In the (T, s) entropy diagram, the two compressions appear on the left (7–8) and in the middle (6–8b) of the diagram, and the expansion (3–4) is at the top right. The heat exchanges correspond to the lines close to the exponentials.

In the (h, P) diagram, the two compressions appear on the left side of the diagram (7–8) and (6–8b), and the expansion (3–4) on the right. The heat exchanges correspond to the plots close to the horizontals.

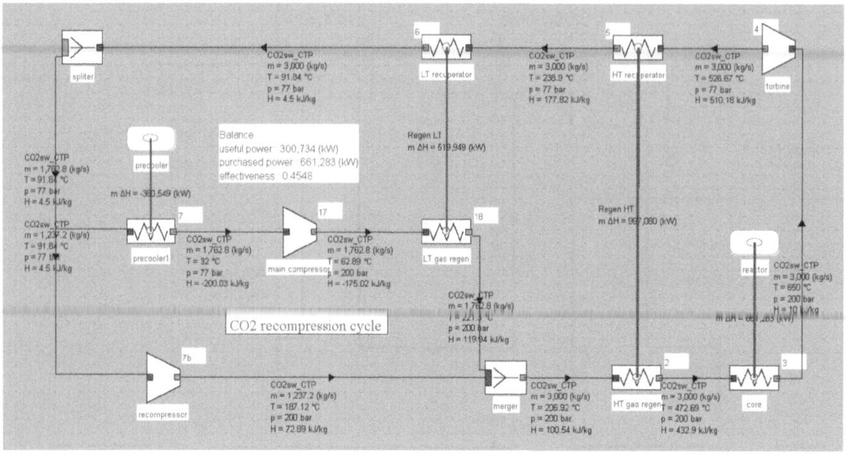

FIGURE 6.2.11 Synoptic view of the CO_2 supercritical recompression cycle.

component	Resource	Product	Exergy efficiency	Irreversibilities	% total	settings
recompressor	84,608.61	77,342.039	91.4115%	7,266.571	3.1738%	
turbine	445,299.934	429,441.76	96.4388%	15,858.174	6.9263%	
main compressor	44,098.72	33,745.097	76.5217%	10,353.623	4.5221%	
merger	896,943.618	896,191.804	99.9162%	751.814	0.3284%	
HT recuperator	545,220.77	0	0	(545,220.77)	(238.1328%)	Regen HT
HT gas regen	0	510,948.759	0	(-510,948.759)	(-223.1640%)	Regen HT
LT gas regen	0	138,410.801	0	(-138,410.801)	(-60.4529%)	Regen LT
LT recuperator	167,384.094	0	0	(167,384.094)	(73.1073%)	Regen LT
core	529,691.03	428,201.04	80.8398%	101,489.99	44.3272%	Tk = 1,200.00 °C
precooler1	29,991.124	0	0	29,991.124	13.0990%	Tk = 20.00 °C
global	529,691.03	300,734.43	56.7754%	228,956.6	1	

FIGURE 6.2.12 Exergy balance of the CO_2 supercritical recompression cycle.

6.2.5 Partial Cooling Cycle

6.2.5.1 Cycle Description

The cycle presented here is a simplified version of the partial cooling cycle (Dostal, 2004), where compression is still two-stage and intercooled with only a portion of the fluid.

After passing through a high-temperature regenerator, then a second low-temperature one and a cooler (Figure 6.2.15), a first compressor compresses the CO_2 from 30 to 77 bar. The pre-compressed CO_2 flow rate is then split into two, with one part being cooled by exchange with the cold source, then compressed in the main compressor, and heated in an LT regenerator.

The remaining flow passes through a third compressor to be compressed to 200 bar before being remixed with the stream exiting the LT regenerator, and then heated in the HT regenerator before returning to the reactor. With the same assumptions as before regarding heat exchanger effectiveness and turbomachinery polytropic efficiency, the efficiency of this cycle is 43.7%, slightly less than that of the recompression cycle, which is also simpler. However, its power output is much higher (494 MW). Its exergy efficiency is also slightly lower.

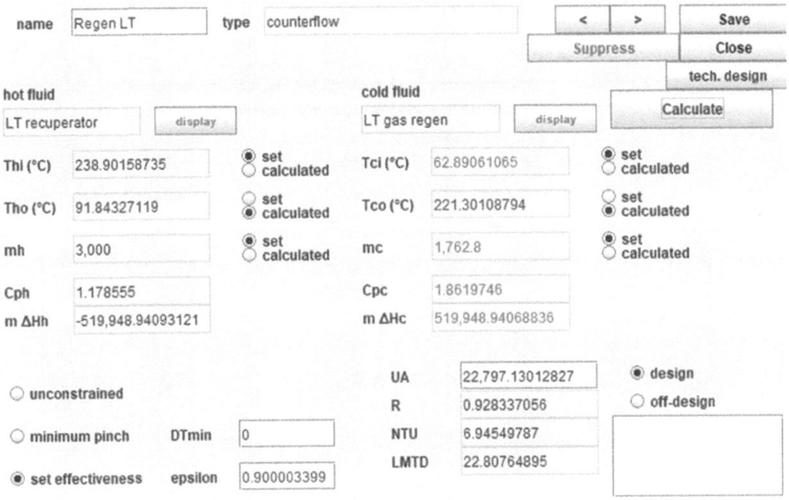

FIGURE 6.2.13 LT regenerator screen of the CO_2 supercritical recompression cycle.

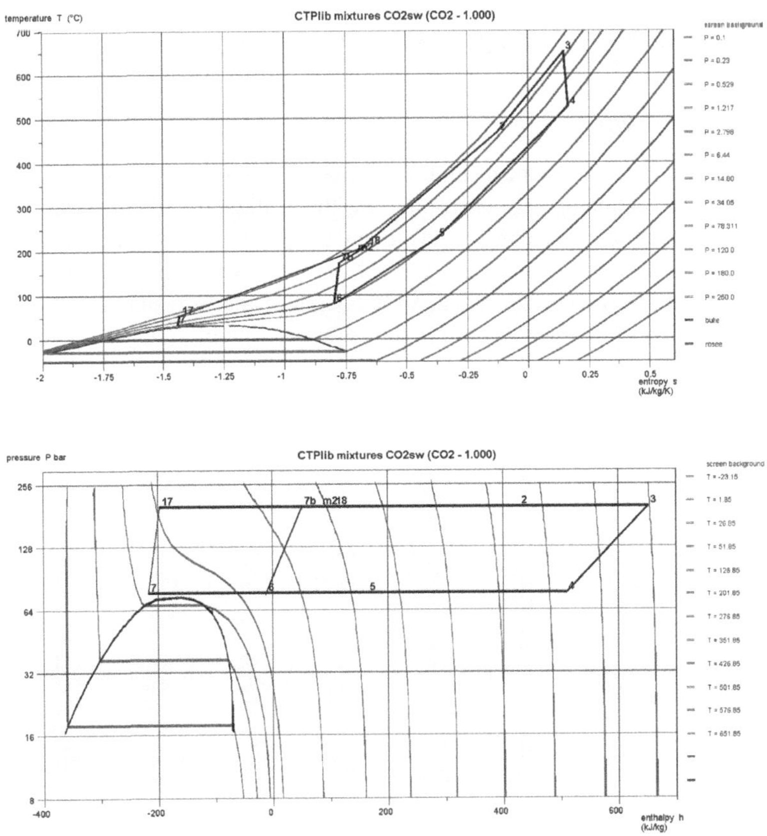

FIGURE 6.2.14 CO_2 recompression cycle in the (T, s) and (h, ln(P)) diagrams.

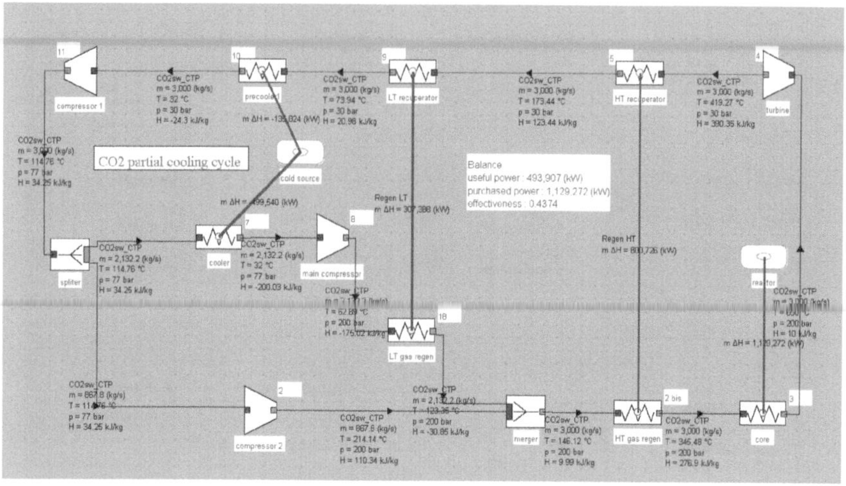

FIGURE 6.2.15 Synoptic view of the partial cooling CO_2 supercritical cycle.

6.2.5.2 Exergy Balance of the Partial Cooling Cycle

The exergy efficiency of the cycle is given in Figure 6.2.16.

6.2.6 Conclusion

Supercritical CO_2 cycles appear quite attractive from a thermodynamic standpoint. The two main technological constraints are the realization of regenerators in the circum-critical zone, particularly to avoid any temperature crossover, and the development of effective turbomachinery for CO_2, an area where there is no established reference. Moreover, these cycles can only operate if the CO_2 state at the precooler outlet is supercritical, which requires a temperature limit of 32°C, which can be difficult to achieve in some countries.

The sCO_2 technology perfectly illustrates the gap between conceptual thermodynamic excellence and industrial reality. More than 50 years after the pioneering work on the subject, and despite intensive development programs, technological barriers persist.

component	Resource	Product	Exergy efficiency	Irreversibilities	% total	settings
compressor 2	66,028.43	60,931.474	92.2807%	5,096.956	1.2412%	
turbine	820,437.837	788,919.26	96.1583%	31,518.577	7.6754%	
compressor 1	175,643.82	158,243.371	90.0933%	17,400.449	4.2374%	
main compressor	53,339.64	40,816.409	76.5217%	12,523.231	3.0497%	
merger	808,383.071	802,312.998	99.2491%	6,070.073	1.4782%	
HT gas regen	0	338,513.778	0	(-338,513.778)	(-82.4348%)	Regen HT
HT recuperator	383,469.192	0	0	(383,469.192)	(93.3823%)	Regen HT
LT recuperator	79,088.103	0	0	(79,088.103)	(19.2595%)	Regen LT
LT gas regen	0	59,775.738	0	(-59,775.738)	(-14.5566%)	Regen LT
core	904,551.819	694,514.827	76.7800%	210,036.992	51.1481%	Tk = 1,200.00 °C
cooler	50,276.852	0	0	50,276.852	12.2434%	Tk = 20.00 °C
precooler1	13,453.539	0	0	13,453.539	3.2762%	Tk = 20.00 °C
global	904,551.819	493,907.371	54.6024%	410,644.448	1	

FIGURE 6.2.16 Exergy balance of the partial cooling CO_2 supercritical cycle.

Without decisive advances in turbomachinery (efficiency, reliability, sealing) and heat exchanger technology, sCO_2 cycles will remain a theoretical promise. The "bottleneck" lies with compressors and turbines, not thermodynamics. The coming years will be crucial: either projects like STEP demonstrate technical and economic viability at an industrial scale, or the technology will remain confined to niche applications where compactness takes precedence over absolute efficiency.

6.3 TRL OF GAS CYCLES

As previously mentioned, the literature on gas-cooled nuclear power cycles is extensive, due to the very high efficiencies that could potentially be achieved. The number of possible technological solutions is considerable, with many competing working fluids. We have focused on helium and supercritical CO_2 cycles, but many others can be considered, notably nitrogen.

Thermodynamic performance is not the determining selection criterion: technological feasibility is even more critical.

The most significant technological constraints arise from high temperatures and allowable pressures, which in turn limit the choice of materials, particularly for turbines and heat exchangers. The compatibility of the working fluid with these materials thus becomes a major criterion.

6.3.1 TRL of Turbomachinery

A major constraint lies in the availability of high-performance turbomachinery for candidate fluids. Currently, the only gas turbines with established technological maturity are open-cycle gas turbines using combustion gases as the working fluid. For compressors, the situation is similar: the only high-efficiency turbocompressors available are those for air. It is estimated that developing new high-performance turbomachines will require at least one to two decades, even in the favorable case where a sufficient market develops.

This explains why the use of nitrogen-based mixtures (which constitute 78% of air) and other gases as working fluids is being seriously considered, to enable easier adaptation of existing machines. An example will be presented in Section 7.1, which covers combined cycles.

One of the main obstacles to the development of these cycles is therefore turbomachinery. There are also constraints regarding heat exchangers, but these are less limiting, except perhaps for the regenerators used in supercritical CO_2 cycles.

As a result, the TRL of helium cycles is currently estimated to be below 7, and that of sCO_2 cycles below 6, whereas for those using nitrogen mixtures, it would be between 7 and 9 (G4NR Handbook p. 823).

For heat exchangers, the TRL for He and sCO_2 would be around 5 or 6, while for nitrogen it would be between 7 and 9.

6.3.2 TRL of HTRs

Section 2.4 provided a comprehensive overview of nuclear reactor types, including the Generation IV concepts selected by the Generation IV International Forum.

This section builds upon that foundation by examining the current status of HTR development and its implications for advanced thermodynamic cycles, beginning with those having the highest technological maturity.

6.3.2.1 Historical Developments on HTRs

Early work on HTRs coupled with gas thermodynamic cycles dates back to the 1960s–1970s. These reactors typically featured fuel packaged as uranium, plutonium, or thorium oxide or carbide kernels coated with pyrolytic carbon and silicon carbide to contain fission products; helium as the coolant to achieve temperatures around 800°C for approximately 40% efficiency; and slow neutron designs with graphite as the moderator.

One of the advantages of these reactors is to significantly exceed temperature levels to which those using water are limited, which allows high conversion efficiencies to be considered.

Let us examine some concepts that have been studied until recently and provide estimates of their TRL.

6.3.2.1.1 Pebble Bed Modular Reactors (PBMRs)

Pebble Bed Modular Reactors are high-temperature thermal neutron reactors with approximately 100 MW output, using weakly enriched uranium fuel pebbles embedded in graphite (serving as a moderator) and helium as a coolant. At TRL 5-7 in 2025, these designs theoretically achieve approximately 48% efficiency, as modeled in this volume.

While early PBMR efforts in South Africa were abandoned in 2002 due to technical and economic challenges, advancements in TRISO fuel technology have revived interest. The Chinese HTR-PM, operational since 2023, employs a similar design with TRISO fuel in graphite spheres and helium cooling, aiming to achieve approximately 40% efficiency with a steam cycle at 750°C and 210 MWe. U.S. projects like X-energy's Xe-100 aim for commercialization by 2030.

6.3.2.1.2 Gas Turbine Modular Helium-Cooled Reactor (GT-MHR)

Building on PBMR principles but with distinct features, the Gas Turbine Modular Helium-cooled Reactor (GT-MHR) is a modular thermal neutron reactor (300 MW units) developed by an international consortium led by General Atomics (USA). At TRL 4-6 in 2025, it uses helium cooling and various fuels (plutonium, enriched uranium, or thorium) in silicon carbide-coated beads.

Unlike PBMR's pebble form, GT-MHR fuel is agglomerated into small cylinders within prismatic graphite elements, enabling better fuel tracking, higher power density, and reduced pressure drops (by a factor of 5–10). Control rods, supplemented by gravity-fed boron particles, manage the chain reaction. In emergencies, passive cooling occurs via conduction to the vessel walls and surrounding ground (the reactor is buried). Developers claim fuel temperatures stay below 1,600°C, even during depressurization.

GT-MHR targets 48% thermal efficiency, with refueling every 18 months (replacing half the fuel). Safety relies on passive mechanisms, though waste management challenges persist, including graphite recycling and actinide reprocessing.

6.3.2.1.3 *Very High-Temperature Reactor (VHTR)*

The Very High-Temperature Reactor (VHTR), inspired by GT-MHR modular designs (thermal neutrons), stands out for its higher temperatures, with a coolant capable of reaching 1,000°C–1,100°C. At TRL 3-5 in 2025, VHTR fuel follows HTR patterns: millimeter beads agglomerated into cylinders within fuel elements, with a helium coolant exiting at approximately 1,000°C.

A priority is to achieve higher fuel conversion rates than in current reactors. With 600 MWe capacity and efficiency exceeding 50%, VHTRs would surpass existing PWRs (33%–36%). Originally designed for highly enriched uranium-thorium mixtures, they now focus on low-enriched uranium and incinerating plutonium and minor actinides from PWR waste.

Work on VHTRs advances refractory materials useful for other systems and opens new markets, such as high-temperature industrial processes. They also benefit gas-cooled fast reactors. A key application could be hydrogen production via sulfur-iodine cycles (900°C–1,000°C) or high-temperature electrolysis, supporting decarbonization in transport and industry.

6.3.2.2 Current Technology Readiness Assessment

When the Generation IV International Forum selected six reactor concepts in the early 2000s (SCWR, VHTR, gas-cooled fast reactors, SFR, LFR, and MSR), the deployment timeline was optimistically set for 2030. More than two decades later, the landscape has evolved significantly, with some technologies advancing faster than anticipated while others face persistent challenges.

The original Generation IV International Forum roadmaps underestimated several critical factors. Materials science limitations at extreme temperatures proved more challenging than expected, while regulatory complexity for novel reactor designs created unforeseen delays. Economic competitiveness requirements in liberalized energy markets imposed stricter criteria than initially anticipated, and public acceptance challenges following Fukushima added another layer of complexity to reactor deployment.

This section evaluates the practical maturity of HTR concepts and their readiness for commercial deployment, starting from the most mature.

6.3.2.2.1 *Operational Breakthroughs (TRL 7-9)*

The Shidao Bay HTR-PM, operational since 2021, represents the sole commercial deployment of Generation IV high-temperature technology.

Its operational parameters reflect practical engineering decisions prioritizing reduced technical risk over maximum thermodynamic performance. The choice of a Rankine steam cycle without reheat rather than direct helium cycles demonstrates how first-of-a-kind technology deployment favors established operational experience, reduced licensing complexity, and proven maintenance procedures.

However, while design specifications claimed thermal efficiencies of 44%–47.6% under optimal conditions (132.4 bar, 535°C–570°C turbine inlet), operational reality tells a different story. Thermodynamic modeling based on actual 2023 test data reveals performance closer to 33.7% efficiency, with only 67 MWe generated.

Although this represents a first prototype, this substantial gap between theoretical potential and achieved performance exemplifies the challenges facing HTR concepts.

6.3.2.2.2 Advanced Development Phase (TRL 4-6)

While less frequently cited in academic literature than more exotic concepts, sodium-cooled fast reactors are the Generation IV technology most likely to achieve rapid commercialization.

China's CFR-600 began construction in 2017 with operation planned for 2024–2025, while Russia's BN-800 has operated since 2016, successfully demonstrating MOX fuel capability under commercial conditions. India's Prototype Fast Breeder Reactor (PFBR) achieved criticality in 2023, marking another milestone for this technology pathway. Several SMR projects employ this technology.

The SFR experience demonstrates that moderate temperature increases (550°C compared to 320°C for PWRs) can achieve significant efficiency gains (38%–42% versus 33%–35%) while maintaining manageable technological complexity. This approach contrasts sharply with higher temperature targets that encounter multiple materials limitations.

6.3.2.2.3 Research Phase Reactors (TRL 2-4)

6.3.2.2.3.1 Very High-Temperature Reactors VHTRs face major materials challenges despite their theoretical advantages of efficiency exceeding 50% and potential applications for supplying heat to industrial processes.

No qualified materials exist for 1,000°C operation in neutron flux environments, and fuel performance limits at extreme temperatures remain unresolved. Heat exchanger materials for gas-to-gas heat transfer at these temperature levels present additional engineering challenges that current alloy development has not addressed.

These limitations have pushed commercial deployment timelines beyond 2045, contingent on materials science breakthroughs being achieved within reasonable timeframes.

6.3.2.2.3.2 Gas-Cooled Fast Reactors Gas-cooled fast reactors encounter major difficulties in combining fast neutrons with gas cooling systems. Inadequate heat removal during accident scenarios creates safety challenges, while complex fuel handling requirements and economic disadvantages compared to SFRs have led most programs to de-emphasize or redirect efforts toward hybrid concepts.

6.3.2.2.3.3 Lead-cooled Fast Reactor As of 2024, LFR technology maturity spans TRL 4-6: TRL 4 for corrosion-resistant materials and pump development in laboratory loops, TRL 5 for integrated testing in lead-bismuth facilities (e.g., CORRIDA, LEAD), and TRL 6 for near-prototypical demonstrators in simulated operational conditions (e.g., MYRRHA in Belgium, ALFRED in Europe). The BREST-OD-300 in Russia is the closest to TRL 7, with construction underway.

6.3.2.2.3.4 Molten Salt Reactor Molten salt reactors have attracted significant private investment since 2015. However, fundamental technical issues persist across multiple domains. Molten salt chemistry control over reactor lifetime remains poorly understood, materials corrosion in high-temperature salt environments presents challenges comparable to those facing VHTRs, tritium management in fluoride-based salts requires complex processing systems, and regulatory framework development lags behind technical progress.

Current projections suggest demonstration reactors might be possible by 2035, but commercial deployment remains highly uncertain given the number of unresolved technical challenges.

6.4 CONCLUSION

This chapter examined thermodynamic cycles where the fluid remains gaseous, focusing on closed-cycle gas turbines and supercritical CO_2 systems. Starting from a simple helium cycle achieving 22.85% efficiency, exergy analysis enabled progressive improvements: a regenerative cycle doubling efficiency to 40.41%, and a cycle with intercooled compression and staged expansion with reheat boosting efficiency to 47.4%. Supercritical CO_2 cycles benefit from reduced compression work above the critical point (78 bar, 31.1°C), with four analyzed variants reaching up to 45% efficiency. Major technical challenges include mastering significant c_p variations near the critical point and the requirement for cold sources below 25°C. Despite their thermodynamic appeal, high-temperature reactor cycles still face barriers related to the development of high-performance turbomachinery (TRL 6-7 for CO_2, below 7 for helium) and material limitations, with only the Chinese HTR-PM having achieved commercial operation.

Further resources on the models presented in this chapter are available on this book's website.[2]

NOTES

1 https://direns.mines-paristech.fr/Sites/Thopt/en/co/createur-fond-diag.html.
2 Available at: https://server.s4e2.com/crc/tcnpp/chapters/chapter6/_resources/index.html.

BIBLIOGRAPHY

"Breakthrough for sCO_2 power cycle as STEP demo completes phase 1 of 10 MW project." (2024). *POWER Magazine*, October.

Brun, K., Friedman, P. & Dennis, R. (2017). *Fundamentals and Applications of Supercritical Carbon Dioxide (SCO_2) Based Power Cycles.* Woodhead Publishing (Elsevier), New York. ISBN 978-0-08-100804-1.

Dostal, V. (2004). *A Supercritical Carbon Dioxide Cycle for Next-Generation Nuclear Reactors.* PhD thesis, Massachusetts Institute of Technology (MIT), January.

Dostal, V., Driscoll, M.J., Hejzlar, P. & Todreas, N.E. (2003). "A supercritical CO_2 *gas turbine power cycle for next-generation nuclear reactors." Proceedings of ICONE-10*, Arlington, Virginia, April 14–18.

El-Sayed, A.F. (2008). *Aircraft Propulsion and Gas Turbine Engines.* CRC Press, Boca Raton, FL. ISBN 978-0-8493-9196-5.

Kugeler, K. & Hoh, J. (2017). *The High Temperature Gas-Cooled Reactor: Lessons Learned and Future Perspectives.* European Commission JRC, EUR 28712 EN, New York. ISBN 978-92-79-71321-3.

Langston, L.S. (1997). "Introduction to gas turbines for non-engineers." *Global Gas Turbine News*, 37(2), 9.

Molière, M., Privat, R., Jaubert, J.-N. & Geiger, F. (2024). "Supercritical CO_2 power technology: Strengths but challenges." *Energies*, 17(5), 1129. DOI: 10.3390/en17051129.

Schobeiri, M.T. (2018). *Gas Turbine Design: Components and System Design Integration*, 2nd revised and enhanced ed. Springer International Publishing, New York. ISBN 978-3-319-58376-1 (print); 978-3-319-58378-5 (eBook). DOI: 10.1007/978-3-319-58378-5.

Span, R. & Wagner, W. (1996). "A new equation of state for carbon dioxide covering the fluid region from the triple point temperature to 1100 K at pressures up to 800 MPa." *Journal of Physical and Chemical Reference Data*, 25, 1509–1596.

Vivier, L. (1965). *Turbines à vapeur et à gaz.* Albin Michel, Paris.

Weston, K.C. (1993). "Dual gas turbine combined cycles." *Proceedings of the 28th Intersociety Energy Conversion Engineering Conference (IECEC'93)*, Boston, Vol. 1, 955–958.

7 Specific Applications and Perspectives
Combined Cycles, Desalination, Hydrogen Production, and Cogeneration

7.1 COMBINED CYCLES

INTRODUCTION

The principle of a combined cycle generally involves operating one or more Brayton cycles in cascade, followed by one or more Rankine cycles, where the hot source of the latter is the cold source of the former. Other configurations exist but are less common. The high-temperature cycles are often referred to as topping cycles, and the low-temperature cycles as bottoming cycles.

In the schematic of the combined cycle of Figure 7.1.1, the Brayton cycle (topping cycle) is represented by solid lines on the left, coupled via an Intermediate Heat Exchanger (IHX) to the nuclear reactor, and via an Heat Recovery Steam Generator (HRSG) to the Rankine cycle, shown in dashed lines on the right (bottoming cycle).

The IHX is used only when the coolant differs from the thermodynamic fluid of the topping cycle, which is generally the case, as their required properties differ, and for radioactive containment reasons.

7.1.1 GENERAL OVERVIEW

The significant advantage of combined cycles is that they achieve the highest efficiency in converting heat into electricity, now exceeding 60% when combining a high-pressure (HP), high-temperature combustion gas turbine and one or more steam cycles.

Despite these high performances, there are currently no examples of combined cycles coupled with a nuclear reactor, as the temperatures reached by nuclear reactors are not yet sufficiently high. However, several are under study. Some of these are hybrid cycles combining nuclear reactors with combustion turbines. These will not be covered in this chapter.

DOI: 10.1201/9781003606024-8

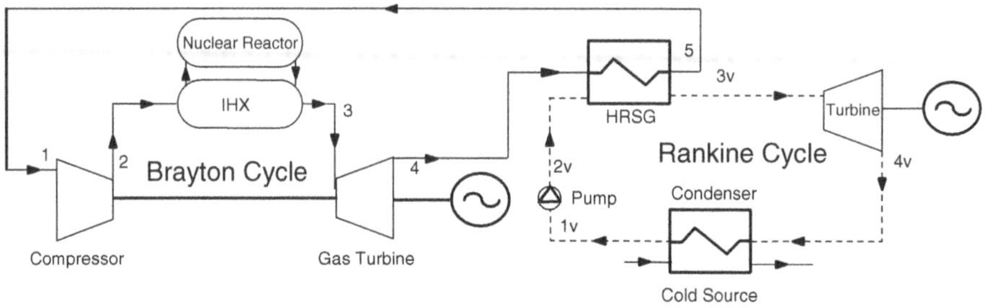

FIGURE 7.1.1 Combined cycle layout.

These systems should be particularly well suited to certain Generation IV nuclear reactors, such as High-Temperature Gas-cooled Reactors or Molten Salt Reactors, which, as we have seen, can reach temperatures of 600°C–1,000°C, or even higher.

In a combined cycle involving one or more steam cycles, the heat exchanger that couples the topping cycle to the bottoming cycle is called an HRSG. We refer to it as a heat exchanger, but in reality, it is often a complex network of heat exchangers that can be challenging to optimize.

Our goal is to familiarize you with powerful optimization methods for Thermodynamic Cycles of Nuclear Power Plants. We will explain in detail how the pinch point method can be applied to determine an efficient configuration of the HRSG for a dual-pressure combined cycle.

7.1.1.1 Heat Recovery Steam Generators

HRSGs are used not only in combined cycles but, more generally, to produce steam from thermal effluents.

Most HRSGs are variations of water tube boilers. The main differences stem from the fact that, on the one hand, the temperature levels of effluents are much lower than those achieved in a boiler, and on the other hand, heat exchange takes place only by convection and not by radiation. Nothing prevents under these conditions the superheater from being positioned upstream of the evaporator, and the HRSG operating sketch and enthalpy diagram are as shown in Figure 7.1.2.

At point 6 on the diagram, you can see a minimum in the temperature difference between the two fluids, called the pinch (see Chapter 3). This point plays a fundamental role in the design of HRSGs, as it represents the smallest temperature difference in the facility. To reduce the cost of equipment, it is preferable that the pinch not be too small. However, the thermodynamic optimization of the complete system requires it to be as small as possible.

Particular attention should be paid to the HRSG design at this pinch, which must exceed a minimum value of about 10°C–15°C, otherwise vaporization cannot occur at the desired temperature.

7.1.1.2 Overall Performance

The enthalpy exchange in a combined cycle can be summarized by the sketch in Figure 7.1.3:

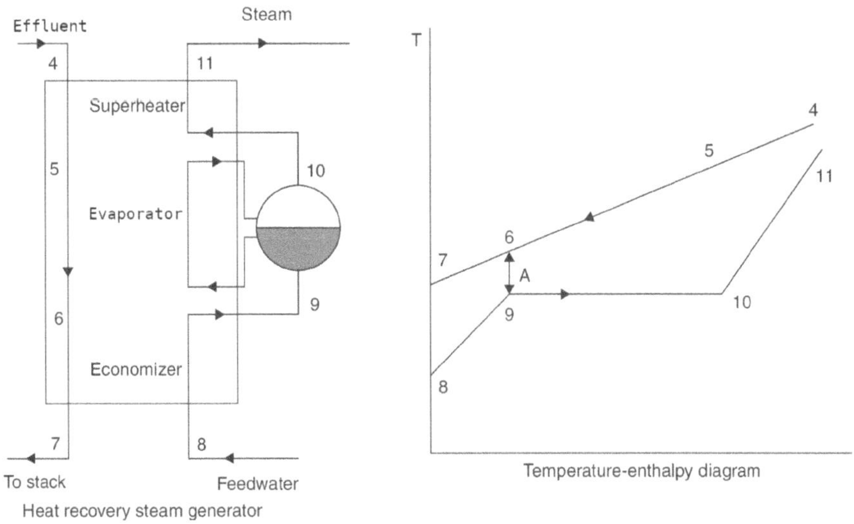

FIGURE 7.1.2 Steam generator exchange configuration p3_5.6.6.

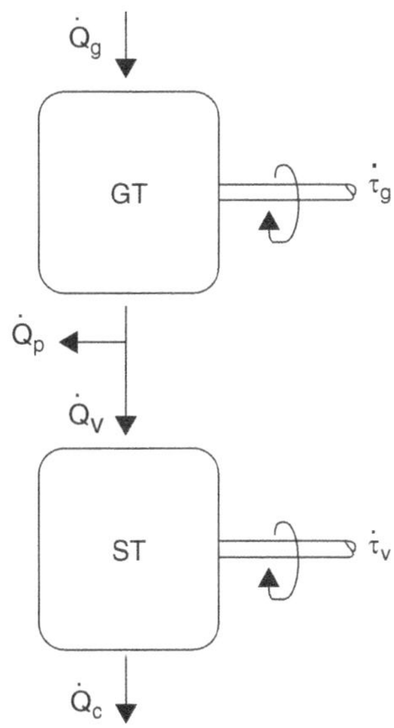

FIGURE 7.1.3 Sketch of a combined cycle.

- The gas turbine cycle GT receives heat Q_g from the hot source, here the nuclear reactor. It provides, on the one hand, useful work τ_g, and, on the other hand, heat $(Q_v + Q_p)$. The first term is the heat supplied to the steam cycle; the second represents losses.
- The steam cycle ST produces useful work τ_v, and the condenser rejects heat Q_c.

Let us call η_g the gas turbine efficiency, η_v that of the steam cycle, η_{cc} that of the combined cycle, and ε the HRSG effectiveness, that is to say the ratio of Qv to Qp + Qv:

$$\varepsilon = \frac{Q_v}{Q_p + Q_v} = \frac{Q_g}{Q_p + Q_v}\frac{Q_v}{Q_g} = \frac{1}{1-\eta_g}\frac{Q_v}{Q_g}$$

$$\eta_{cc} = \frac{\tau_g + \tau_v}{Q_g} = \eta_g + \eta_v \frac{Q_v}{Q_g} = \eta_g + \varepsilon\left(1-\eta_g\right)\eta_v$$

$$\eta_{cc} = \eta_g + \varepsilon\left(1-\eta_g\right)\eta_v \tag{7.1.1}$$

The combined cycle efficiency equals the sum of the gas turbine efficiency and the product of its complement to 1 by the HRSG effectiveness and the steam cycle efficiency.

For example, with $\eta_g = 0.29$, $\eta_v = 0.32$, and $\varepsilon = 0.83$, we obtain $\eta_{cc} = 0.48$.

7.1.1.3 Combined Cycle Optimization

We have seen that overall performance is given by equation (7.1.1): $\eta_{cc} = \eta_g + \varepsilon\left(1-\eta_g\right)\eta_v$.

This expression shows that it is equally important to optimize the steam cycle and the recovery steam generator, and thus its effectiveness ε. Difficulties arise because the problem is highly constrained and conflicts may exist between these two objectives.

The optimization of such a combined cycle is based on the reduction of its internal irreversibilities, which can be grouped into two broad categories:

- Mechanical irreversibilities that take place in the compressor and turbines;
- Thermal irreversibilities related to temperature differences in the heat exchangers.

Much has already been done to limit the mechanical irreversibilities. It is today a purely technological problem that only machine manufacturers are able to tackle, and on which they are indeed working diligently.

We therefore focus in what follows only on the reduction of thermal irreversibilities, i.e., on the optimization of plants whose turbine outlet temperature is fixed. These irreversibilities result from temperature differences between the hot and cold parts of the cycle.

In some cogeneration plants [Combined Heat and Power (CHP)] studied in Section 7.4, problems arise in a similar manner, especially if steam requirements at medium and HP are important.

In a combined cycle plant, the stream of hot gases exiting the gas turbine must be cooled by water of the steam recovery cycle. In a single-pressure cycle, water enters the heat exchanger in the liquid state at about 30°C after being compressed by the feedwater pumps downstream of the condenser. It is heated to the boiling temperature corresponding to its pressure (economizer), then evaporated at constant temperature, and superheated before being expanded in the steam turbine. The diagram in Figure 7.1.2 shows the heat transfer within the heat exchanger between the reactor coolant and water. It shows that if we set, for technical reasons, a minimum pinch value (temperature difference between both fluids) between points 6 and 9 on the one hand, and between points 4 and 11 on the other hand, heat exchanges actually take place with much larger differences in nearly all of the heat exchanger. This stems from the need to vaporize water, which induces a very significant "plateau" at constant temperature.

Losses in the HRSG and those in the exhaust are linked. Their reduction is therefore an important issue. The ideal heat exchange corresponds to the case where the curve of gas cooling and that of water heating would be parallel. The irreversibility would then be minimal. This is not feasible with water, and the single-pressure cycle has strong internal irreversibilities. We will see in Section 7.1.5 that Kalina cycles attempt to reduce this constraint.

To improve the cycle performance, we use multiple steam cycles at different pressure levels (two, three, or even four). Figure 7.1.4 shows the value of using multi-level pressure: with some simplifying assumptions and a choice of scales, we can superimpose on an entropy chart gas turbine and steam power plant cycles. In all three cases, the gray surface represents work provided for the same heat input in the gas turbine. The rectangle in dashed lines is the Carnot cycle.

The left figure corresponds to a single gas turbine, the middle to a single-pressure combined cycle, and the right to a three-pressure combined cycle. The gain provided by the increased number of pressure levels is very clear, but designing the HRSG is a complex optimization problem that does not arise in conventional boilers.

The optimization of such cycles is a complex problem because, to achieve better cooling of the hot gas stream, there are many degrees of freedom regarding the pressure levels, the corresponding flow rates, and the placement of heat exchangers (in series or in parallel). Figure 7.1.5 gives an example of an industrial three-pressure HRSG, with the complete combined cycle.

This optimization problem is relatively new. It did not arise in old power plants, in which very large irreversibilities occurred for technical and economic reasons related to the thermal resistance of steel boilers and sulfur content of fumes. There is no proven method to solve it.

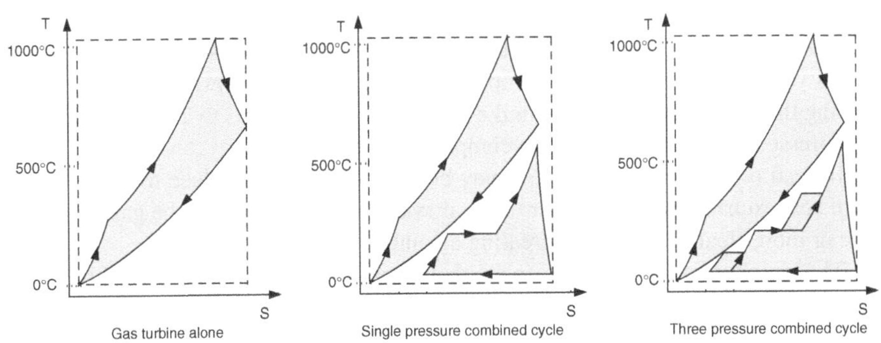

FIGURE 7.1.4 Comparison of work provided by a gas turbine and combined cycles.

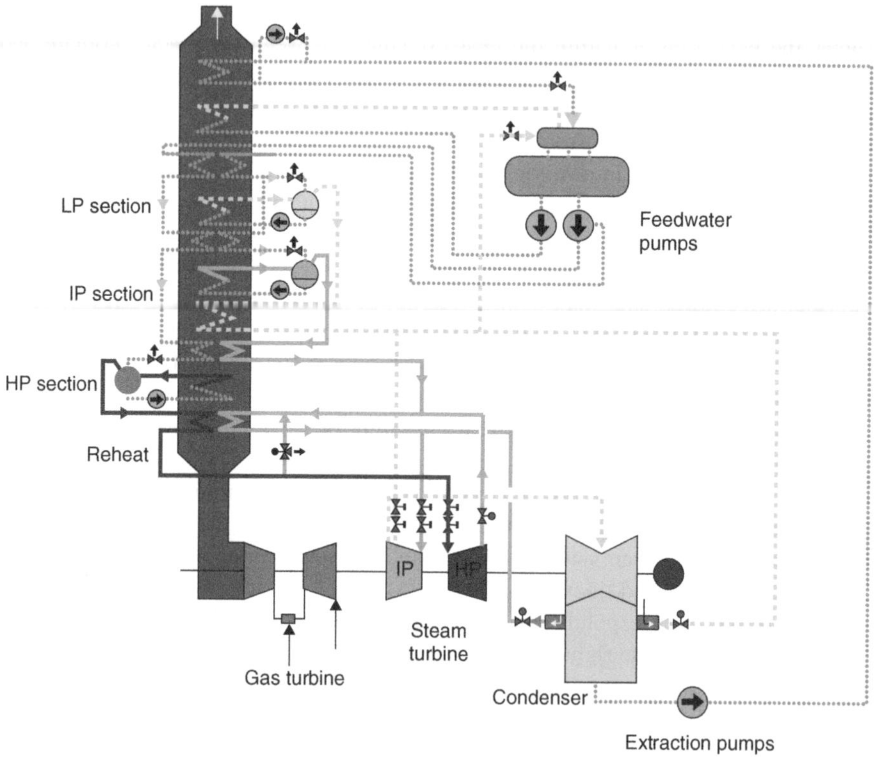

FIGURE 7.1.5 Three-pressure HRSG.

The optimization method introduced in Chapter 3 is the extension to power generation or cogeneration plants of the pinch method developed in chemical engineering to optimize the configuration of very large exchanger networks (such as those of a refinery).

We will later see how this method can be practically implemented.

7.1.2 Single-Pressure Combined Cycle

To illustrate our point, we will revisit the regenerative helium cycle studied in Section 6.1:

As you can observe in this example (Figure 7.1.6), the low-pressure (LP) helium exiting the regenerator is cooled in the precooler from 247.3°C to 28°C before being compressed, in order to reduce the compression work.

Instead of operating this way, it may be advantageous to replace the regenerator with heat exchangers that transfer the heat available at the outlet of the gas turbine to one or more steam cycles, thus creating a combined cycle.

The simplest configuration is one with a single steam cycle comprising an economizer, an evaporator, and a superheater. This is called a single-pressure steam cycle.

Knowing the condensation temperature and thus the pressure, three parameters of the steam cycle can be freely chosen: the HP of the cycle, the superheat temperature, and the flow rate.

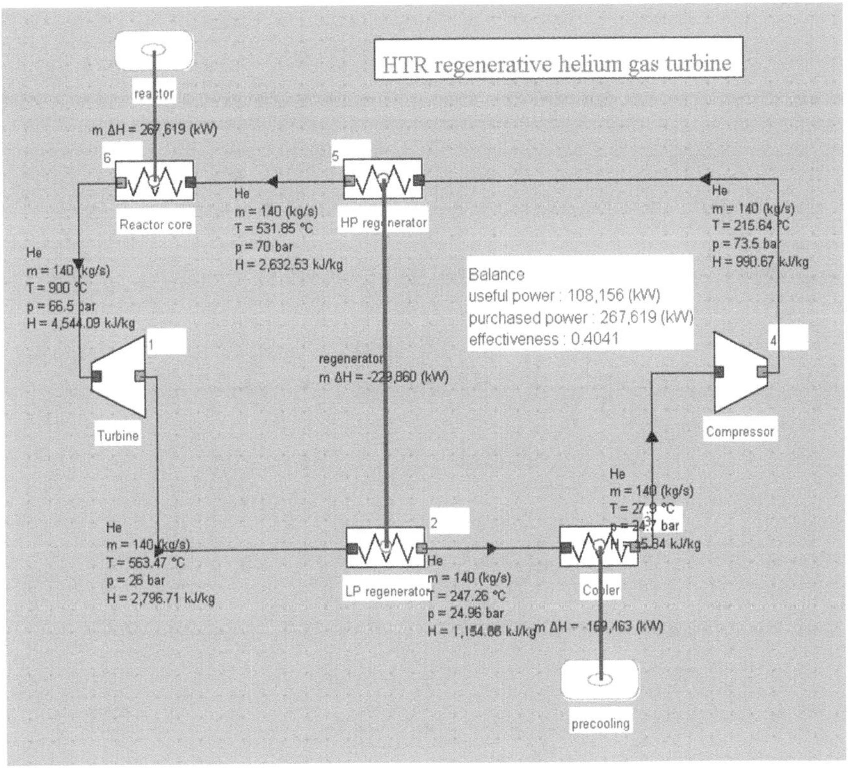

FIGURE 7.1.6 Synoptic view of the regenerative gas turbine cycle.

Assuming a superheat temperature of 500°C, approximately 50°C less than the temperature of the helium at the outlet of the turbine, one possible solution results in a pressure of 120 bar and a water flow rate of 82.9 kg/s. The cycle is represented in the synoptic view of Figure 7.1.7. The power output is more than doubled, and the efficiency increases from 40.41% to 45.62%.

The influence of the thermal power rejected at the precooler requires careful analysis. In this example, although it has been reduced by a third, it is not zero, as a single-pressure level is insufficient to cool the helium adequately.

However, as noted in Section 6.1.2, cooling the helium has the positive effect of reducing the compression work, but as the outlet temperature of the compressor decreases, the heat that the nuclear reactor must supply increases. In the case of this combined cycle, this results in a significant increase in mechanical power output but also a slight decrease in overall efficiency.

With a precooler, the mechanical power output is 227 MW with an efficiency of 45.62%. Without this heat exchanger, these values are 158.4 MW and 49.65%, respectively. It is thus possible to modulate the power output delivered by the combined cycle by approximately 45% by adjusting the cooling upstream of the compressor, but at the cost of an efficiency reduction of about 10% (Figure 7.1.8).

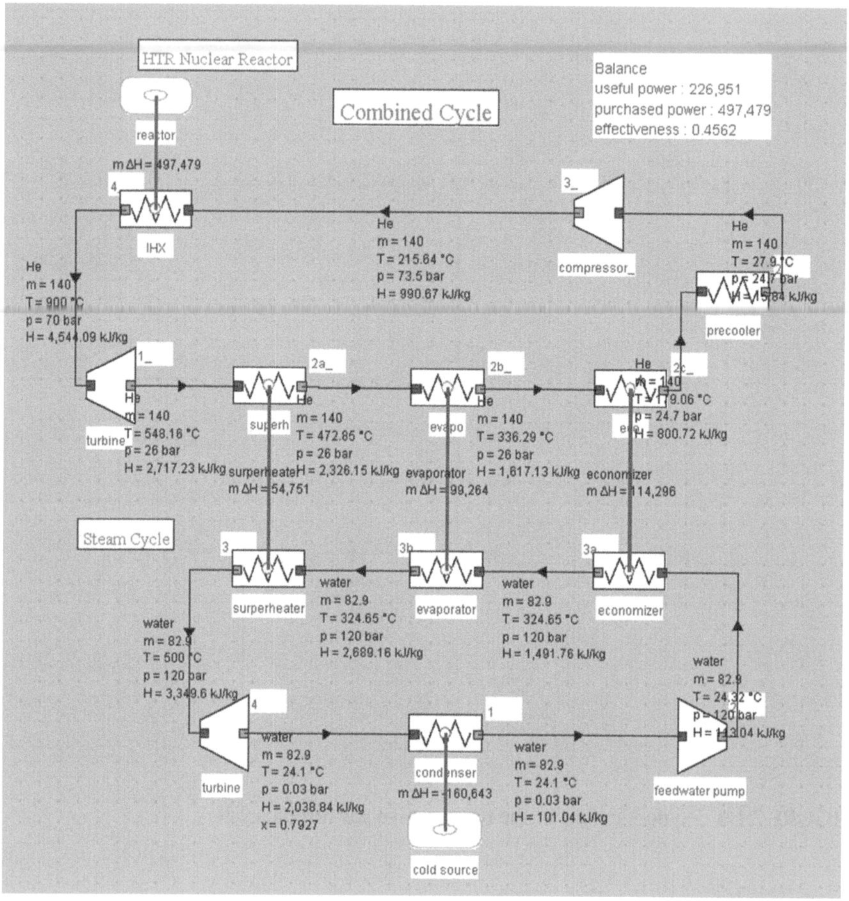

FIGURE 7.1.7 Synoptic view of the combined cycle.

It should be noted that the parameterization of this combined cycle has not been fully optimized. It would be possible to adjust the three degrees of freedom of the steam cycle to find the set of values maximizing either the power output or the efficiency. We will not do so here.

7.1.2.1 Exergy Balance of the Combined Cycle

The exergy balance of this cycle is provided in Figure 7.1.9. Its exergy efficiency (62%) has significantly increased compared to that of the regenerative cycle alone (53%).

The main irreversibilities occur in the IHX (41.7%). Other losses are distributed fairly evenly among the various components. In descending order, they occur in the economizer, compressor, steam turbine, gas turbine, evaporator, superheater, and condenser.

Irreversibilities in the turbomachines account for 29% of the total. Manufacturers continue their efforts to reduce these, and significant progress has already been made in recent decades. Consequently, the potential for further improvement is increasingly limited.

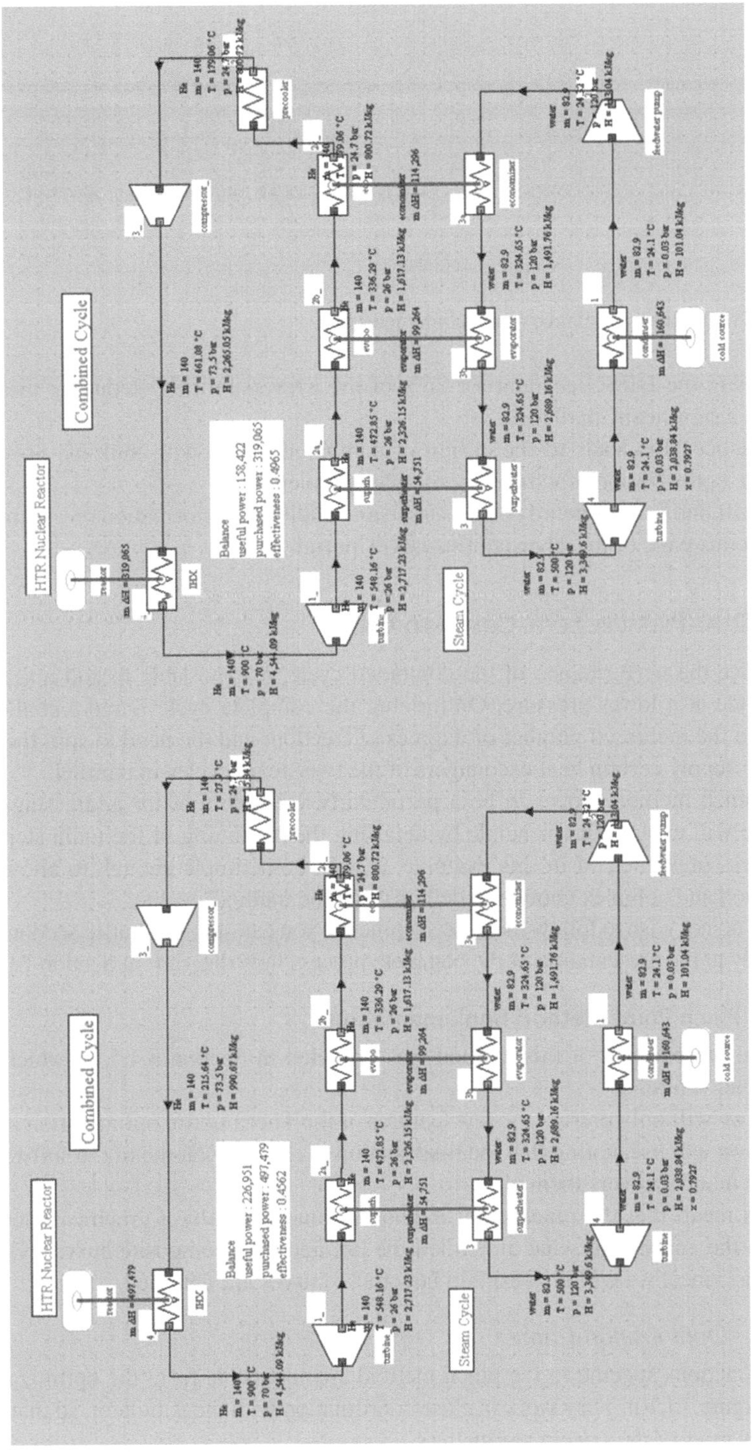

FIGURE 7.1.8 Comparison of the synoptic views of the combined cycle.

component	Resource	Product	Exergy efficiency	Irreversibilities	% total	settings
compressor_	136,476.67	126,147.356	92.4314%	10,329.314	6.0218%	
turbine_	264,202.736	255,760.45	96.8046%	8,442.286	4.9217%	
feedwater pump	994.57	994.561	99.9991%	0.00882	0.0000%	
turbine	117,964.2	108,662.12	92.1145%	9,302.08	5.4230%	
economizer	0	37,919.18	0	(-37,919.18)	(-22.1062%)	economizer
eco	55,080.696	0	0	(55,080.696)	(32.1111%)	economizer
evaporator	0	50,585.7	0	(-50,585.7)	(-29.4906%)	evaporator
evapo	56,180.802	0	0	(56,180.802)	(32.7525%)	evaporator
superh	34,253.715	0	0	(34,253.715)	(19.9693%)	surperheater
surperheater	0	30,682.245	0	(-30,682.245)	(-17.8872%)	surperheater
IHX	398,482.84	306,749.934	76.9795%	91,732.906	53.4787%	Tk = 1,200.00 °C
condenser	2,217.487	0	0	2,217.487	1.2928%	Tk = 20.00 °C
precooler	23,179.34	0	0	23,179.34	13.5132%	Tk = 20.00 °C
global	398,482.84	226,951.33	56.9539%	171,531.51	1	

FIGURE 7.1.9 Exergy balance of the combined cycle.

Losses in the HRSG account for 26% of the irreversibilities. Reducing these is therefore a significant challenge.

In practice, this leads to the design of combined cycles with multiple pressure levels, the optimization of which is a complex problem.

You will find further details on this issue and additional information on combined cycles on this page of the Thermoptim-UNIT portal.[1]

7.1.3 Dual-Pressure Level Combined Cycle

To improve the performance of the combined cycle, it is possible to add a second steam cycle at a lower pressure. Optimizing the complete cycle is more challenging due to the increased number of degrees of freedom and the need to split the gas stream to supply certain heat exchangers of the two steam cycles in parallel.

The pinch method proves to be a particularly valuable tool for addressing this issue. We will explain the principle by detailing the beginning of the main steps of the process implemented in this example, as it is both simple enough to allow for explanation and complex enough to demonstrate the method's value.

If you do not wish to follow these developments, you can skip this entire section and go directly to the presentation of the complete process flow diagram in Section 7.1.3.4.

7.1.3.1 Pinch Point Method Implementation

A general presentation of this method was provided in Section 3.4.5, to which we recommend you refer.

Here, we will only present the few features of the Thermoptim optimization interface that we will use, noting that the final section of the first volume of the software's reference manual covers them all.

Let us recall that the pinch point method is fundamentally a graphical method based on the analysis of what are called the hot and cold composite curves. Given their importance, we will also explain how these curves are obtained.

7.1.3.1.1 Optimization Frame

All the functions specific to the pinch method are accessible from the optimization frame (Figure 7.1.10). They work in close coordination with the simulator, so that you can easily modify the system parameters.

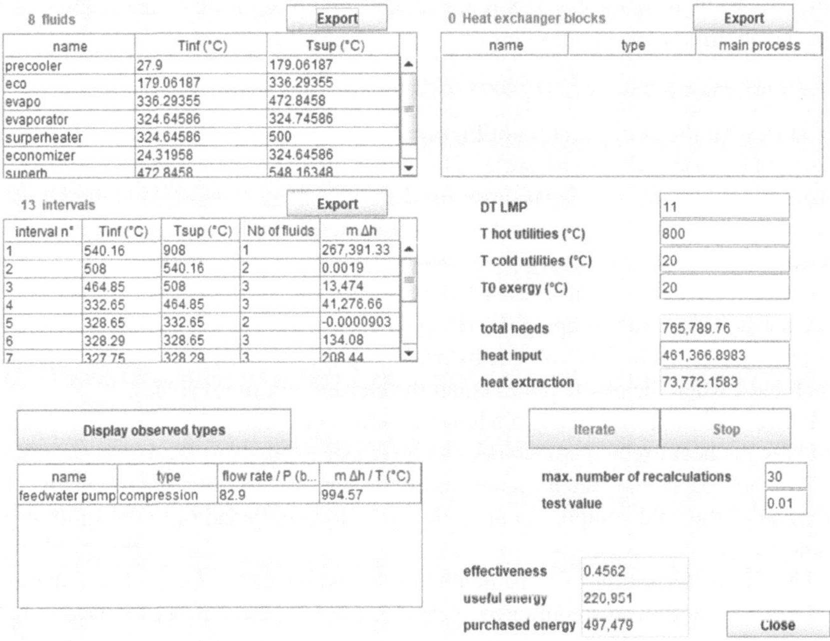

FIGURE 7.1.10 Optimization screen.

To open the optimization screen, type Ctrl+M or select "Optimization tools" in menu "Special" of Thermoptim professional or pinch version.

It comprises three main tables:

- The fluid table contains all the exchange streams in which the "pinch method fluid" checkbox is selected. These are the fluids which are processed by the variant of the Problem Table Algorithm implemented in Thermoptim.
- The interval table contains the list of intervals which are built by the Problem Table Algorithm variant.
- The heat exchanger block table contains the exchange blocks which can be defined to facilitate stream matching in heat exchangers.

In the lower right part of the screen appears the "Iterate" button which automatically recalculates Thermoptim.

Here, we will mainly use two menu lines:

- *Update and minimize target* which updates the simulator, calculates the Problem Table Algorithm, and sets up the Composite Curves.
- *Plot the Carnot Factor curves* that plot the exergy composite curves.

To implement the pinch point method, it is first necessary to specify the minimum pinch values chosen for the different types of heat exchange processes. Typically, these are set at 16 K for gases, 8 K for liquids, and 6 K for fluids undergoing phase changes (boiling and condensation).

7.1.3.1.2 Optimization of the LP Steam Cycle Flow Rate

We start with the configuration of Figure 7.1.11, where an LP steam cycle has been added below the HP steam cycle. The labels of the various components in the schematic differ slightly from those used for the single-pressure combined cycle, but this has no practical significance, as the structure and parameterization remain the same.

Initially, we will determine the LP flow rate that leads to the best performance for the overall cycle, while keeping the HP cycle unchanged. As explained in Chapter 3, the pinch point method allows this by analyzing the composite curves: the optimum is achieved when the pinch values equal the chosen minimum values.

In a second step, we will explain how the composite curves are practically obtained and how the heat exchanger network can be constructed.

This approach may seem counterintuitive, but explanations regarding the construction of the cold composite curve are easier to understand once the LP flow rate is known.

The parameter settings for the helium cycle and the HP steam cycle are the same as those for the previous combined cycle. For the LP steam cycle, we have initially chosen a flow rate of 1 g/s, a pressure of 6 bar, and a superheat temperature of 275°C.

With such a low LP flow rate, the performance summary of the two combined cycles is almost identical, which allows us to verify that the model is correctly parameterized.

We now increase the LP cycle flow rate to 20 kg/s, which is done by double-clicking on its process and recalculating it. In the optimization interface, we iterate until the performance summary stabilizes. The efficiency increases and is now 48.25%.

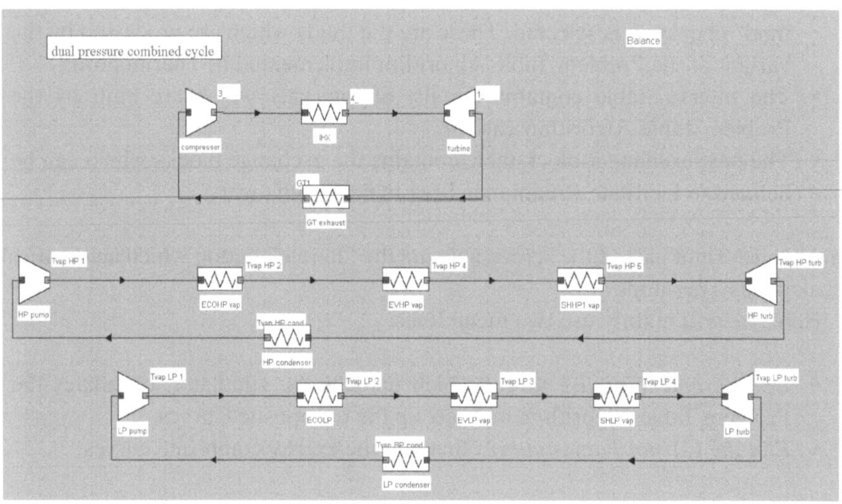

FIGURE 7.1.11 Initial layout.

The pinch point method is implemented by pressing Ctrl+U or activating the "Update and minimize target" menu option. At this stage, no additional heat input is required, meaning the LP flow rate can be further increased. By trying 25 kg/s, an additional heat input of slightly over 4 MW is needed.

By reducing the LP flow rate to 23 kg/s, a valid parameterization is obtained, leading to an efficiency of 48.76%.

The exergy composite curve allows for simple visualization of the pinch points. They occur where the two upper or lower curves are closest. The difference between the two curves appears at the bottom. It aligns with the smooth, decreasing curve representing the locus of minimum pinch points.

Two pinch points are visible. Clicking on the energy demand curve at the pinch points reveals their values, here around 160°C and 325°C. Additionally, it can be seen that the helium is not fully cooled. A third steam cycle could be inserted (Figure 7.1.12).

7.1.3.1.3 Construction of Composite Curves

Now, let us address how the composite curves are practically constructed. We start with the following simplified schematic, which corresponds to the single-pressure combined cycle with an added LP cycle, but without including the condensers and feedwater pumps.

We have deliberately simplified the schematic to make our presentation as educational as possible (Figure 7.1.13). For the same reason, we have assumed that the minimum pinch values equal zero. This reduces the number of intervals to consider.

Figure 7.1.14, showing the enthalpy diagram of the three superimposed fluids, helps us understand how the composite curves are constructed and how the heat exchanger network can be determined.

The ordinate represents temperatures in °C, and the abscissa represents the enthalpies involved.

For the cooling of helium, it is a simple straight line, from the turbine outlet temperature of 548.16°C to the compressor inlet temperature, assumed to be 27.9°C.

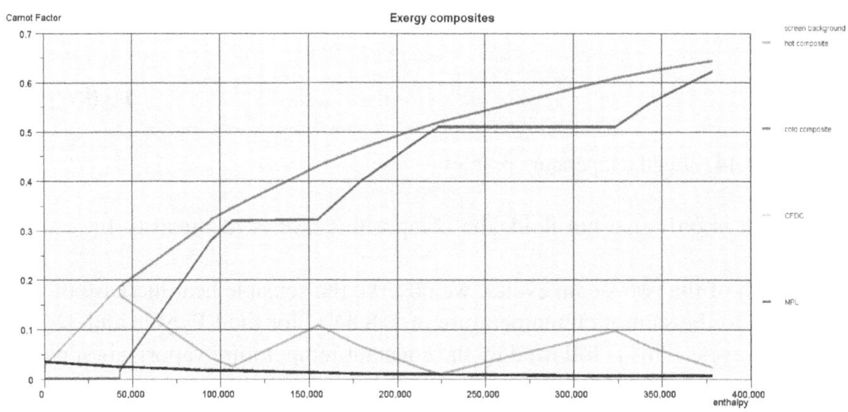

FIGURE 7.1.12 First composite curves.

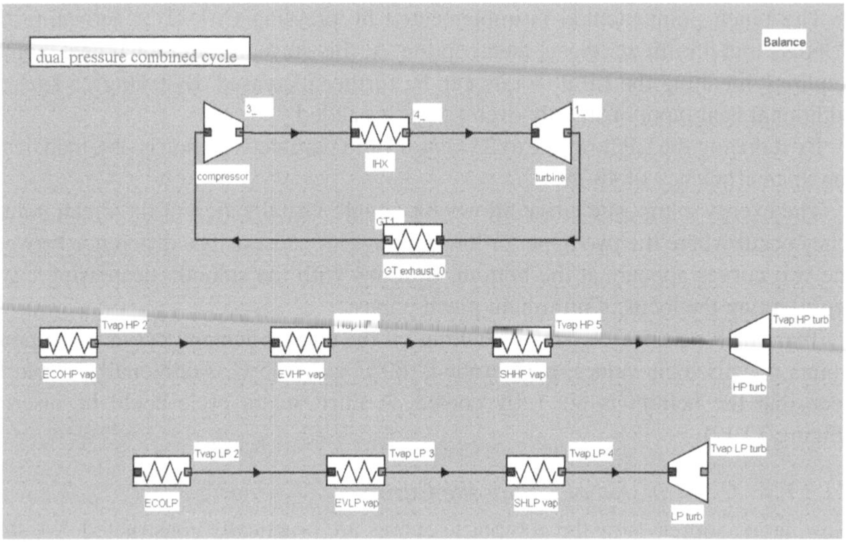

FIGURE 7.1.13 Simplified schematic.

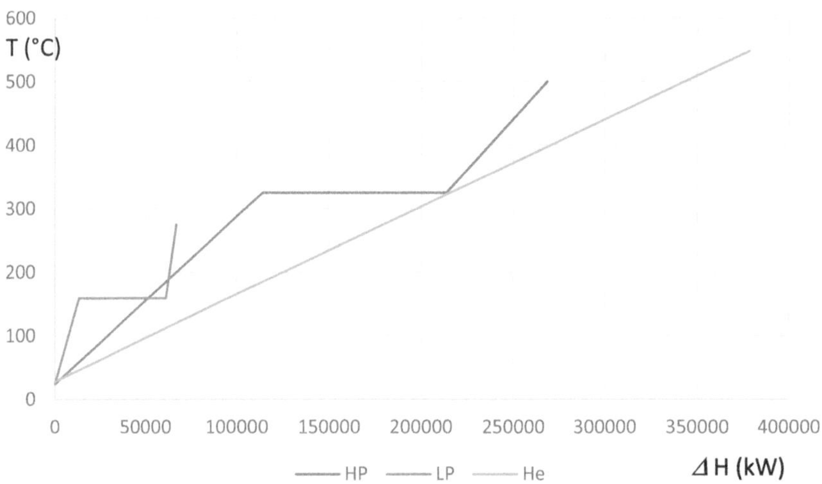

FIGURE 7.1.14 Fluid temperature profiles.

Since there is only one hot fluid, the composite curve is identical to the enthalpy diagram.

For each of the two steam cycles, we observe the sensible heat increase of water, from 24°C to the saturation temperature of 158.84°C for the LP cycle and 324.65°C for the HP cycle. This is followed by the constant-temperature vaporization plateaus and then the superheating phases.

The cold composite curve is easily derived from this. The interval boundaries are the temperatures at which slope changes occur: 500°C (end of HP superheating),

324.65°C (HP saturation temperature), 275°C (end of LP superheating), 158.84°C (LP saturation temperature), and finally 24°C.

Table 7.1.1 compiles all the relevant information. From left to right, it includes the interval number, the upper and lower temperatures, the names of the fluids involved, the temperature difference between the boundaries, and the sum of their thermal capacities (mC_p), which represents the inverse of the slope of the corresponding curve in the enthalpy diagram.

The plot of the two composite curves can then be drawn, and it matches the pattern highlighted earlier, with the ordinate representing temperatures rather than the Carnot factor (Figure 7.1.15).

The pinch points occur in the zones where the two composite curves are closest, namely at the outlets of the two economizers.

We have shown how the cold composite curve can be determined. Once both composite curves are established, the pinch point method considers the set of intervals corresponding to these two curves. In our case, this involves adding two intervals: one at high temperature to represent the cooling of helium at the turbine outlet, and another to represent the cooling of helium below the LP saturation temperature, in parallel with the two economizers, before only these two fluids remain.

TABLE 7.1.1
Construction of the Simplified Composite Curves

Interval	T_i (°C)	$T_i + 1$ (°C)	Streams	$T_i - T_{i+1}(K)$	$\sum(mC_p)(kW/K)$
1	500	324,65	SHHP	175,35	312
2	324,65	324,65	EVHP	0,1	99,264
3	324,65	275	ECOHP	49,65	381
4	275	158,84	ECOHP, SHLP	116,16	431
5	158,84	158,84	ECOHP, EVLP	0,1	47 956
6	158,84	24	ECOHP, ECOLP	130,94	478

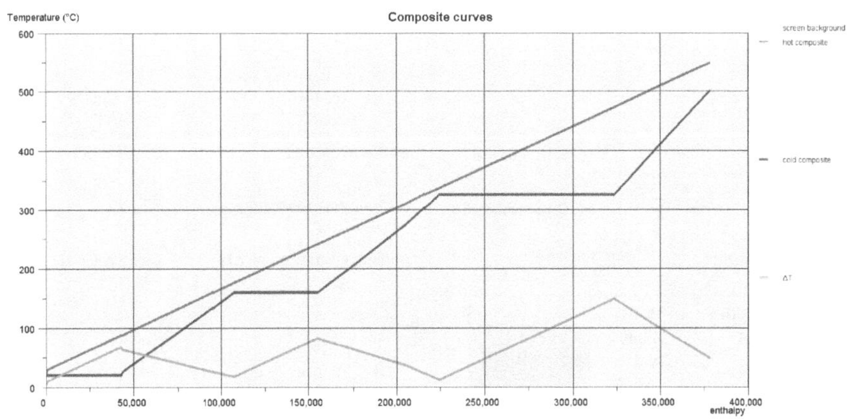

FIGURE 7.1.15 Temperature composites.

7.1.3.1.4 Creation of the Heat Exchanger Network

Let us now examine how the heat exchanger network can be constructed.

Heat integration is a valuable guide for designing the overall architecture of the exchanger network by choosing the best pairs of streams exchanging heat.

Note first that for a pinch to exist, there must be, as shown in Figure 7.1.16:

- Either appearance of a new cold stream (need);
- Or disappearance of a hot stream (availability);
- Or simultaneity of both events: onset of a cold stream and disappearance of a hot stream.

Here, we are clearly in the first case, the new cold fluid appearing being the evaporation of steam (Figure 7.1.17).

As the most constrained zone for fluid matching in heat exchangers is located at pinches, one must start by analyzing intervals at their level, knowing that the heat capacity rates of the fluids matched in each exchanger must comply with the following inequations (in absolute values):

- Just above the pinch: $\dot{m}\ c_{ph} < \dot{m}\ c_{pc}$;
- Just below the pinch: $\dot{m}\ c_{ph} > \dot{m}\ c_{pc}$.

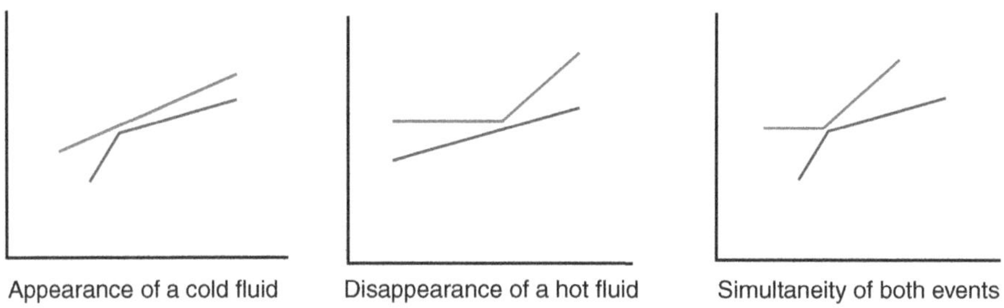

FIGURE 7.1.16 Different types of pinches.

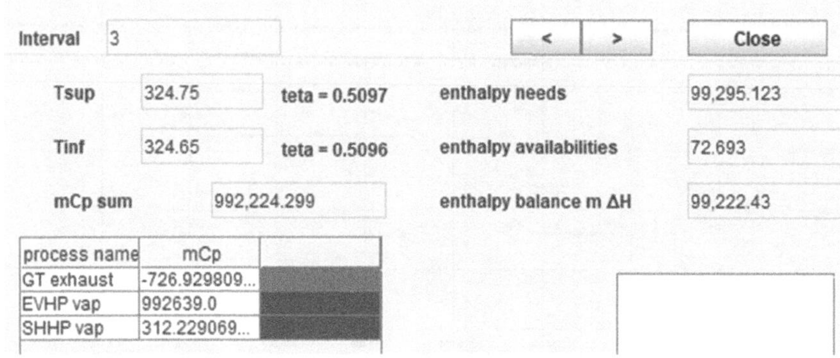

FIGURE 7.1.17 Interval 3 streams.

The upper pinch (interval 3) is created by the appearance of the HP evaporator. At that level, existing fluids are given in Figure 7.1.17:

- Just above, the constraints are met if the HP evaporator is matched with the gas (the constraint writes 726.9 < 992,638), the superheater being moved upstream (for the latter, the constraint writes 0 < 312.2) (Figure 7.1.18);
- Just below (Figure 7.1.19), the constraint is also verified (726.9 > 398.85).

However, a third fluid appears in the following interval: the LP superheater. We shall see later that, even if the pinch constraint is verified below the HP evaporator, it will be necessary to parallelize the LP superheater and the HP economizer.

Obviously, the gas has to be divided into two parts, one above and one below the pinch. This will also be true for the lower pinch (interval 5), which is created by the appearance of the LP evaporator.

Above the second (cold) pinch the streams of Figure 7.1.19 exist.

As previously, the constraints can be met by matching the LP evaporator with the gas at that level (762.92 < 479,557) and moving the other exchangers upstream (0 < 380.53, and 0 < 50.42).

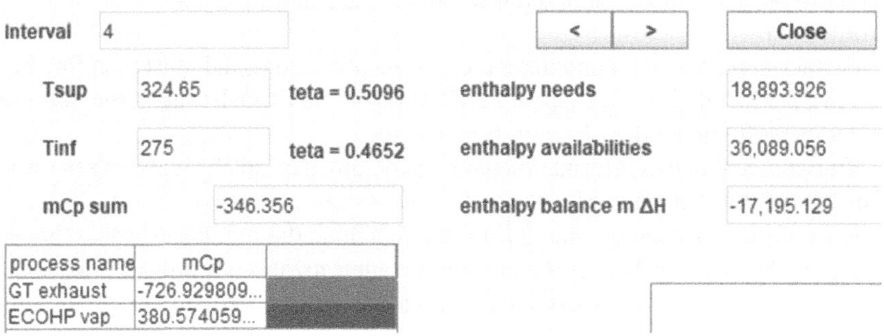

FIGURE 7.1.18 Interval 4 streams.

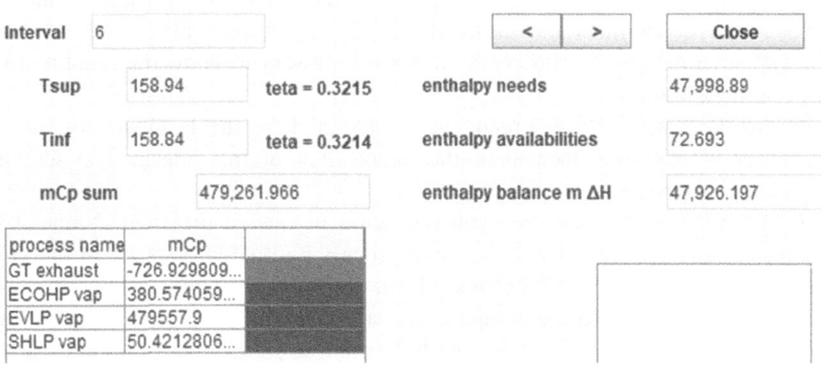

FIGURE 7.1.19 Interval 6 streams.

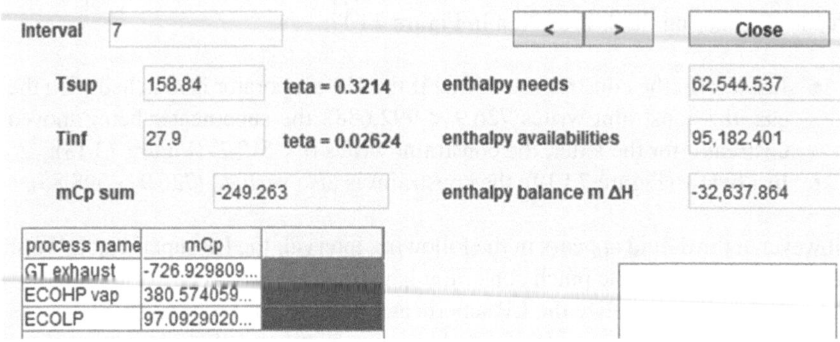

FIGURE 7.1.20 Interval 7 streams.

Below the pinch, the situation is that of Figure 7.1.20. The LP evaporator and superheater have disappeared, replaced by the LP economizer. Here heat capacity rate constraints can only be met if the flue gas stream is split into two parts, as there is only one hot stream and the constraints are that each cold stream capacity flow rate must be lower than that of a hot stream: the HP and LP economizers have to be parallelized.

Furthermore, the HP economizer crosses the pinch zone and must therefore be cut into two parts, which we will name ECOHP vap 1 and ECOHP vap 2, the intermediate temperature being the LP saturation one.

You can see that this information has been obtained by solely analyzing the stream distribution at the pinch level.

Knowing that, it becomes possible to start matching the streams in heat exchangers.

Using the heat exchanger blocks of the Optimization frame of Thermoptim enables you to do that progressively, without having in a first step to decide on the detailed placement of the heat exchangers.

We have seen that the need for parallelization occurs only below the lower pinch, and that other heat exchangers can be put in series. We will then divide the HRSG into two main blocks. To do that, begin by creating two intermediate points, one inside the HP economizer and one inside the flue gas stream, and split these two units into two streams. For the time being it is not necessary to know the exact temperatures of these points.

Go to the optimization screen and double-click in the headband of the heat exchanger blocks zone located at the upper right of the screen. The screen of Figure 7.1.21 appears.

It is similar to a node, the controls being of course different. Name it (for instance "block 1") and select the main process by double-clicking in the corresponding field. Choose "GT exhaust" from among the list of exchange processes proposed. Its inlet and outlet temperatures are displayed. Select then the different elements which have to be matched with the main stream, either by double-clicking in the headband of the table, or by clicking on the button named "add an element." Start by the HP superheater and the HP evaporator. Set then the checkboxes located in the upper left part of the screen: "series," "upper constraint" as the upper temperature of the main stream is set (here the inlet one), and "active" to indicate that the block is selected.

Now that the different streams are selected, you can adjust the block enthalpy balance by clicking the button named "adjust main process." Thermoptim calculates the total enthalpy corresponding to the different elements which have been selected and sets that of the main stream to that value. If "upper constraint" is selected, the minimum temperature of the stream is modified (the outlet one for a hot stream, and the inlet one for a cold one). If "lower constraint" is selected, this is done conversely. Then click "test the block" to verify that the block can be built. You get the result in Figure 7.1.22.

Intermediate temperatures of the main stream are then calculated: 472.85°C at the superheater outlet, and 336.29°C at the evaporator outlet.

You can now build the two heat exchangers by proceeding as follows. Click on "divide main process," which creates two new processes and two additional blocks (Figure 7.1.23). By accessing each of these blocks, you can create the corresponding

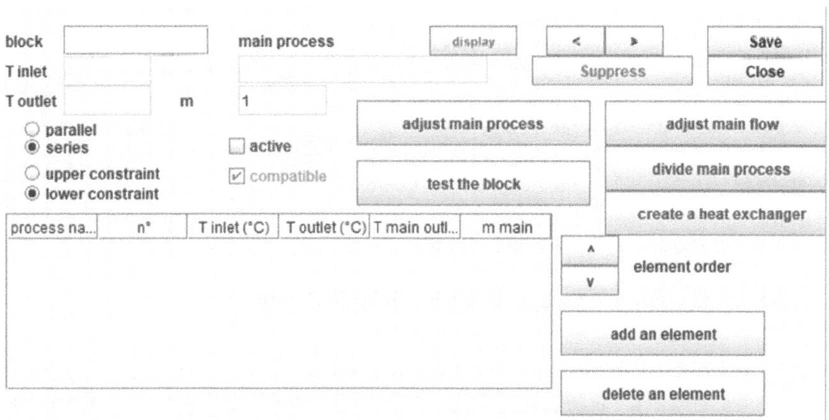

FIGURE 7.1.21 Default heat exchanger block.

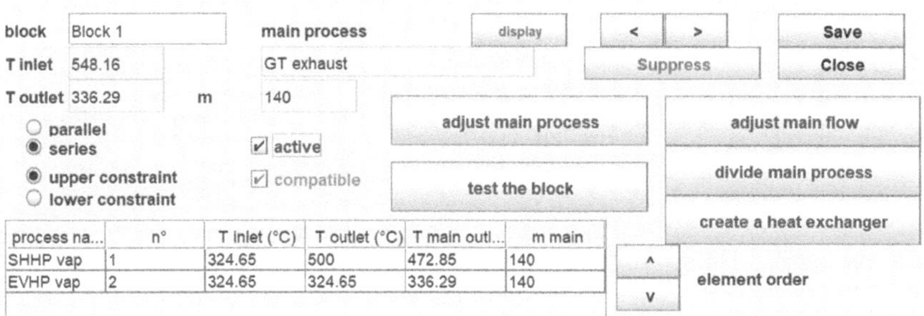

FIGURE 7.1.22 Heat exchanger block with settings.

heat exchanger in the simulator and parameterize it by setting the parameters related to the steam cycle and recalculating the helium outlet temperature.

The transfer of this heat exchanger to the process flow diagram editor is not automated. To do so, you need to use the interface between these two environments.

Start by updating the element table and deselecting all elements, and then select only the new processes. Next, update the diagram editor. The processes are created but not yet connected to the rest of the diagram. Disconnect the upstream of the old GT_exhaust process and connect the two new processes in the correct order, and then reconnect the GT_exhaust process. You can then create the two heat exchangers by linking the paired processes and assigning them the same names as in the simulator (Figure 7.1.24).

The block above the upper pinch being built, you can operate in a similar manner to build that between the two pinches.

Start by renaming GT_exhaust to GT_ext, and then select it as the main process of the new Block 2.

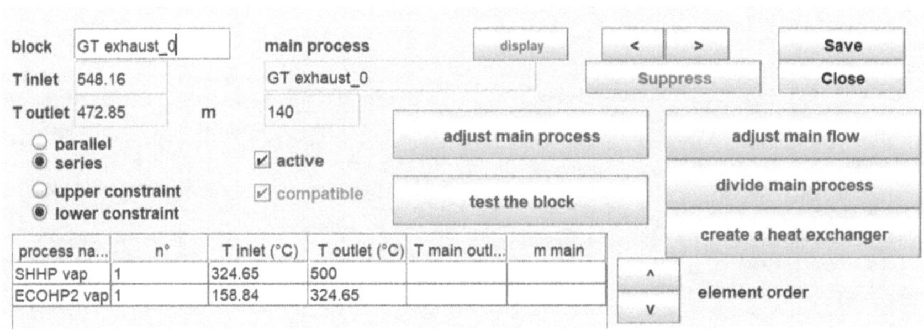

FIGURE 7.1.23 Heat exchanger block with divided GT_exhaust.

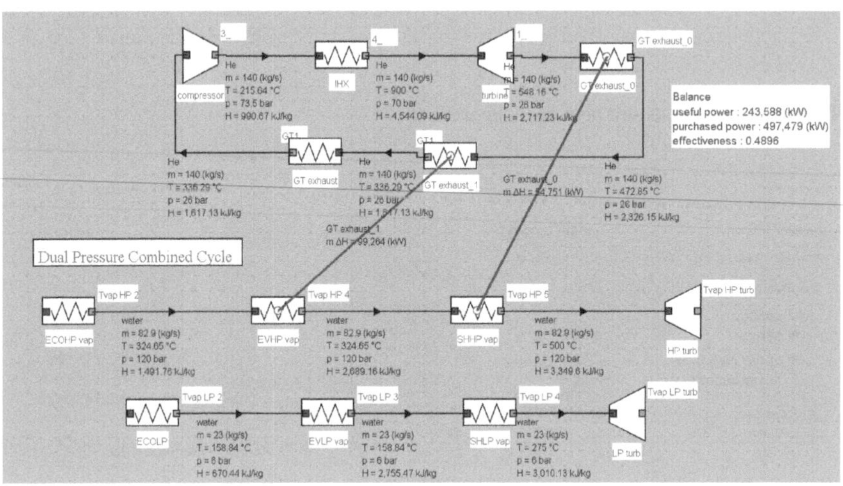

FIGURE 7.1.24 Partial synoptic view.

In this case, a problem arises when you adjust the block in series, the LP superheater temperature being too high for it to be placed downstream of the HP economizer. It is therefore necessary to parallelize this block.

Ultimately, the only possible solution is to start by parallelizing the LP superheater and HP economizer (Figure 7.1.25).

Divide the main process by creating the nodes, then build the heat exchangers in a manner similar to before, and transfer the new elements to the diagram editor, removing any excess links if necessary.

The continuation of the heat exchanger network proceeds without difficulty. Downstream of the two previous heat exchangers, the LP evaporator is located, which exchanges heat with the entire helium stream.

The last two heat exchangers, corresponding to the remaining two economizers, must be parallelized (Figure 7.1.26).

The principle of constructing the heat exchanger network is thus explained, in this simplified case where the minimum pinch values, condensers, and feedwater pumps were not taken into account to limit the number of intervals.

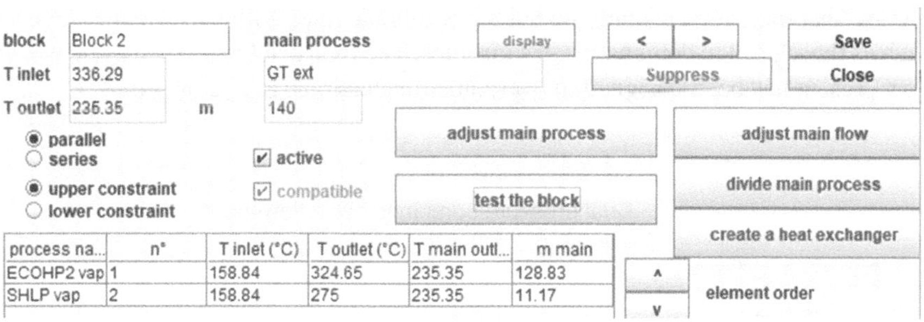

FIGURE 7.1.25 Block 2 screen.

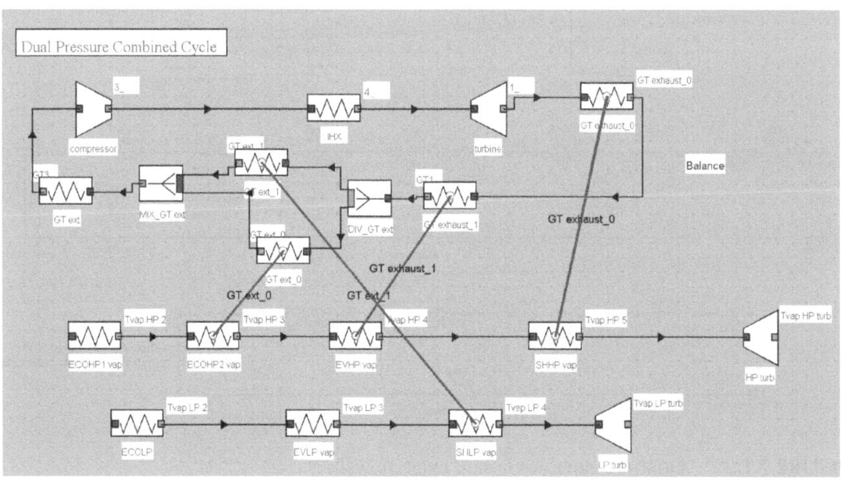

FIGURE 7.1.26 Optimized layout.

7.1.3.2 Presentation of the Complete Process Flow Diagram

We will now present the results obtained for a more realistic cycle, without detailing the intermediate steps.

The result obtained is that of Figure 7.1.27.

In this case also, the precooler has a similar influence to that observed for the single-pressure level cycle, though less significant.

By adjusting the precooler, it is possible to increase the mechanical power output from 208.4 to 235.6 MW, an increase of 13%, but at the cost of a slight decrease in efficiency from 48.73% to 47.26%.

We have thus modeled and parameterized a high-efficiency helium combined cycle that could be coupled to a nuclear reactor of the High-Temperature Reactor (HTR) or Molten Salt Reactor type. It should be noted that this cycle has not been fully optimized. To do so, sensitivity tests on its efficiency or power output would need to be conducted by adjusting the six existing degrees of freedom, which would represent significant work, justified only once the characteristics of the nuclear reactor and the IHX are known with certainty.

Remember, as mentioned in Section 6.3 Technology Readiness Level of Gas Cycles, that there are currently no helium turbomachines with performance levels as high as those used in combustion gas turbines. The industrial combined cycle we will now present has been designed to overcome this technological constraint.

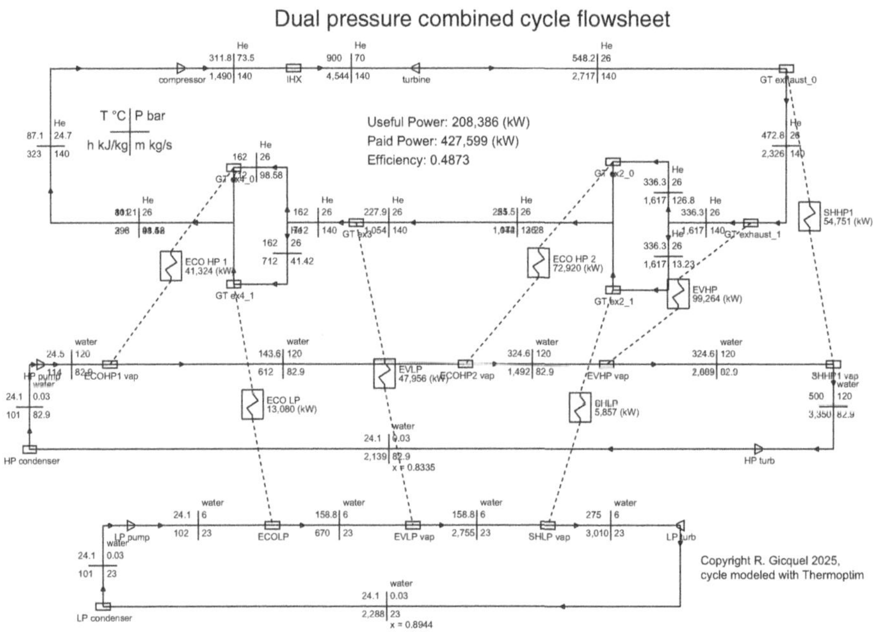

FIGURE 7.1.27 Dual-pressure combined cycle flowsheet.

7.1.3.3 The Areva GT-MHR Based Combined Cycle

The Gas Turbine Modular Helium Reactor (GT-MHR[2]) was the result of optimization work performed by General Atomics in 1985. Since 2000, Areva, which was involved in this work, changed the basic concept (helium Brayton cycle) to make the design easier (Figure 7.1.28) and the adaptation to cogeneration more immediate (Gauthier et al, 2006; Gosset et al., 2005).

Considerations governing the selection of cycles are numerous. Obviously, intrinsic cycle efficiency is fundamental, but technological feasibility is equally important. This has led Areva to choose for HTR-VHTR a combined cycle using a helium–nitrogen mixture (20%–80% wt) instead of pure helium, because of the considerable experience gained over several decades with air gas turbines. In 2002, Areva optimized with Thermoptim a combined cycle associated with an HTR (Figure 7.1.28).

The gas circuit (a mixture of helium and nitrogen) comprises a heat exchanger IHX that allows the transfer of energy from the coolant fluid (helium) warmed in the nuclear reactor. A temperature difference of about 50°C between the hot fluid inlet and cold fluid outlet is desirable to maintain the surface area of this exchanger within a reasonable range of values.

Before passing through the exchanger IHX, the gas is first compressed by a compressor to a pressure between 55 and 70 bar, the latter value of 70 bar being a maximum for reasons of material strength. At the exchanger outlet, the gas is expanded to about 40 bar in a turbine and exits at a still relatively high temperature (about 600°C). It is then cooled in a heat exchanger and redirected to the compressor. The bottoming thermodynamic cycle selected utilizes the enthalpy available in this exchanger to operate a steam cycle and thereby obtain a combined cycle.

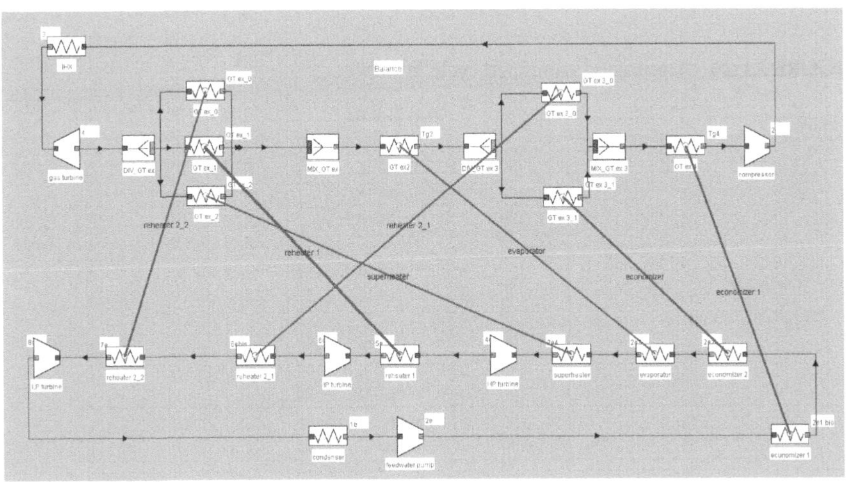

FIGURE 7.1.28 Optimized combined cycle.

Gas temperature at the IHX heat exchanger inlet (thus at the compressor outlet) is 300°C and pressure 57,75 bar. Gas temperature at the IHX exchanger outlet is 800°C and pressure 55 bar. This results from a temperature differential of 50°C selected for the IHX exchanger and pressure drops of about 5%.

The steam circuit consists of three successive turbines (HP, intermediate pressure, and LP). Superheated steam available at the entrance of the HP turbine passes through an economizer that heats the liquid, an evaporator that vaporizes it and a superheater. Two reheats are provided at the entrance of the Intermediate Pressure (IP) and LP turbines. Its flowsheet is shown in Figure 7.1.29.

The optimization problem involves 14 fluids and 16 intervals. It leads to the composite curves shown in Figure 7.1.30.

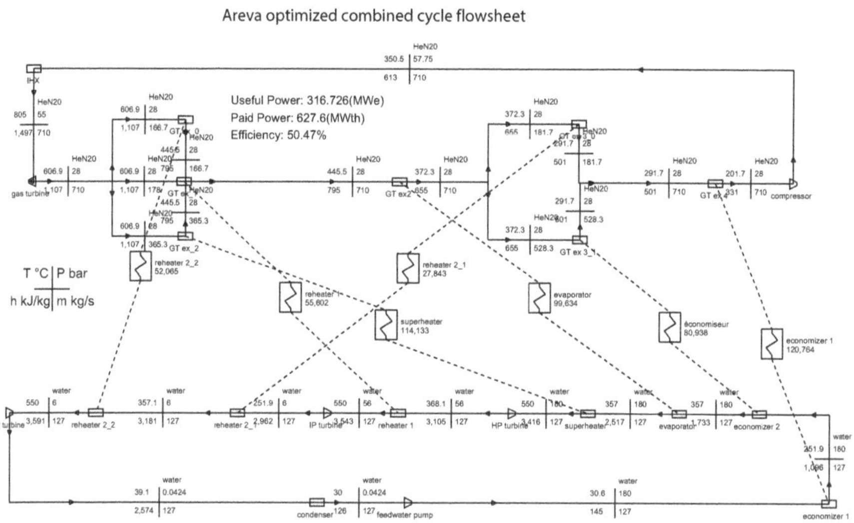

FIGURE 7.1.29 Optimized combined cycle flowsheet.

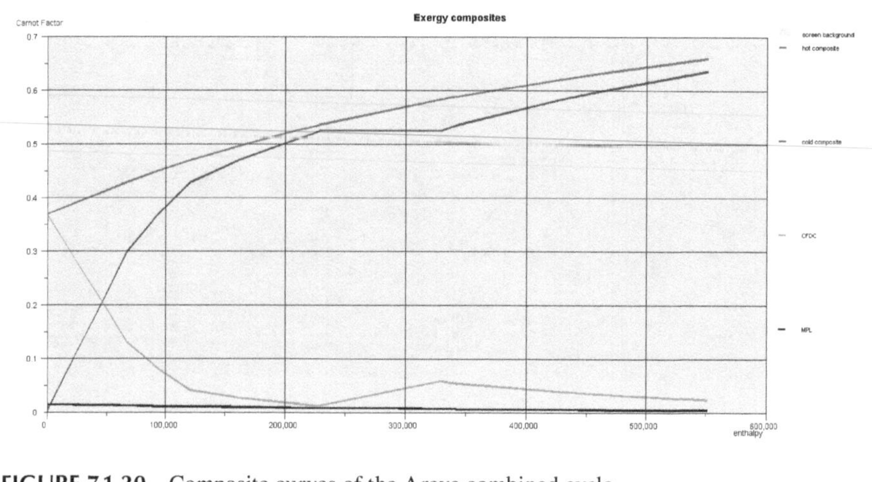

FIGURE 7.1.30 Composite curves of the Areva combined cycle.

Variations of this cycle can provide superheated steam at 110 bar using water at 120°C for some cogeneration applications, for capacities from 50 to 300 MW. The efficiency announced by Areva in electricity production is only 47%, but the objective was actually 50%, without any particular technological development except the IHX intercooler, with Mitsubishi committing to manufacture all turbomachinery (gas and steam) on the basis of current knowledge.

The exergy balance of this cycle can easily be established (Figure 7.1.31). Its exergy efficiency is 65.6%. Half of the irreversibilities occur in the IHX.

Variations of this cycle adapted to CHP have also been optimized by Areva with Thermoptim. We will discuss these in Section 7.4.

7.1.3.4 Dual-Pressure Level Cycle with a He–N_2 Mixture

It is possible to take the schematic of the dual-pressure level combined cycle optimized with helium in the topping cycle and replace this fluid with a He–N_2 mixture. The result obtained is shown in Figure 7.1.32.

The main differences between the He–N_2 mixture cycle and the pure helium cycle stem from the thermophysical properties of the working fluids, which affect compression work, heat transfer characteristics, and overall cycle performance. As a result, efficiency is slightly lower.

7.1.4 Supercritical CO_2 Combined Cycle

We have seen that supercritical CO_2 (sCO_2) cycles are generating significant interest worldwide. Naturally, combined cycles incorporating them have been proposed. We will consider two single-pressure level cycles: one combining two sCO_2 cycles, and the other combining an sCO_2 cycle with an ammonia cycle.

These cycles are quite similar to the first one we studied, with the main difference being that the topping cycle retains its regenerator. Under these conditions, the

component	Resource	Product	Exergy efficiency	Irreversibilities	% total	settings
feedwater pump	2,411.84	2,290.745	94.9791%	121.095	0.0776%	
HP turbine	41,489.828	39,463.82	95.1169%	2,026.008	1.2980%	
IP turbine	78,566.393	73,848.96	93.9956%	4,717.433	3.0224%	
LP turbine	143,054.324	129,202.83	90.3173%	13,851.494	8.8744%	
gas turbine	285,281.08	276,637.76	96.9702%	8,643.32	5.5376%	
compressor	200,015.43	190,991.456	95.4884%	9,023.974	5.7815%	
MIX_GT ex	671,711.963	671,711.941	100.0000%	0.02235	0.0000%	
MIX_GT ex 3	558,927.946	558,927.934	100.0000%	0.01186	0.0000%	
economizer 2	0	40,202.194	0	(-40,202.194)	(-25.7568%)	economizer
GT ex 3_1	41,675.63	0	0	(41,675.63)	(26.7008%)	economizer
GT ex 4	52,502.498	0	0	(52,502.498)	(33.6374%)	economizer 1
economizer 1	0	33,673.497	0	(-33,673.497)	(-21.5740%)	economizer 1
evaporator	0	53,270.955	0	(-53,270.955)	(-34.1297%)	evaporator
GT ex2	56,772.075	0	0	(56,772.075)	(36.3728%)	evaporator
reheater 1	0	33,165.982	0	(-33,165.982)	(-21.2488%)	reheater 1
GT ex_1	35,146.007	0	0	(35,146.007)	(22.5174%)	reheater 1
reheater 2_1	0	13,669.5	0	(-13,669.5)	(-8.7578%)	reheater 2_1
GT ex 3_0	14,336.29	0	0	(14,336.29)	(9.1850%)	reheater 2_1
reheater 2_2	0	30,957.354	0	(-30,957.354)	(-19.8338%)	reheater 2_2
GT ex_0	32,910.421	0	0	(32,910.421)	(21.0851%)	reheater 2_2
superheater	0	66,162.06	0	(-66,162.06)	(-42.3888%)	superheater
GT ex_2	72,143.712	0	0	(72,143.712)	(46.2211%)	superheater
IHX	483,091.706	399,776.293	82.7537%	83,315.414	53.3786%	Tk = 1,000.00 °C
global	483,091.706	327,007.843	67.6906%	156,083.863	1	

FIGURE 7.1.31 Optimized combined cycle exergy balance.

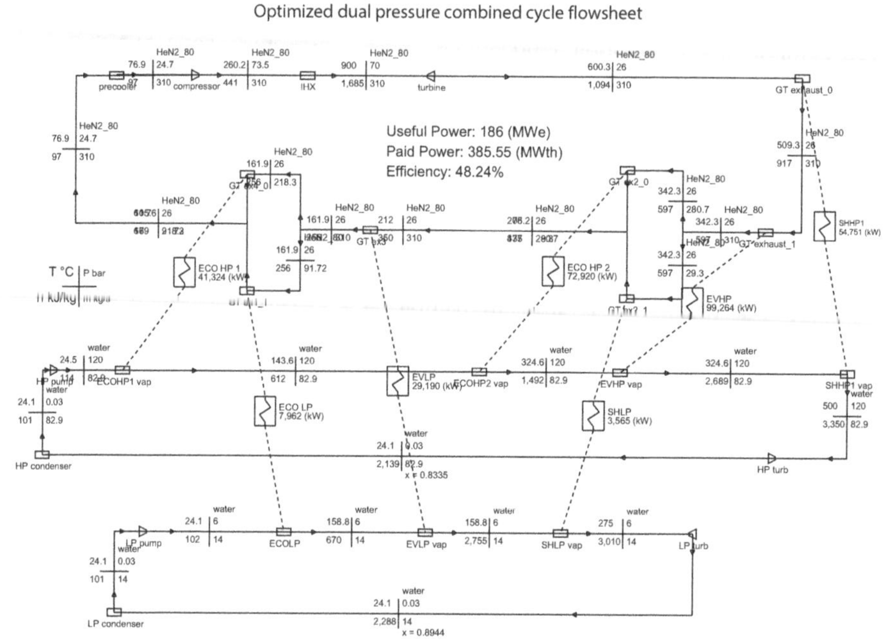

FIGURE 7.1.32 Optimized combined cycle flowsheet.

temperature level of the gas stream to be cooled is much lower than before, making steam bottoming cycles unsuitable, whereas an ammonia cycle may be appropriate. However, it results in a lower efficiency than the first option.

In both cases, the topping cycle is the supercritical CO_2 cycle studied in Section 6.2.2, but with slightly different parameters.

Since the structure of the combined cycles is similar to that of the single-pressure helium cycle, we will not detail them, limiting ourselves to providing an example of the results achieved.

7.1.4.1 sCO_2–sCO_2 Cycle

The first solution involves using a Rankine cycle with CO_2 as the bottoming cycle.

We were inspired for this example by the cycle presented in Chapter 21 of the G4NR, in Figure 21.1.3, p. 763, which involves two supercritical CO_2 cycles, and we modeled it in Thermoptim.

To make the comparison with the regeneration cycle easier, we have not represented the primary circuit with helium, which incurs a significant consumption to overcome the pressure drops. It would be easy to complete the model.

In the supercritical CO_2 topping cycle (Figure 7.1.33), a flow of 3 t/s of CO_2 at 200 bar enters the reactor at a temperature of about 400°C and leaves it at 607°C. It is then expanded to the pressure of 54 bar in a turbine. Regeneration takes place between expanded CO_2 and that which flows out of the compressor.

At the exit of the regenerator, the CO_2 temperature (about 200°C) is much lower than in the previous combined cycles. CO_2 then enters the heat exchanger with the

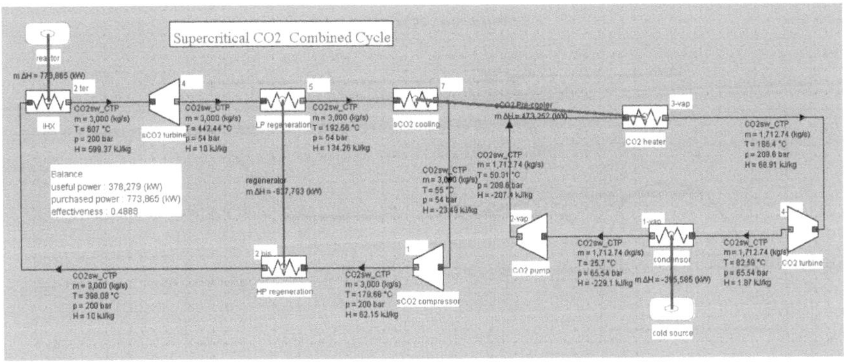

FIGURE 7.1.33 sCO_2–sCO_2 combined cycle.

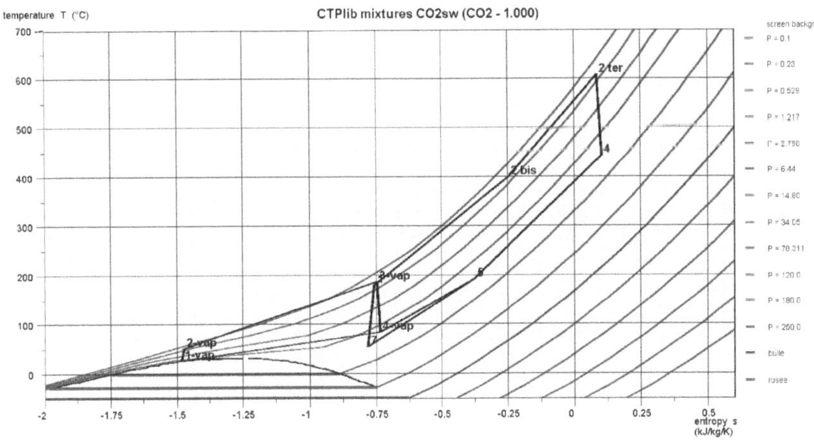

FIGURE 7.1.34 sCO_2–sCO_2 combined cycle in the (T, s) diagram.

bottoming cycle, which it leaves at 55°C, before being drawn into the compressor, which brings it to 200 bar.

The bottoming cycle is a second simple supercritical CO_2 cycle between pressures of about 66 bar and 210 bar. The effectiveness of the coupling heat exchanger is high (about 97%).

The efficiency is higher than the regeneration cycle. However, as the gain compared to a recompression sCO_2 cycle is not truly significant, the benefit of using a combined cycle is not clearly demonstrated.

The plotted cycle in the (T, s) and (h, P) diagrams is given in Figures 7.1.34 and 7.1.35.

7.1.4.2 sCO_2–NH_3 Cycle

In the case where the bottoming cycle uses ammonia, the heat exchanger is a boiler comprising the three classic components: an economizer, an evaporator, and a superheater.

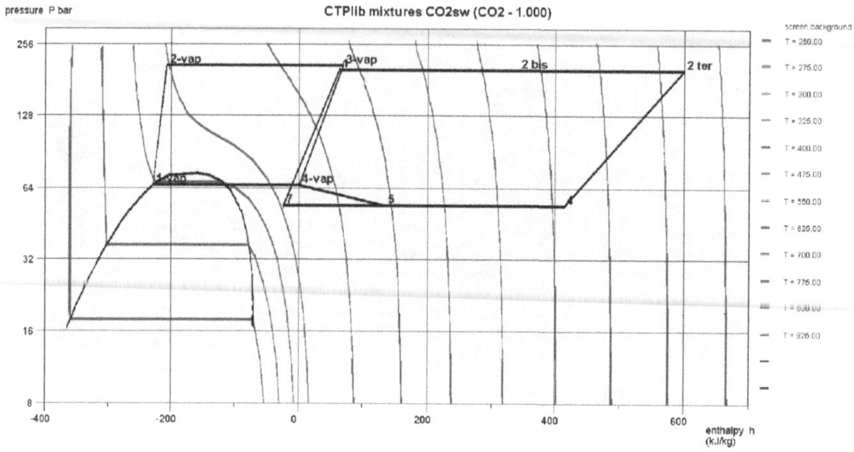

FIGURE 7.1.35 sCO_2–sCO_2 combined cycle in the (h, P) diagram.

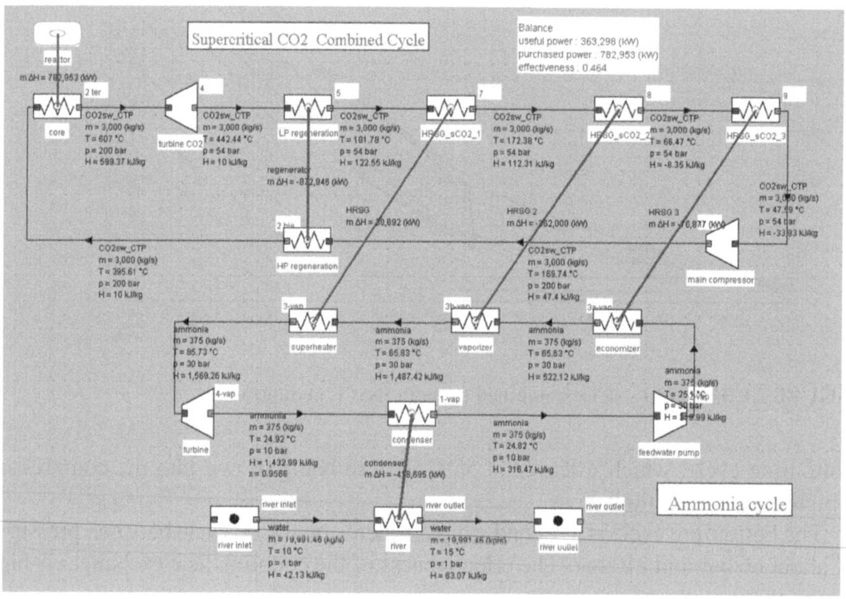

FIGURE 7.1.36 sCO_2–NH_3 combined cycle.

The ammonia cycle is a simple Organic Rankine Cycle (ORC), condensing at about 25°C. The HP and LP of this cycle are derived from the properties of the fluid: 30 and 10 bar (Figure 7.1.36). Finding the flow rate of ammonia that leads to the best performance is not a particular problem. Here we find 375 kg/s for 3 t/s of CO_2.

The efficiency of the combined cycle is also significantly higher here than that of the regenerative cycle, but slightly less than the first option.

7.1.5 Kalina Cycle

Although Kalina cycles are currently used only for very specific applications, their concept is highly interesting, and nothing excludes their potential application in the nuclear field, which justifies their inclusion in this section.

As we discussed at the beginning of this chapter, the optimization of a combined cycle is a particularly complex problem that leads to attempting to minimize irreversibilities stemming from temperature heterogeneity between the two cycles. When the recovery cycle uses a pure substance like water, the existence of a vaporization plateau induces losses that can be reduced only with difficulty by multiplying pressure levels. To overcome this constraint, Kalina has proposed using as working fluid a water–ammonia mixture, which has a large temperature glide.

An elementary Kalina cycle (Figure 7.1.37) consists of three main elements:

- A recovery steam generator (HRSG);
- A turbine;
- A distillation and condensation system.

The complexity of the cycle results from two causes:

- To benefit from a significant temperature glide in the temperature range corresponding to the hot gas or nuclear reactor cooling, the ammonia quality of the mixture should be quite high (typically 50% NH_3, 50% H_2O).
- To be condensed at a temperature above that of the surroundings and at LP, the ammonia quality of the mixture should be low (typically 20% NH_3, 80% H_2O).

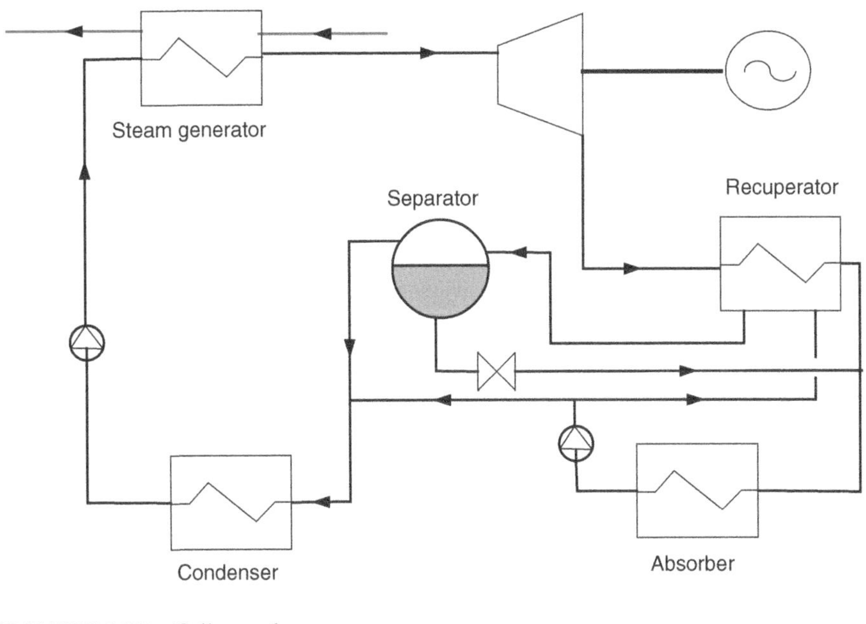

FIGURE 7.1.37 Kalina cycle.

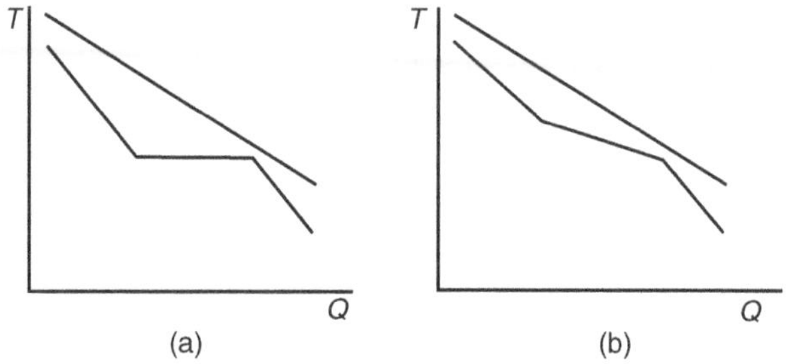

FIGURE 7.1.38 Internal irreversibilities in a Hirn (a) and Kalina (b) cycle.

The temperature-enthalpy diagram of a Kalina cycle is given in Figure 7.1.38. It can produce mechanical power from a heat source at variable temperatures, such as effluent or exhaust gas from a gas turbine.

As shown in Figure 7.1.38b, the existence of this temperature glide reduces the average temperature difference between the gas stream cooling in the HRSG and the working fluid as compared to what occurs in a Hirn cycle (Figure 7.1.38a). Internal irreversibilities due to the temperature gradient are reduced.

One is therefore led to change the mixture composition between the hot zone (HRSG) and cold zone of the cycle, which is achieved in the distillation and condensation system.

In the distillation and condensation system, the rich mixture leaving the turbine is cooled in a recuperator and then mixed with a weak solution of NH_3 to raise the condensation temperature. The mixture (basic solution) is condensed in the absorber, recompressed, and directed to the recuperator. A portion of the flow is used to dilute the rich solution leaving the separator, while the main stream is distilled at the recuperator outlet. The vapor is mixed with the extraction from the basic solution, condensed, and compressed before entering the HRSG.

An example of a Kalina cycle modeled with Thermoptim is presented in the Thermoptim-Unit portal at this address:

https://direns.minesparis.psl.eu/Sites/Thopt/en/co/cycle-kalina.html

In 2025, the Kalina cycle remains a promising but underutilized technology, with successful commercial applications in areas such as geothermal energy and industrial heat recovery.

Current research focuses on improvements in efficiency and flexibility, particularly with second-generation cycles and the use of alternative fluids to ammonia, which presents multiple constraints, notably safety concerns. However, complexity, initial costs, and competition with established technologies such as the Rankine or ORC cycles have so far limited its large-scale adoption.

7.2 DESALINATION OF SEAWATER

INTRODUCTION

This section focuses on seawater desalination, which consumes large amounts of heat and can therefore constitute a viable market for nuclear reactors.

The limitation of drinking water resources in many parts of the world has led mankind for centuries to seek methods for producing fresh water from seawater and brackish water. Note that the issue of desalination is similar in many respects to that of concentrating aqueous solutions, with the exception that the useful product differs: we are interested in the distillate rather than the concentrate. In this chapter, we present the main desalination techniques currently implemented industrially, without providing an exhaustive treatment.

One of the most important parameters for characterizing the water to be treated is its salinity, usually expressed in g/L. This value varies according to different seas around the globe, from 35 g/L for the Atlantic to 50 g/L and higher in the Persian Gulf.

7.2.1 Boiling Point Elevation

An important feature to consider when modeling saltwater is the phenomenon known as boiling point elevation.

As the solute concentration increases, the saturated vapor pressure of the solute–solvent mixture decreases, in accordance with Raoult's law. Consequently, at a given pressure, the boiling point increases slightly compared to that of the pure solvent. We call this temperature difference the **boiling point elevation**, ΔT_{eb}. Physically, the presence of the solute impedes the evaporation of the solvent, which can only occur at a higher temperature.

It can be shown that ΔT_{eb} is given by the following law:

$$\Delta T_{eb} = i\ x\ \frac{K_{eb}}{\rho} \tag{7.2.1}$$

- i is the solute van't Hoff coefficient, representing the number of elementary particles (ions, etc.) formed in the solution.
- x is the solute mass concentration.
- K_{eb} is the pure solvent ebullioscopic constant.
- ρ is the solvent density.

Law (7.2.1) indicates that the boiling point elevation is proportional to x. In practice, if we can assume that ρ and K_{eb} are constant—an assumption valid over a small temperature range—it suffices to know several values of the boiling point elevation to identify it according to (7.2.2):

$$\Delta T_{eb} = Kx \tag{7.2.2}$$

The external class SalineWater used in the Thermoptim model of saline water accounts for the boiling point elevation.

The EvapoConcentrator.java external class, which accounts for the boiling point elevation, will be used in the models shown in Figures 7.2.1 and 7.2.2.

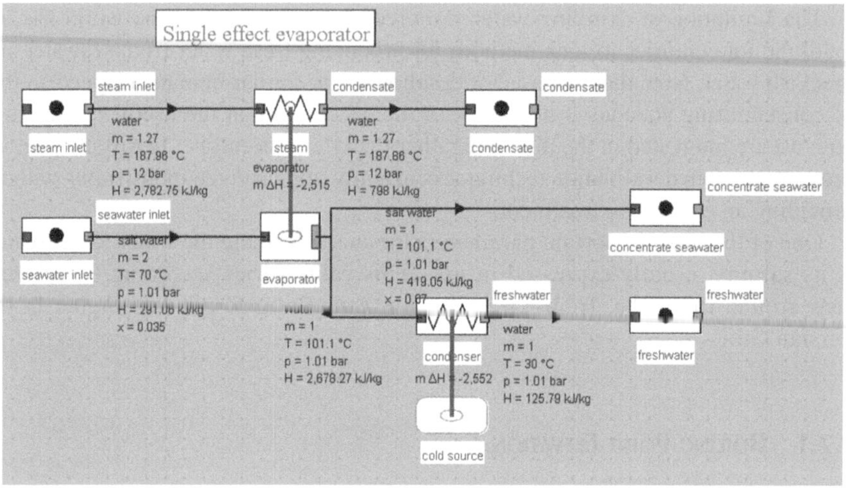

FIGURE 7.2.1 Single-effect evaporator.

7.2.2 Single-Effect Distillation

Simple distillation is the most straightforward method for desalinating seawater. It involves vaporizing the water by supplying high-temperature heat and then condensing the steam produced.

A single-effect desalination cycle can be easily modeled in Thermoptim using class EvapoConcentrator.java (Figure 7.2.1).

In this example, 2 kg/s of seawater with a salinity of 35 g/L, preheated to 70°C, enters the evaporator to produce 1 kg/s of freshwater. The thermal power involved is 2,513 kW in the absence of losses, and 2,552 kW must be rejected at the condenser to cool the distilled water to 30°C.

Another way to present the results is to note that 1.27 kg/s of steam is required per 1 kg/s of distillate. The steam-to-distillate ratio (SDR) is thus 1.27, and its inverse is the Gain Output Ratio (GOR $\approx$ 0.79). The specific energy consumption (SEC) is about 820 kWh per ton of distillate. The Performance Ratio, defined as Performance Ratio = 2,326/SEC kJ/kg, is $\approx$ 0.79, close to the GOR for a single-effect evaporator.

7.2.3 Multiple Effect Distillation

To recover part of the available enthalpy in the distilled vapor, it is possible to arrange several distillation units in cascade. This is referred to as a multiple effect distillation (MED) cycle. However, the increase in seawater concentration at the evaporator outlet leads to a rise in evaporation temperature due to boiling point elevation, and it is therefore necessary to gradually lower the pressure in successive units.

It is also possible, as explained below, to use mechanical vapor compression (MVC).

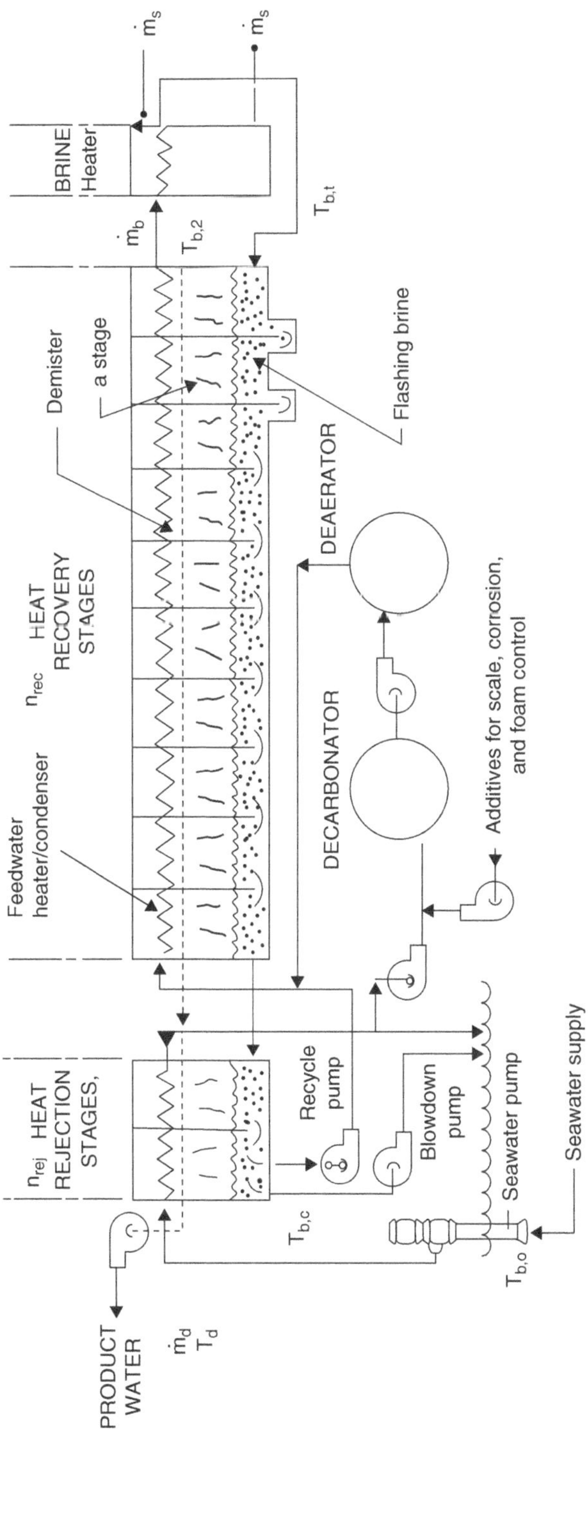

FIGURE 7.2.2 Multistage flash desalination system.

7.2.4 Multistage Flash Desalination Cycle

In a multistage flash (MSF) cycle, seawater is preheated by heat exchange with distilled steam. To reduce the temperature differences between the two fluids, a counter-flow arrangement is used, with several chambers (stages) in series as shown in Figure 7.2.2 (Lior, 2000). In each stage, salt water is expanded by flashing, causing partial vaporization.

Seawater at low temperature is heated in the lowest pressure chamber, then passes into the next stage, and so on. A high-temperature heat input is supplied by live steam.

The warm seawater is expanded in the first stage, with the distilled steam generated being condensed as described above. The salt-concentrated residue is then routed to the next stage at lower pressure, and so on.

The synoptic view in Figure 7.2.3 shows that the steam input required to evaporate 1 kg/s of water SDR is 0.627 kg/s, that is to say, half of that of the simple cycle.

An example of a flash chamber model is available in the model library.

7.2.5 Reverse Osmosis Desalination

Let us consider two media consisting of two mixtures of the same solvent–solute pair but of different concentrations, separated by a semipermeable membrane. Experience shows that:

- if the two media are at the same pressure, a solvent transfer called **osmosis** takes place.

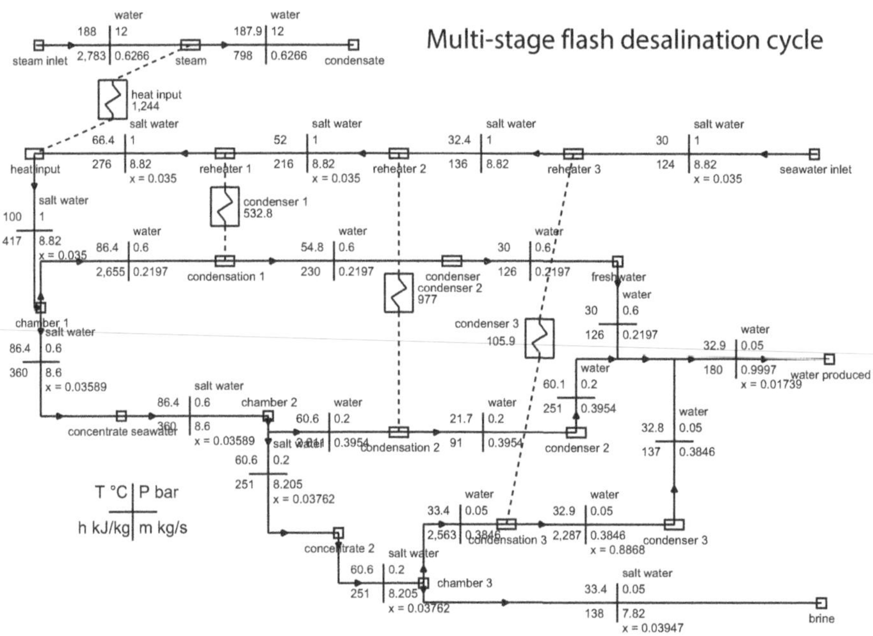

FIGURE 7.2.3 Multistage flash desalination cycle.

- if a certain pressure, called osmotic pressure π, is applied to the more concentrated medium, this transfer can be canceled.
- if the pressure is greater than π, the solvent migrates from the more concentrated solution to the other: this is called **reverse osmosis**.

Desalination plants using reverse osmosis operate on this principle (Figure 7.2.4): by applying a pressure higher than the osmotic pressure of salt water in a chamber equipped with a semipermeable membrane, it is possible to collect on the other side of the membrane a solution of very low salinity, called permeate.

The energy consumed is only the compression work of the initial solution, and is therefore much lower than that brought into play (as heat) in most devices studied thus far.

The van't Hoff law (7.2.3) states that the osmotic pressure exerted by the solute equals the pressure it would exert in the perfect gas state in the same volume and at the same temperature. If the solute dissociates into ions, the osmotic pressure is multiplied by the number of ions present. in equation (7.2.3) X is the concentration of solute.

$$\pi = n_i \, X R T \qquad (7.2.3)$$

For seawater, the magnitude of π is 25–30 bar.

ΔP being the pressure difference across the membrane, we can show that the flow of solvent through the membrane J_e is given by (7.2.4).

$$J_e = A(\Delta P - \Delta\pi) \qquad (7.2.4)$$

A is called the water permeability of the membrane. It is a characteristic parameter of the membrane, which depends on temperature according to an Arrhenius type law (7.2.5).

$$A = A_0 \exp\left[\frac{E}{R}\left(\frac{1}{298} - \frac{1}{T}\right)\right] \qquad (7.2.5)$$

Even though there is a preferential transfer of the solvent, a small fraction of the solute also crosses the membrane. The flow of solute J_s is given by (7.2.4). It is proportional to the concentration difference.

$$J_s = B\Delta X \qquad (7.2.6)$$

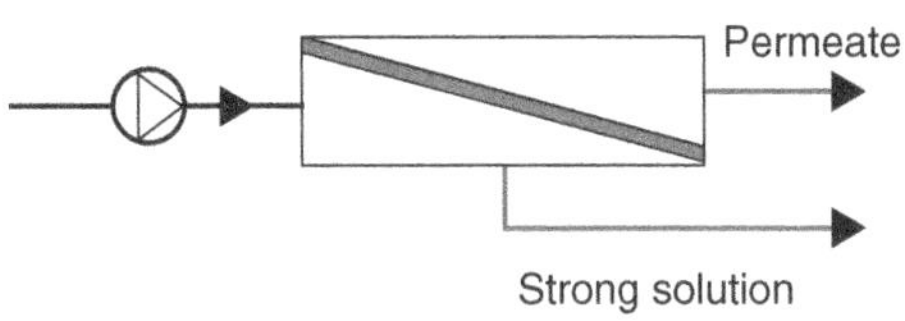

FIGURE 7.2.4 Reverse osmosis unit.

B is called the salt permeability of the membrane. This parameter depends on the membrane but not on temperature.

We define the conversion rate as the ratio of flow through the membrane to the feed flow rate, and the retention rate as the ratio of the concentration difference between the initial solution and permeate to the initial solution concentration.

Figures 7.2.5 and 7.2.6 show the Thermoptim diagram and component screen for modeling a Dow Chemical FilmTech SW30-4040 membrane used to desalinate seawater. The values of A and B were estimated from the manufacturer's documentation.

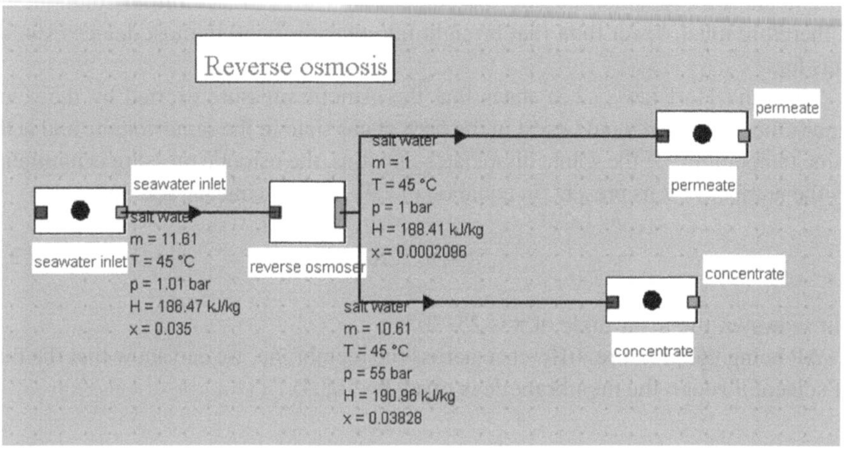

FIGURE 7.2.5 Thermoptim diagram of a reverse osmosis unit.

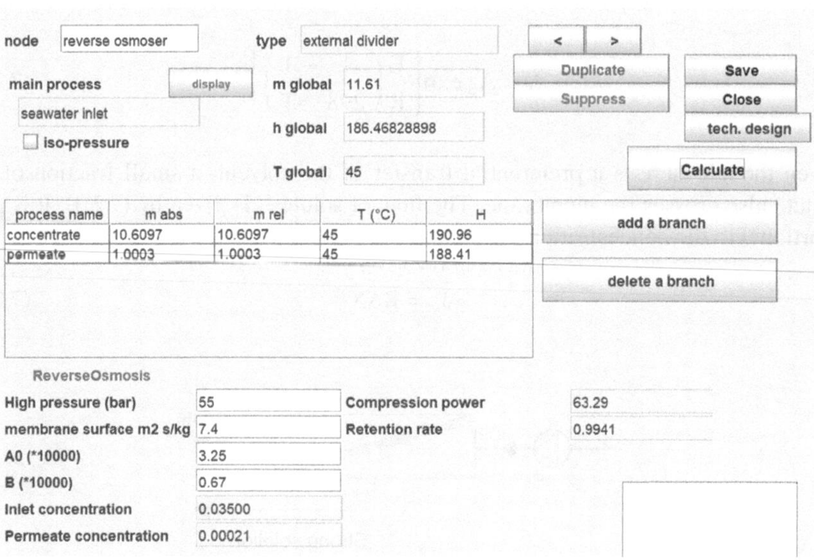

FIGURE 7.2.6 Reverse osmosis component screen.

The compression power required to produce 1 kg/s of fresh water equals 63.3 kW, or nearly twice that required by MVC as we will see. However, it is possible to recover some of that power by expanding the concentrate in a turbine connected to the compressor.

This reverse osmosis desalination model is available in the modeling library.

7.2.6 Mechanical Vapor Compression MVC

In most situations suitable for heat pump applications (small temperature difference between a "free" cold source and heating needs), MVC presents an efficient alternative if the cold source is a gas or vapor. It is then possible to avoid the evaporator, and possibly the condenser, thus achieving high performance.

Indeed, compression of a vapor (steam or process gas) raises its pressure and temperature, and thus its enthalpy. If the compressed vapor can be recovered either in a process or in terms of energy, the compression operation can be very economically attractive.

The two most conventional MVC applications are distillation and concentration by evaporation. It can also be used for waste heat recovery. However, its cycle is then similar to that of an HRSG as discussed in Section 7.1.

7.2.6.1 Evaporative MVC Cycle

In an evaporative MVC cycle (Figure 7.2.7), the idea is to raise the solvent vapor enthalpy level so that it can be directly used to provide heat to the evaporator.

The basic scheme is shown in Figure 7.2.7, and its cycle is plotted on the entropy diagram of Figure 7.2.8: at point 0, the product to be concentrated enters two heat recovery exchangers where it is preheated (0–1) by cooling the concentrated product (1′–7) and condensate (5–6).

In the evaporator, the product to be concentrated is partially evaporated (1–2) by exchange with its own vapor, which condenses (4–5).

At the bottom of the evaporator, the concentrate is extracted at point 1′, while the solvent vapor exits at point 2 to be recompressed to point 3, gaining sufficient enthalpy to serve as the cycle's hot source. To ensure large heat exchange coefficients (two-phase flow) in the evaporator, steam is frequently desuperheated from 3 to 4

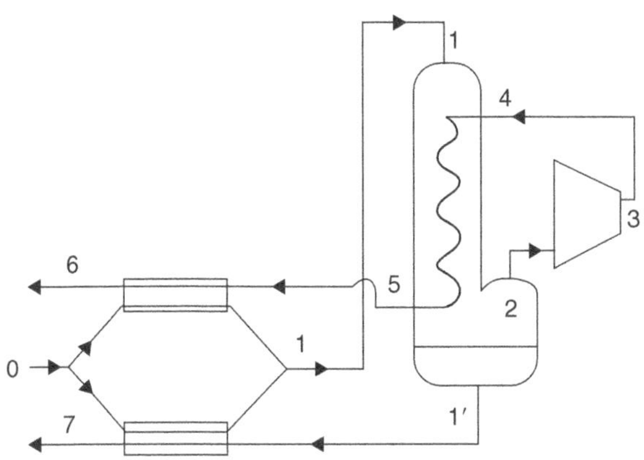

FIGURE 7.2.7 Evaporative MVC cycle.

via a separate heat exchanger, cooled by desuperheating water, even if this requires steam addition at point 4.

The energy advantage of this operation is that by providing a small amount of additional enthalpy h_{23} (but mechanical in nature), it is possible to recover the solvent vapor condensation enthalpy h_{45}. As an indication, if the solvent is water, which has a very high heat of vaporization (almost 10 times that of oil), h_{23} is typically between 3% and 9% of h_{45}.

Figure 7.2.9 presents a modification of the single-effect evaporator in Figure 7.2.1 to which mechanical steam compression has been added for most of the heat supply. The input steam flow is divided by a factor greater than 3, while the compressor power is only 200 kW. The ratio (steam evaporated)/(heating steam) is 2.2.

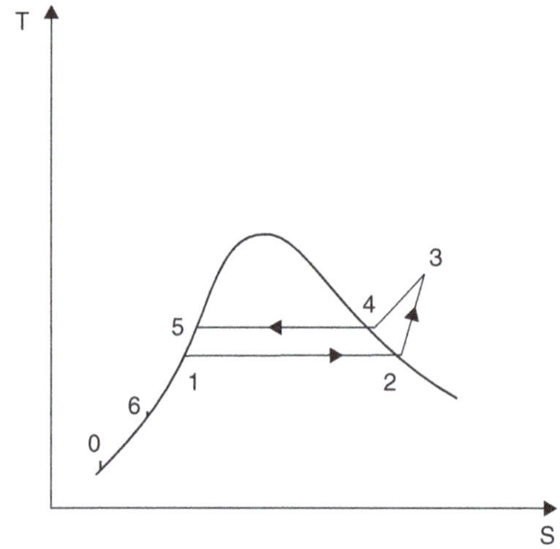

FIGURE 7.2.8 MVC cycle.

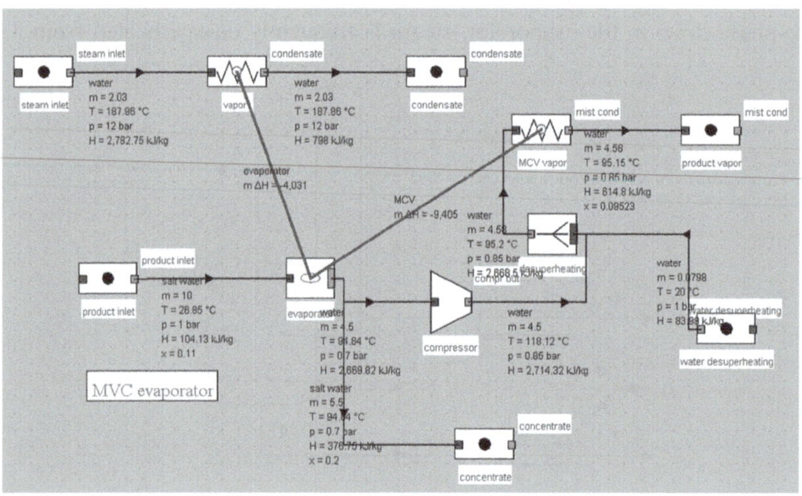

FIGURE 7.2.9 MVC evaporator.

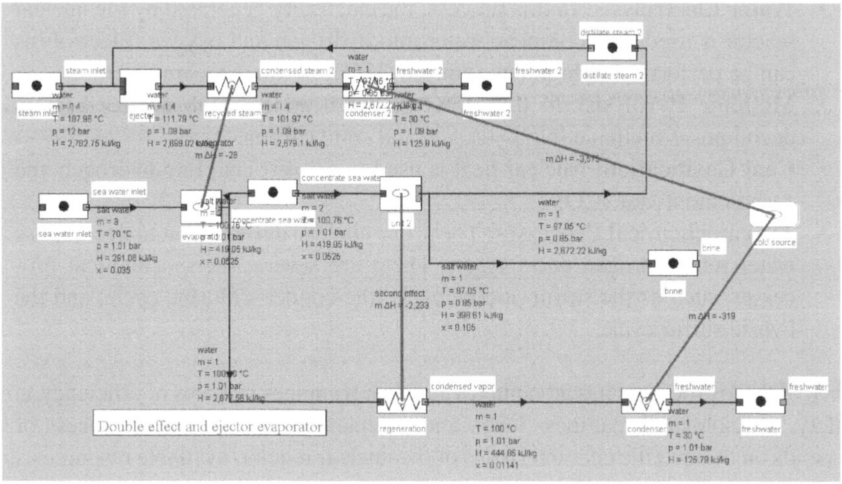

FIGURE 7.2.10 Ejector evaporator.

7.2.6.2 Desalination Ejector or Thermocompression Cycle

Another broadly applicable solution is thermal vapor compression (TVC): available steam is used as the motive fluid in a steam ejector to entrain the vapor leaving the last effect; the mixture is recompressed and routed as heating steam to the evaporator (first effect).

Figure 7.2.10 shows the layout of such a cycle. Unlike MVC, which relies on electrical work and no motive steam, TVC requires a non-negligible motive-steam flow but uses a low-cost, robust ejector instead of a compressor.

We will see in Section 7.4, which deals with cogeneration, how such a system can be coupled to a pressurized water reactor (PWR)-type nuclear power plant.

7.3 HYDROGEN PRODUCTION, METHANE REFORMING, AND ELECTROLYSIS

Introduction

In this section, we study the large-scale production of hydrogen by developing two technologies that can be coupled with nuclear reactors: methane reforming and High Temperature Electrolysis HTE.

Hydrogen can be used as a clean fuel for transportation or for electricity production. To produce it on a large scale from a nuclear reactor, various processes can be employed. The main processes are as follows.

7.3.1 Hydrogen Production Processes

- **Methane Reforming:** This process uses the heat produced by the nuclear reactor to convert methane (CH_4) into hydrogen and carbon dioxide (CO_2). Reforming can be achieved through steam (Steam Methane Reforming), partial oxidation, or autothermal reforming.

- **Water Electrolysis:** In this process, the electricity generated by the nuclear reactor is used to decompose water into hydrogen and oxygen. Electrolysis can be performed at high temperature (HTE) or low temperature.
- **Methane Pyrolysis:** In this process, the heat from the nuclear reactor decomposes methane into hydrogen and solid carbon.
- **Coal Gasification:** Nuclear heat is used to convert coal into hydrogen and carbon monoxide (CO).
- **Thermochemical Processes:** These use nuclear-derived heat to decompose water into hydrogen and oxygen. There are several thermochemical processes, such as the sulfur–iodine cycle, the copper–chlorine cycle, and the hybrid sulfur cycle.

Each of these processes has advantages and disadvantages in terms of efficiency, cost, safety, technological readiness level, and complexity. The choice of process often depends on the specific characteristics of the nuclear reactor, available resources, and market requirements for the produced hydrogen.

In what follows, we will limit ourselves to developments related to the first two processes, understanding that methane reforming models are relatively complex and this topic can therefore be considered optional within the framework of this book.

We will not address thermochemical cycles, which are even more complex and for which the technological readiness level is low.

7.3.2 Reforming

It is possible to produce hydrogen from another fuel. This operation is called reforming. There are five main reforming techniques: Steam Cracking, Partial Oxidation, Autothermal Reforming, Thermal Cracking, and Plasma Reforming.

7.3.2.1 Steam Cracking

Catalytic steam cracking is an endothermic process that transforms hydrocarbon molecules into CO, CO_2, and H_2 by reacting them with a sufficient amount of water. Generally, the equation for the steam cracking of fuel C_nH_m is written as:

$$C_nH_m + nH_2O \rightarrow nCO + \left(n + \frac{m}{2}\right)H_2$$

The steam cracking of light hydrocarbons takes place at temperatures between 700°C and 1,000°C, in the presence of nickel-based catalysts. The feed must first be desulfurized. The synthesis gas produced contains carbon monoxide, which must then be converted, as we will see later, by a reaction known as the shift reaction.

7.3.2.2 Partial Oxidation

Partial oxidation is an exothermic reaction that occurs in the gas phase, with or without a catalyst:

$$C_nH_m + \frac{n}{2}O_2 \rightarrow nCO + \frac{m}{2}H_2(\Delta H < 0)$$

This is an incomplete combustion of the hydrocarbon with rapid kinetics, out of equilibrium.

7.3.2.3 Autothermal Reforming

Autothermal reforming combines steam cracking and partial oxidation. In this way, the heat required for steam cracking is provided by partial oxidation. It allows conversion rates close to 100% at 650°C, but the reaction is relatively difficult to control.

7.3.2.4 Thermal Cracking

Thermal cracking involves decomposing the hydrocarbon molecule at high temperatures (1,000°C–2,000°C) according to the reaction:

$$C_nH_m \rightarrow nC + \frac{m}{2}H_2(\Delta H < 0)$$

This reaction is also difficult to control.

7.3.2.5 Plasma Reforming

Plasma reforming is a process that cracks the hydrocarbon molecule in the presence of an electric arc at a temperature close to 2,000°C.

7.3.2.6 Conversion of CO to CO_2

In the first three reforming techniques mentioned, the reaction products contain carbon monoxide, which may need to be eliminated. The water–gas shift reaction is often used to convert CO to CO_2:

$$CO_2 + H_2 \leftrightarrow CO + H_2O \ (\Delta Hv = 41 \text{ kJ/mol})$$

This is a catalytic reaction carried out in one or more reactors, with slow kinetics that are strongly influenced by temperature. It generally converts most of the CO.

7.3.2.7 Modeling of a Reformer in Thermoptim

In the following, we limit ourselves to the study of methane reforming without the elimination of carbon monoxide. The process therefore involves steam cracking, which we assume to be partial. Indeed, in the presence of water and at high temperature, the water dissociation reaction must sometimes be taken into account.

The basic equation for the reforming of a fuel such as methane is:

$$CH_4 + H_2O \rightarrow CO + 3H_2 \ (\Delta H = 206{,}140 \text{ kJ/kmol}) \tag{7.1}$$

This equation allows the conversion of the fuel into hydrogen but requires water and a heat input, and produces carbon monoxide.

For simplicity, we assume in this example that natural gas contains only methane, ethane, carbon dioxide, and nitrogen. The consideration of other reactive or inert components does not pose any particular problem, provided that we assume that the other reactants react completely, like ethane.

The reformer receives humidified natural gas, which reacts according to the three reactions:

$$CH_4 + H_2O \rightarrow CO + 3H_2 \quad (\Delta H = 206{,}140 \text{ kJ/kmol}) \tag{7.3.1}$$

$$C_2H_6 + 2H_2O \rightarrow 2CO + 5H_2 \quad (\Delta H = 347{,}000 \text{ kJ/kmol}) \tag{7.3.2}$$

$$CO + H_2O \leftrightarrow CO_2 + H_2 \quad (\Delta H = -41{,}200 \text{ kJ/kmol}) \tag{7.3.3}$$

The ethane reaction can be considered complete, while the other two are in equilibrium, and their constants being given by:

$$\ln(Kp_{1a}) = 31.0152 - \frac{28{,}357.7}{T} + \frac{610{,}573}{T^2} \tag{7.3.4}$$

$$\ln(Kp_2) = -3.57414 + \frac{3{,}642.48}{T} + \frac{292{,}593}{T^2} \tag{7.3.5}$$

Calling F° the molar flow of natural gas, x the molar flow of CO from reaction (7.3.1), y the molar flow rate of CO2 from (7.3.3), w/2 the molar flow of CO from reaction (7.3.2), a and d the mole fractions of CH4 and CO2 in the natural gas, and SC the number of moles of H2O per mole of natural gas, we can write:

$$F_{CO} = x + 2w - y$$

$$F_{CO_2} = y + d.F°$$

$$F_{H_2} = 3x + 5w + y$$

$$F_{CH_4} = a.F° - x$$

$$F_{H_2O} = SC.F° - x - 2w - y$$

$$F_{tot} = (1 + SC).F° + 2x + 4w$$

$$K_{1a}(T) = P^2.\left(F_{H_2}{}^3.F_{CO}\right)/\left(F_{CH_4}.F_{H_2O}.F_{tot}{}^2\right) \tag{7.3.6}$$

$$K_2(T) = \left(F_{CO_2}.F_{H_2}\right)/\left(F_{CO}.F_{H_2O}\right) \tag{7.3.7}$$

The two equilibrium equations (7.3.6) and (7.3.7) form a nonlinear system that is difficult to solve. However, it can be noted that reaction (7.3.7) is a quadratic form in x and y, which allows one of these two variables to be formally expressed as the solution of a second-degree equation in terms of the other.

If x is expressed in terms of y in this way and the value obtained is reinjected into the other equation, we obtain an implicit formulation in y, which can easily be solved numerically. Only one of the two solutions of the quadratic equation must be chosen, and the equation in y has only one real positive solution. The search range for its value is not easy to find, especially because of the existence of a singularity in the function, corresponding to the zero of F_{CH_4}.

Since x is an increasing function of y, one solution is to seek an upper interval bound y_{max} by solving equation $y = f(aF°)$ from (7.3.7). The lower bound can be chosen proportionally to y_{max} : $y_{min} = 0.9\ y_{max}$. The solution of (7.3.6) is then easily obtained by dichotomy between these two bounds (Figure 7.3.1).

The model that can be retained under these conditions is as follows:

- species composition is determined by solving the equations above: from the molar flow of humidified fuel input, we determine the compositions and molar flow rates of different species at the outlet, the only parameter to consider being the reactor temperature.
- the heat required by the reaction can then be calculated and set by a thermocoupler.
- this heat is supplied by a combustion chamber located directly upstream of the thermocoupler.

We present here a very theoretical process flow diagram intended simply to show how the reformer calculation can be carried out. A high-temperature nuclear reactor cooled by helium with approximately 300 MWth capacity is used to provide the necessary heat for the reformer to operate. This cycle does not represent the network of heat exchangers that preheat the natural gas and steam entering the reformer. We will see in the next section dedicated to a high-temperature electrolyzer that a

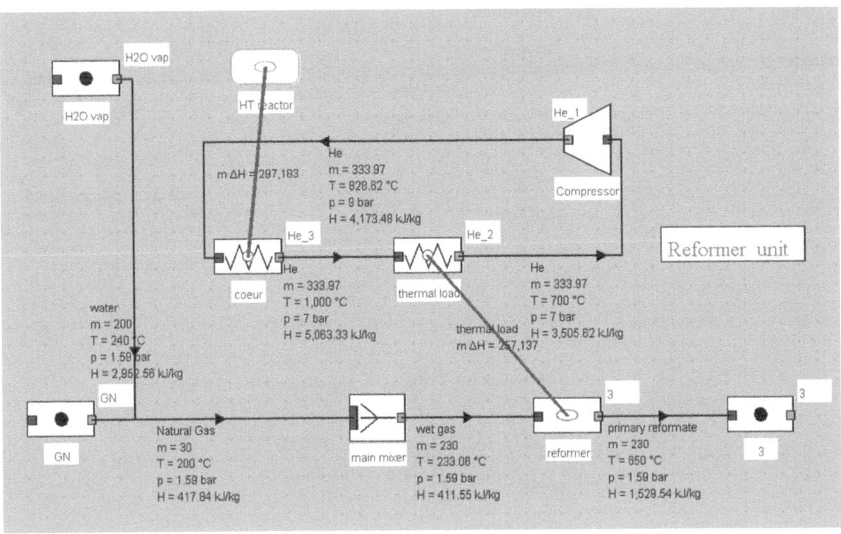

FIGURE 7.3.1 Process diagram of the reformer.

similar issue exists, and we will provide some indications on how the network of heat exchangers can be partially constructed, without however going to completion.

In Thermoptim, the reformer is represented by an external process connected to a thermocoupler. Its class is Reformer.java.

The screens of the reformer component and its thermocoupler are shown in Figures 7.3.2 and 7.3.3.

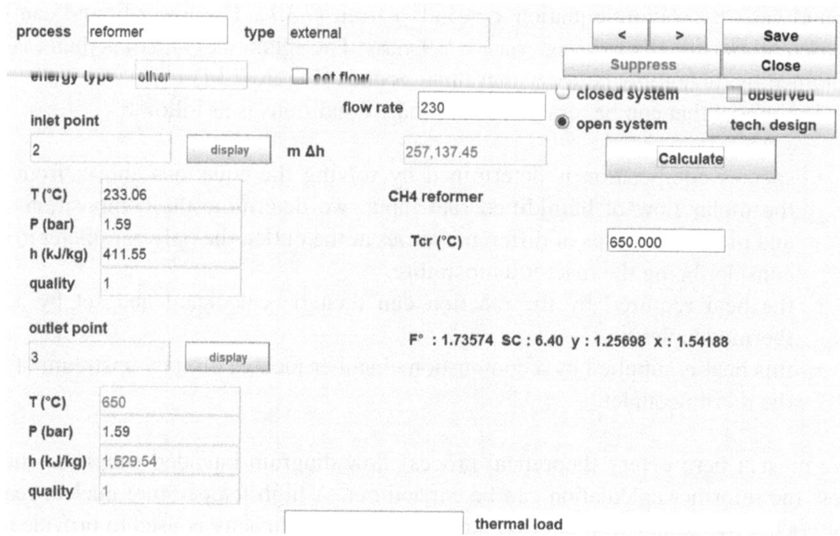

FIGURE 7.3.2 Screen of the reformer component.

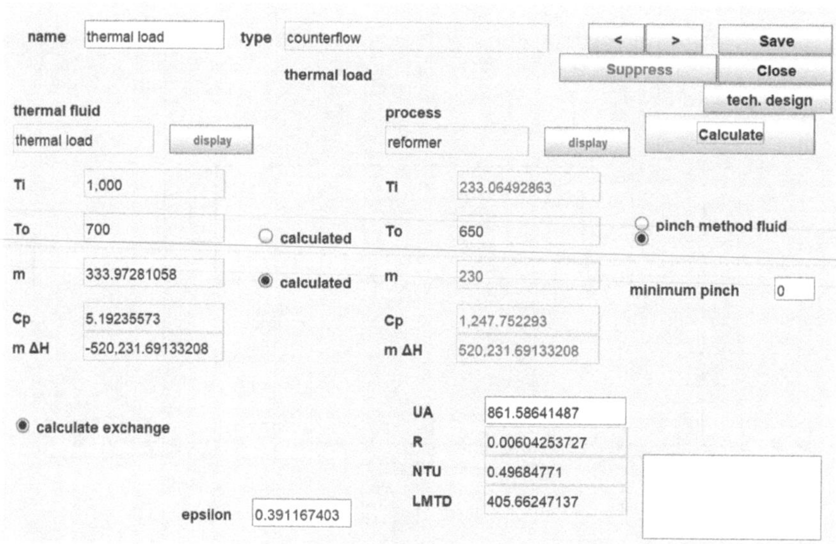

FIGURE 7.3.3 Reformer thermocoupler screen.

Inlet flows are determined by components located upstream of the process and the combustion chamber. As they are expressed in mass units, the molar flow rate F°, the value of SC, and the values of x and y are displayed on the process screen.

The thermal power supplied by the thermocoupler is determined by the process, and the thermocoupler calculates the required helium flow rate.

The gas compositions obtained are shown in Figures 7.3.4 and 7.3.5.

7.3.3 Electrolyzers

An electrolyzer performs an electrochemical reaction that produces hydrogen and oxygen from water through electricity consumption via the endothermic reaction:

$$H_2O \rightarrow H_2 + \frac{1}{2}O_2$$

An electrolyzer is composed of two electrodes, the anode and cathode, separated by an electrolyte (Figure 7.3.6).

Water enters at the cathode where an electrochemical reduction induced by the electric field occurs within the cell: $H_2O + 2e^- \rightarrow H_2 + O_2^-$. At the anode, the O_2^- ions are oxidized into oxygen and release electrons: $2O_2^- \rightarrow O_2 + 4e^-$.

7.3.3.1 High-Temperature Water Electrolysis

HTE is a process that uses electricity to decompose water into hydrogen and oxygen at high temperatures, typically between 800°C and 1,000°C.

The principle of nuclear HTE electrolysis is similar to that of low-temperature electrolysis, but it has several advantages. First, the high temperature reduces the amount of electrical energy required to decompose water, making the process more energy efficient. In addition, the heat produced by the nuclear reactor can be used directly to heat the water, reducing production costs.

component name	molar fraction	mass fraction
CH4 methane	0.1243921	0.1113837
C2H6 ethane	0.008112529	0.01361569
CO2	0.001352088	0.003321271
N2	0.001352088	0.002114077
H2O	0.8647912	0.8695652

FIGURE 7.3.4 Humidified natural gas, LHV: 6,217 kJ/kg.

component name	molar fraction	mass fraction
CO	0.03018695	0.06006241
CO2	0.07799918	0.2438392
H2	0.3919335	0.05612304
CH4 methane	0.003366579	0.003836472
H2O	0.4954514	0.6340248
N2	0.001062407	0.002114077

FIGURE 7.3.5 Primary reformate, LHV: 7,531 kJ/kg.

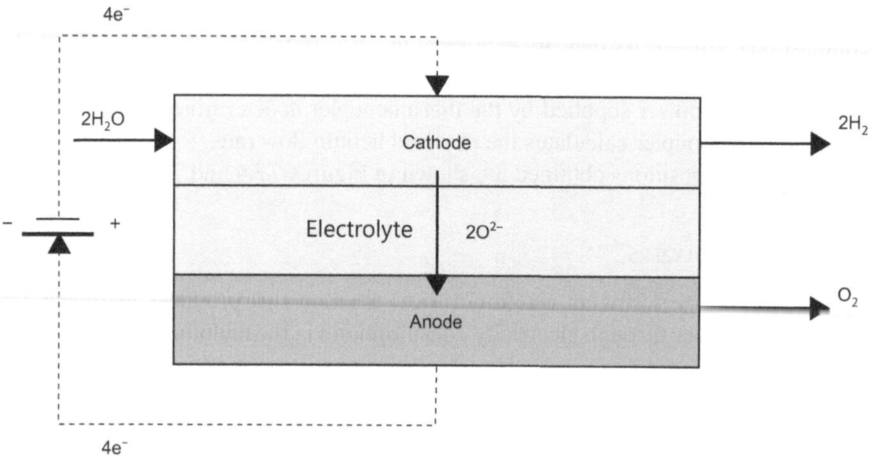

FIGURE 7.3.6 Sketch of an electrolyzer.

The HTE electrolysis process generally uses solid electrolytes, such as cerium oxides or zirconates, which are stable at high temperatures. The electrodes are often made of refractory materials such as nickel or cobalt.

However, HTE electrolysis is still in the development phase and is not yet widely used on an industrial scale. Research is underway to improve the energy efficiency of the process and reduce production costs.

Electrolyzers are currently used to produce pure hydrogen for a number of industries such as electronics, pharmaceuticals, and food. Electrolysis also produces commercially valuable oxygen gas that has many industrial applications. Technologies vary according to operating temperature and electrolyte.

The alkaline process was developed first and is the most mature method of hydrogen production, with the disadvantage of having poor performance due to low conversion efficiency and high energy consumption.

Recently, research and development has focused on polymer electrolyte membrane cells using a solid polymer electrolyte that is more compact and offers higher efficiency and current density.

However, at low temperatures, the performance of electrolyzers is quite low due to thermodynamic barriers.

The production of hydrogen by HTE therefore appears to be a promising avenue, so much so that a great deal of research is currently underway on this subject, particularly on solid oxide electrolysis cell (SOEC) electrolyzers, based on solid oxide ion conductor technology, such as that used for solid oxide fuel cells (SOFC).

7.3.3.2 Modeling of a High-Temperature Electrolyzer in Thermoptim

A high-temperature electrolyzer receives a mixture of water vapor and hydrogen (typically with a molar composition of 80%–20%) at 30 bar and 800°C. It produces two fluids: oxygen and the initial mixture enriched with hydrogen. At the outlet of the electrolyzer, the hydrogen mixed with water vapor is collected. The separation does not pose any particular problem: it is sufficient to condense the water to obtain pure hydrogen.

The Thermoptim model includes a classic mixer upstream, used to create the mixture ([H_2], [H_2O]) at the entrance of the electrolyzer, and an external splitter downstream, both connected by a process-point playing a passive role. A thermocoupler is used to provide heat to the component.

The structure of the model is shown in Figure 7.3.7 and the screen of the electrolyzer is shown in Figure 7.3.8.

The model we develop is very simple: it takes as parameters the molar fraction of hydrogen at the outlet α or the electricity supplied Pelec.

The balance of species is easy to establish. Electricity supplied electrolyzes water and produces hydrogen and oxygen, while heat provided by the high-temperature thermocoupler raises the temperature of the device (Figure 7.3.9).

Two calculation methods are possible:

- either determine Pelec knowing α;
- or determine α knowing Pelec.

In the first mode, the calculation is as follows: the model determines the inlet molar fraction and molar flow rate from the external divider main flow, deduces the molar flow of hydrogen electrolyzed, and establishes its balance.

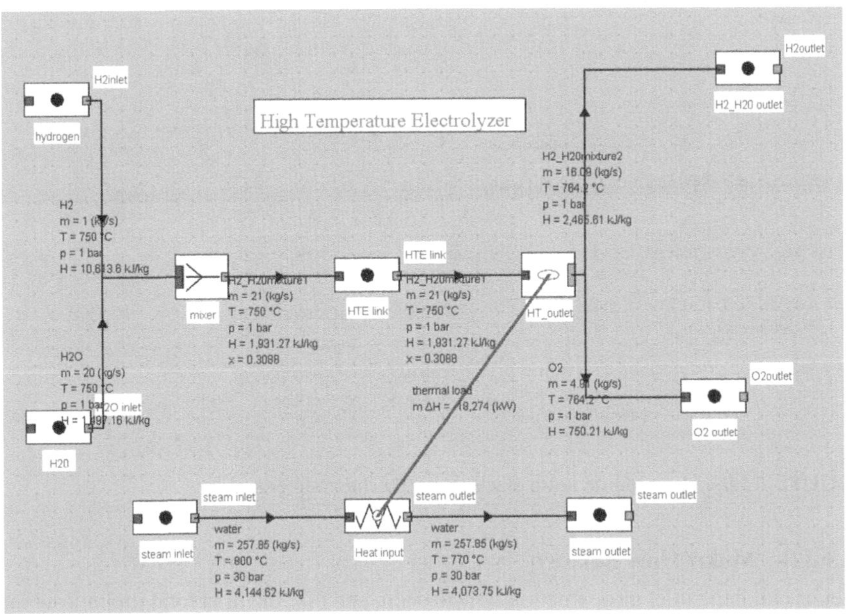

FIGURE 7.3.7 Diagram of the HTE component with connections.

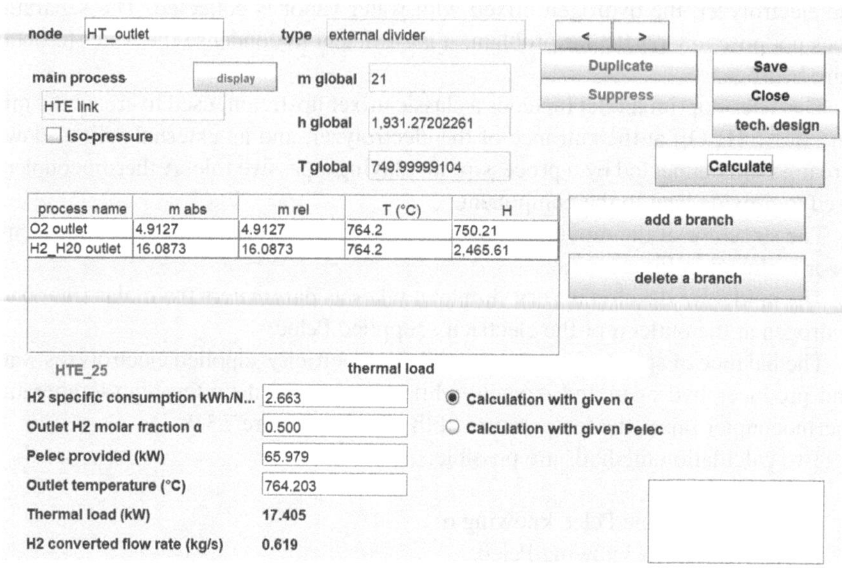

FIGURE 7.3.8 Screen of the HTE component.

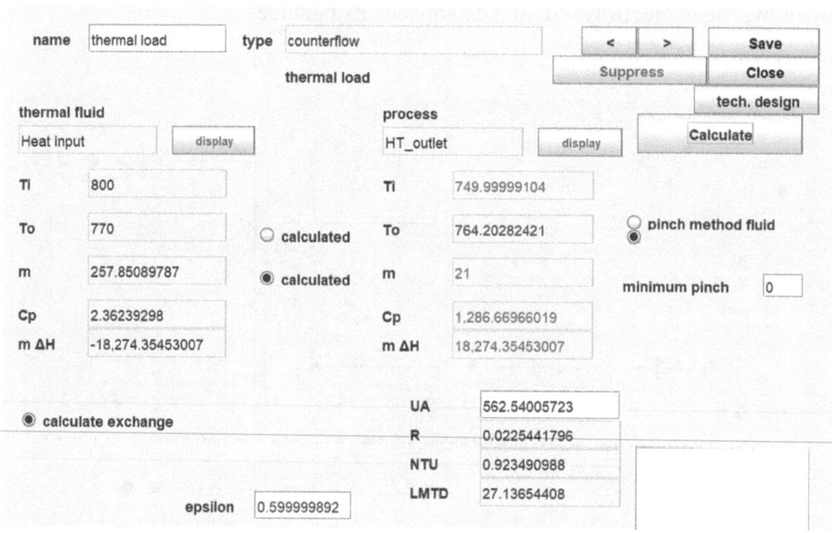

FIGURE 7.3.9 High-temperature electrolyzer thermocoupler screen.

7.3.3.3 Molar Flow Balance

Let us call β the inlet mole fraction of hydrogen, and $\dot{n}_{mol}$ the inlet total molar flow rate.

The molar flow of hydrogen electrolyzed $\dot{m}H_2$ is:

$$\dot{n}_{H_2} = (\alpha - \beta)\, \dot{n}_{mol}$$

The composition of the mixture $\frac{[H_2]}{[H_2O]}$ being known, its molecular weight can be determined. As its molar flow rate is equal to the inlet molar flow rate, the mass flow is known.

The molar flow rate of oxygen is $\frac{\dot{n}H_2}{2}$. Its mass flow rate is deduced directly.

7.3.3.4 Energy Balance

The overall energy balance of the water dissociation reaction is described by the First Law of thermodynamics, based on the relationship between enthalpy (ΔH), Gibbs energy (ΔG), and entropy (TΔS):

$$\Delta H = \Delta G + T\Delta S$$

Where:

- ΔH (Reaction Enthalpy) represents the total energy (electric + thermal) required for the reaction. It is the minimum total energy that must be provided.
- ΔG (Gibbs Energy) represents the part of the energy that must be provided in the form of electrical work. It is the minimum electric energy required.
- TΔS (Thermal Energy) represents the energy that can be provided in the form of heat. It is this component that makes the reaction endothermic.

Regressions giving the theoretical values have been performed:

$$T\Delta S_0 = 61.491\ \ T/1{,}000 - 6.2334;$$

$$\Delta G_0 = -55.578\ \ T/1{,}000 + 248.08;$$

7.3.3.5 Calculation of ΔG = Pelec

The cell voltage is a linear function of the HTE temperature:

$$\text{Ucell} = aT + b$$

The total current is given by Faraday's law:

$$I_{total} = \frac{\dot{n}_{H_2O} \times 2 \times F}{\eta_F}$$

The electric power is then given by:

$$\text{Pelec} = \text{Ucell}\ I\_\text{total}$$

An efficiency can be expressed as $\eta = \Delta G_0/\text{Pelec}$.

7.3.3.6 Calculation of $T\Delta S$ = Qabs

The heat provided by the thermocoupler is given by Qabs = $\dot{n}H_2\, T\Delta S_0$ where $T\Delta S_0$ is calculated by the regression given above.

In practice, we consider that the heat provided must slightly exceed this value, and we increase it by 5%.

The enthalpy balance is given by the following.

7.3.3.7 Calculation of ΔH

$$\Delta H = Pelec + Qabs$$

The "specific consumption" in kWh/Nm^3 allows calculation of the amount of electricity to be supplied ΔH by:

$$\Delta H = \dot{n}H2\ MH_2 \frac{3{,}600}{1{,}000.0.08988}$$

ΔH_0 being the theoretical heat of reaction (246 kJ/mol), the heat released is:

$$Q = \Delta H - \dot{m}H2\ \Delta H_0$$

In the second mode, the sequence of calculations is slightly modified: knowing Pelec, $\dot{m}H_2$ can be determined, from which we deduce α. Q is determined in the same way.

7.3.3.8 Results

With the settings above (1 kg/s hydrogen and 20 kg/s of water at the electrolyzer inlet), the gas compositions obtained are shown in Figures 7.3.10 and 7.3.11.

This model can be used to represent an electrolyzer integrated into a system, and particularly to study the thermal integration of the process in order to minimize the external heat input.

The external class is named HTE_25.java.

An alternative model configuration without the thermocoupler could supplement this approach. Here, the inlet mixture would be preheated by the thermal energy (Qabs)

component name	molar fraction	mass fraction
H2	0.3088358	0.04761905
H2O	0.6911642	0.952381

FIGURE 7.3.10 Humidified hydrogen at the inlet, flow rate 21 kg/s, LHV: 5,712 kJ/kg.

component name	molar fraction	mass fraction
H2	0.5	0.1006372
H2O	0.5	0.8993628

FIGURE 7.3.11 Humidified hydrogen at the outlet, flow rate 16.09 kg/s, LHV: 12,072 kJ/kg.

accounting for the TΔS thermodynamic term in the electrolysis model, with the outlet temperature of the electrolyzer component being the same as the reference case.

This model has some limitations:

- Simplified linear voltage relationship (reality is more complex)
- No consideration of degradation
- Constant efficiency (actually varies with current)
- Ignores auxiliary consumption

7.3.3.9 Coupling a High-Temperature Electrolyzer to a Nuclear Reactor

As we indicated, the heat needed for evaporating the water entering the electrolyzer can be provided by different sources, notably nuclear reactors.

7.3.3.9.1 Overall Sketch of the Device

As shown in Figure 7.3.12, given the high temperatures at which the electrolyzer operates, it is indeed necessary to achieve as much regeneration as possible between the incoming flows (water vapor and any hydrogen) and the outgoing flows (humidified hydrogen and oxygen).

The temperature to which the vapor must be brought depends on many factors, including the type of nuclear reactor and the pressure in the electrolyzer. Recent studies show that a pressure of around 30 bar would be favorable for the process, but this needs to be confirmed.

If, considering the unavoidable pressure losses, a water vapor pressure of 30 bar is adopted upstream of the regeneration exchangers of the electrolysis device, the boiling temperature of water is 242.54°C.

To minimize the input from the reactor, it is generally preferable to superheat the steam not with nuclear heat, but by cooling the humidified hydrogen and oxygen produced by the electrolyzer.

The synoptic of Figure 7.3.13 provides an example of an electrolyzer device receiving 12.2 kg/s of steam at 30 bar and 26.85°C. It uses the HTE_25 electrolyzer model presented above and available in the model library.

This steam flow is shared between two regeneration exchangers, with the upper part cooling the humidified hydrogen and the lower part cooling the produced oxygen.

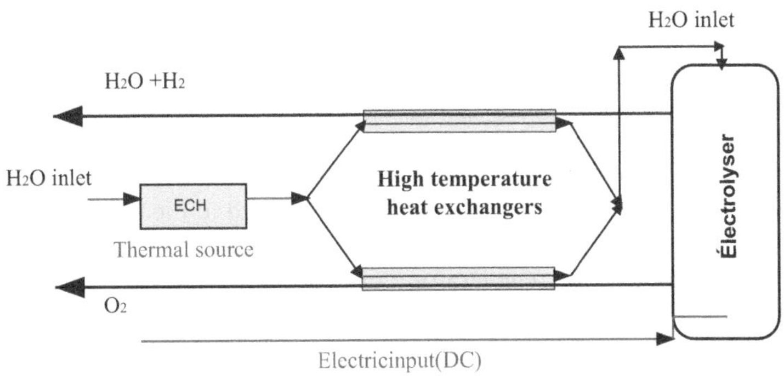

FIGURE 7.3.12 Electrolyzer heat exchanger partial layout.

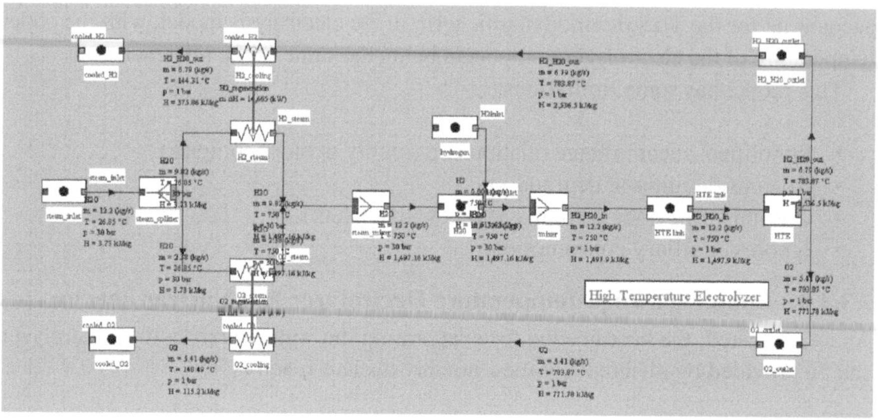

FIGURE 7.3.13 Synoptic view of the electrolyzer cycle.

The two flows at 750°C are directed to the entrance of the electrolyzer. In this example, there is no input of hydrogen, so we have entered a very low flow value in the corresponding process.

At the outlet of the electrolyzer, we find the hot products, which are cooled in the regeneration exchangers.

So far, we have not represented the coupling to the nuclear reactor. The need to provide heat at temperatures around 800°C limits the possibilities to a few reactor types: HTR, VHTR, or Molten Salt Reactor.

In practice, this coupling problem can prove to be complex, and the optimization must then concern the complete system, both the thermodynamic cycle of the nuclear reactor and the electrolysis device. The thermal integration method of Thermoptim that we used in Section 7.1 can be an appropriate tool for this.

7.3.3.9.2 Coupling with an HTR Reactor

To go further while keeping it simple, let us examine what could be achieved by coupling a high-temperature electrolyzer to an HTR reactor.

In Section 6.1.2, we presented a model of the thermodynamic cycle of a helium regenerative gas turbine. We will start from this model, which will serve as our reference.

At the outlet of the reactor, the helium stream at 900°C is split into two, with several possible configurations. One stream is directed to the main turbine, while the second can either be used solely to supply the thermocoupler and then be mixed with the helium exiting the regenerator, or expanded in a second turbine and then mixed with the stream exiting the main turbine before entering the regenerator.

The two examples we present correspond to these two variants (one and two turbine(s)). In both cases, we adjusted the flow rate through the reactor so that the electrical power output roughly matches that required to perform the electrolysis.

The performances are close: approximately 200 MW of nuclear heat is required to produce the 72.75 MW of electrical power and the 20.15 MW of thermal energy needed for the operation of the electrolyzer.

The synoptic views obtained are shown in Figures 7.3.14 and 7.3.15.

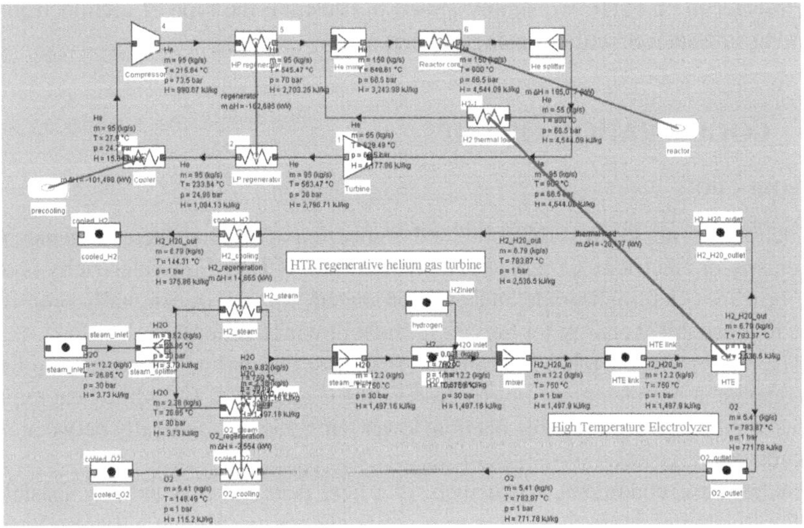

FIGURE 7.3.14 Synoptic view of the electrolyzer cycle coupled to the HTR reactor, variant with a single turbine.

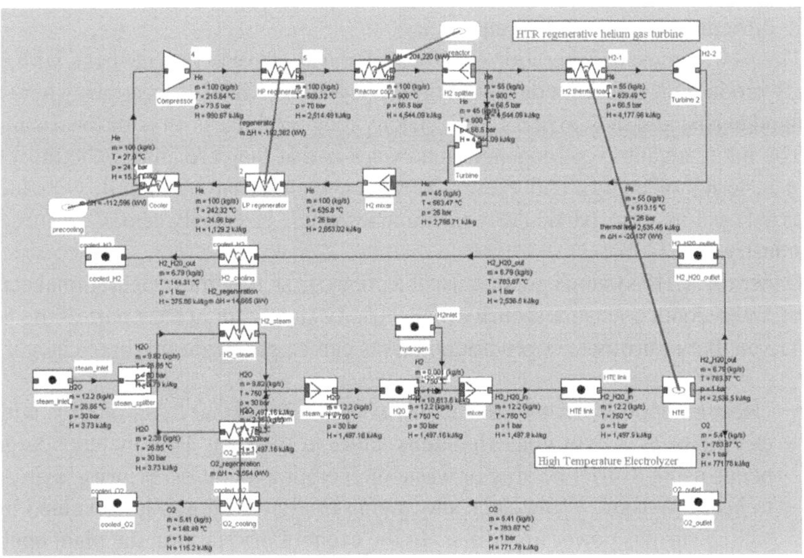

FIGURE 7.3.15 Synoptic view of the electrolyzer cycle coupled to the HTR reactor, variant with 2 turbines.

It should be noted that we have not modeled the entire network of heat exchangers necessary to vaporize the makeup water required to compensate for the hydrogen and oxygen produced. In this model, water is represented by the ideal gas "H_2O," which is a valid approximation only at high temperature.

Lacking data on a realistic thermodynamic cycle for coupling a high-temperature electrolyzer to an HTR, we simply aimed to illustrate the type of architecture that could be considered, without going further.

7.4 COGENERATION OR CHP

INTRODUCTION

We define **cogeneration** as the combined production of thermal energy and mechanical energy or electricity Combined Heat and Power (CHP). Since electricity is often the form in which high-grade energy is produced, we will occasionally omit reference to mechanical energy for brevity in the following sections.

The basic principle of cogeneration is that most thermal power generation facilities operate at high temperatures (between 300°C and 1,000°C), while heat requirements in industry or for heating occur at lower temperatures, generally between 80°C and 300°C.

Under these conditions, when there is a heat demand, it is entirely feasible to exploit this temperature difference to produce electricity through a thermodynamic cycle. For this purpose, the hot source of the thermodynamic cycle is the nuclear reactor and the cold source is the heat demand.

It is also theoretically possible to produce heat at high temperature at facilities primarily generating electricity, but this generally proves unfavorable as heat is much more difficult to transport than electricity.

This is why most cogeneration facilities primarily serve to provide heat, with electricity production as a secondary function. They are installed at locations where heat demand exists, and they export the electricity they produce as an additional output.

The main advantage of cogeneration cycles is that they are among the most efficient in terms of energy and exergy. However, their economic viability should be assessed case-by-case, particularly in countries where electricity prices are attractive to industry.

Generally, CHP systems pursue two objectives: first, to achieve operational economies, and second, to ensure security of electricity supply for at least part of the facility. Given these purposes, cogeneration plants can be grouped into three categories:

- "Heat and power" installations where heat is the primary commodity, with electricity as a byproduct that adds value to the fuel. This includes large plants using district heating or waste incineration facilities. Priority is given to heat provision, while electricity, being easily transportable, is valued by selling surplus power to utilities. In the event of grid failure, the plant operates in islanding mode.
- "Total energy" systems seeking to ensure electrical autonomy, with heat as the byproduct. These are generally off-grid plants and ships.

- Non-autonomous facilities, undersized for economic reasons, which receive supplemental electricity from utilities and supplemental heat from conventional boilers. The installation operates only when electricity prices are high and heat demands are significant. This type of installation is quite common and often leads to the best financial results for the operator.

Technically, CHP systems are customarily classified into two families, depending on the thermodynamic cycle employed:

- Boiler and steam turbine systems, which are widespread, as the benefits of this configuration have been recognized for over a century. They can use a wide variety of fuels, including coal or waste.
- Internal combustion engine systems, which employ either gas turbines or reciprocating engines (especially diesel and gas engines). Heat is recovered from exhaust gases as well as coolants and lubricants. Only liquid and gaseous fuels can be used in these engines.

This classification does not, however, cover all scenarios. There are also systems such as Stirling engine cogeneration, which is an external combustion engine. Finally, when a plant produces heat, motive power, and cooling simultaneously, it is called trigeneration or combined cooling, heating, and power generation.

Here we focus exclusively on the first category, which is most suitable for nuclear reactors.

After an initial section covering cogeneration generalities, we will focus more specifically on systems involving nuclear reactors. We present a first example corresponding to an HTR and then develop two variants of a cogeneration system intended for seawater desalination using a PWR.

7.4.1 Performance Indicators

A cogeneration plant produces both heat and electricity. To describe its performance in both regulatory and technical terms, we introduce several indicators, defined below.

Let Q_c denote the heat supplied to the cogeneration machine, that is, the heat released by the nuclear reactor; Q_u the useful heat; and τ the electricity produced. In what follows, these different energies are expressed in the same units, usually kWh or MJ.

We define:

- **Mechanical efficiency** as the ratio $\eta_m = \frac{|\tau|}{Q_c}$.

  It characterizes the performance of the facility as an electricity generator. The highest mechanical efficiencies are obtained in conventional NPPs where $Q_u = 0$.

- **Overall efficiency** as the ratio $\eta_g = \frac{|\tau + Q_u|}{Q_c}$.

  It characterizes, in energy terms, the overall efficiency of the facility.

- **Exergy efficiency** as the ratio $\eta_x = \frac{\left|\tau + \left(1 - \frac{T_0}{T}\right) Q_u\right|}{Q_c}$.

  where T_0 is the ambient temperature and T is the temperature at which heat is supplied. Through the introduction of the Carnot factor, this characterizes the temperature level at which heat is provided.

- **Heat-power ratio** as the ratio $CF = \frac{Q_u}{|\tau|}$.

It represents the distribution of energy between heat and electricity.

These indicators, as defined above for a given operating point, are likely to vary depending on operating conditions, including ambient temperature. To estimate their average values over a long period, such as the heating season or a year, they are calculated from the cumulative values of the variables considered.

7.4.2 Boilers and Steam Turbines

In a boiler and steam turbine cogeneration plant, there is either a single turbine, called back-pressure, or two turbines, called extracting and condensing (Figures 7.4.1 and A2.1.3).

In back-pressure turbines, well suited when heat demands vary little, steam produced in the boiler at an initial pressure generally between 30 and 50 bar is expanded to a pressure (called back-pressure) of approximately 2–6 bar and temperatures of 130°C–160°C. This steam is then used directly in industrial processes or district heating.

In extracting and condensing turbines, all the steam passes through the HP section, which behaves like a back-pressure turbine. A fraction is then extracted to feed industrial processes or district heating, while the remainder is expanded in an LP section and finally condensed either by air (Figure 7.4.1) or by water cooling.

This type of turbine can largely decouple electricity production from heat production, and is therefore very well suited for cogeneration facilities used for space heating. In summer, extraction is minimal and electricity production maximal; in winter, the opposite occurs.

The overall plant efficiency of back-pressure facilities is, however, slightly higher than that of extracting and condensing units, because in the latter, part of the heat is

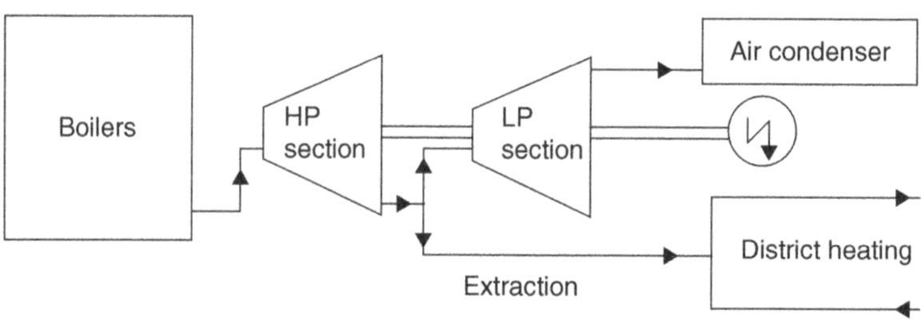

FIGURE 7.4.1 Boiler and steam turbine cogeneration unit.

discharged to the condenser. These configurations are widely used for applications where heat demands are significant, such as district heating networks and heavy industries.

The fraction of the load directed to supply the district heating network can be modulated to some extent, and where appropriate, the LP turbine section or heating network can be bypassed. Only a detailed study of the off-design behavior of the various system components, accounting for the respective prices of heat and electricity, can determine the economic optimum.

Under these conditions, the heat-power ratio CF can vary from one plant to another or during operation. For smaller units (a few hundred kilowatts), values of 8–15 are typical. For larger installations, this ratio decreases to values between 3 and 5. Overall efficiency is generally excellent, exceeding 80% even for large heat-power ratios. Mechanical efficiency is relatively low, between 6% and 22%.

For those particularly interested in this type of application, we recommend studying the Diapason S47 session on district heating.[3] Although it pertains to a combustion gas turbine cogeneration installation, it will enable study of the issues related to this type of installation, which could be implemented with a high-temperature nuclear reactor.

A more general overview of cogeneration installations can be found on this page of the Thermoptim-UNIT portal.[4]

7.4.3 Nuclear Cogeneration

Although not yet widespread, cogeneration facilities associated with nuclear reactors have existed for many years, and many are currently under development.

This 2017 IAEA publication provides an overview of the opportunities for cogeneration from nuclear energy.[5]

For example, in Switzerland, the 970 MW(e) Gösgen NPP has been supplying process steam to a neighboring cardboard factory since 1979. The process steam (13.7 bar, 220°C) is produced in a tertiary steam cycle using live steam from the NPP. It is then transported via pipelines over a distance of 1,750 m to the cardboard factory. After condensation, it returns as hot water at 100°C to the PWR. A maximum process steam extraction of 22.2 kg/s is possible, representing a thermal power of approximately 54 MW or about 2% of the total thermal power of the NPP.

Below, we present three types of nuclear cogeneration installations.

7.4.3.1 Project Antares of Areva Framatome

Figure 7.4.2 shows the different cogeneration options feasible with a high-temperature nuclear reactor. On the right side of the diagram, the temperature levels where heat demands arise and the corresponding market applications are indicated.

In Section 7.1.3.3, we presented the Antares project of Areva Framatome. Variants of this cycle adapted to cogeneration have also been optimized by Areva using Thermoptim. They provide superheated steam at 110 bar from water at 120°C for certain cogeneration applications, for power outputs from 50 to 300 MW.

An example of a Thermoptim diagram for such a cogeneration installation is given in Figure 7.4.3, and the flowsheet in Figure 7.4.4.

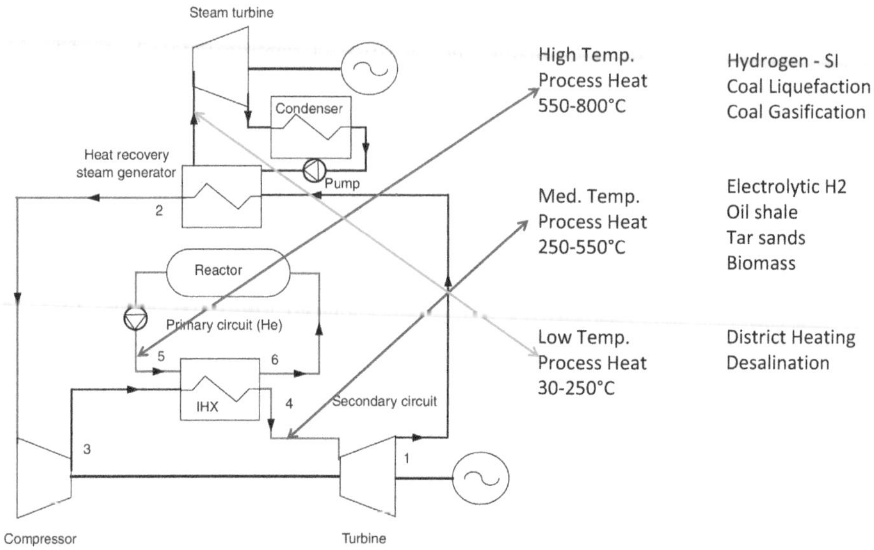

FIGURE 7.4.2 Diagram of the Antares cogeneration concept.

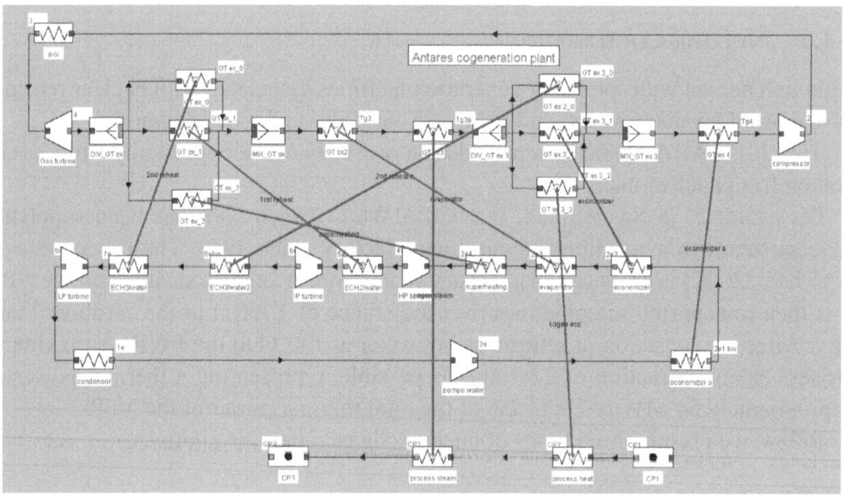

FIGURE 7.4.3 Antares cogeneration process flow diagram.

Compared to the combined cycle configuration, the He–N_2 flow had to be increased to provide the heat required by the cogeneration steam (50 kg/s of water at 110 bar and 110°C heated to the temperature of 318.4°C, i.e., 110 MWth). As a result, the turbine and gas compressor power is approximately 22 MWe higher. The mechanical conversion efficiency decreases from 50.5% to 44.4%, but the overall efficiency, including cogeneration, equals (336.253 + 46.927 + 63.006)/757.08 = 58.9%.

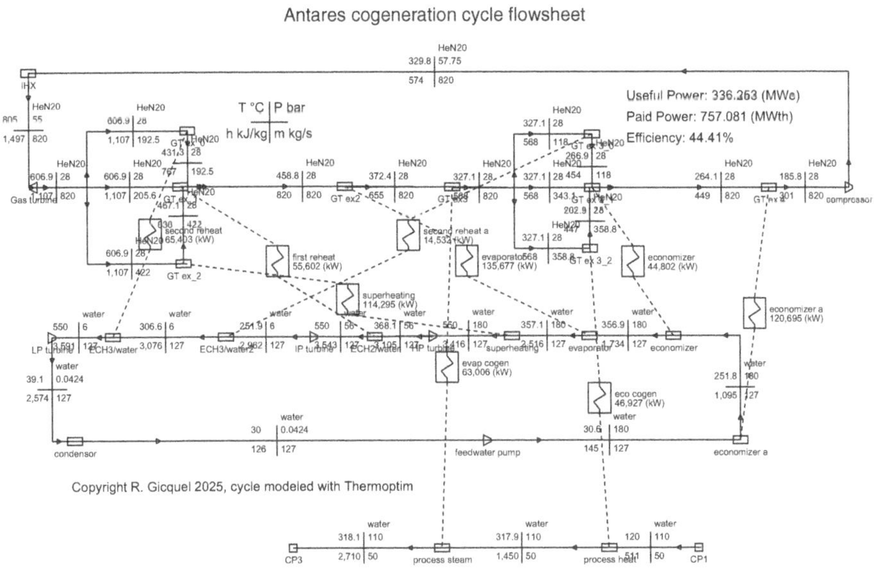

FIGURE 7.4.4 Antares cogeneration flowsheet.

7.4.3.2 Nuclear Cogeneration Installation Intended for Seawater Desalination

In Section 7.2, we studied an example of a seawater desalination cycle using TVC. Two variants of nuclear cogeneration systems employing such a device are presented here.

This cycle was chosen partly because of its good performance, which could be further improved by increasing the number of effects, and partly for educational reasons, since it remains simple enough to be easily understood. In a real industrial project, however, many additional parameters would have to be considered, both on the nuclear reactor side and on the desalination system side.

As we have seen, thermocompression increases the enthalpy level of the vapor by raising its pressure and, consequently, its phase-change temperature. Such a device is used to upgrade the vapor produced by an evaporation–concentration boiler, which then serves as the heat source for a first effect. The Thermoptim diagram of Figure 7.4.5 illustrates the operation of the desalination plant.

In the first unit of interest, 300 kg/s of seawater at 35 g/L and 70°C enters the boiler, heated by steam at about 104°C. It produces 100 kg/s of fresh water in vapor form and 200 kg/s of concentrated brine at 52.5 g/L and 100.4°C. This concentrated brine then enters a second boiler, heated by pure steam from the first boiler, which condenses. It yields 100 kg/s of brine at 105 g/L and 100 kg/s of distilled vapor at 0.85 bar and 97°C.

This distilled vapor enters an ejector whose motive fluid is 259 kg/s of hot water at 11.1 bar and 184.5°C supplied by the PWR reactor. At the ejector outlet, 359 kg/s of steam at 1.16 bar and 103.8°C serves as the heat source for the first boiler. The nearly condensed steam is then partly returned to the reactor (259 kg/s) and partly condensed to produce fresh water.

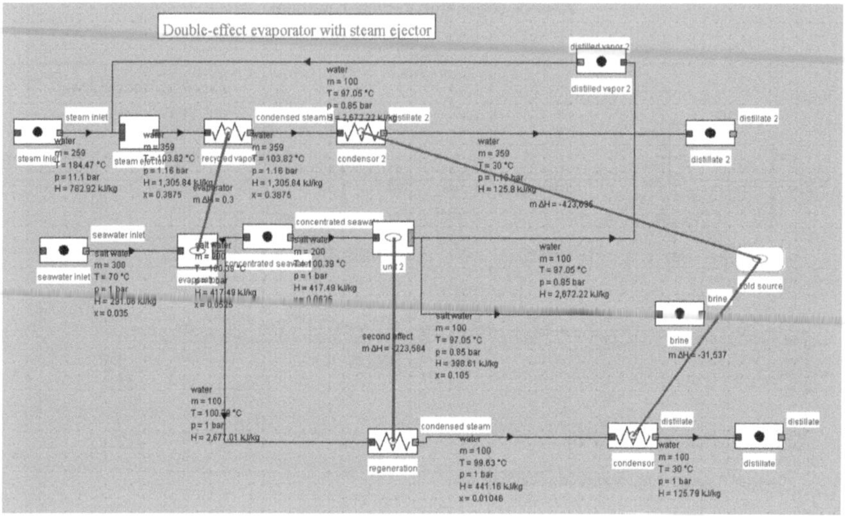

FIGURE 7.4.5 Double-effect ejector evaporator fed with pressurized water.

The total distillate production is therefore 200 kg/s.

The flowsheet of Figure 7.4.6 shows how the desalination unit is coupled with the PWR reactor, which thus operates in cogeneration mode.

The two processes on the right side of the diagram represent:

- At the top, the motive flow to the ejector of 259 kg/s of medium-pressure hot water representing the liquid phase exiting the separator.
- At the bottom, the return to the feedwater tank of an equivalent flow of LP, nearly condensed steam. Note that the diagram is missing both a booster pump to raise the pressure of this flow and a condenser so that this pump can draw liquid water.

As can be seen, the extraction from the reference nuclear cycle is made from a diverted flow of the feedwater tank, which has a nearly negligible impact on the performance of this cycle. The Thermoptim performance summary for this cogeneration plant are as follows: Useful power: 1,658.7 MW; Heat input: 5,229 MW; Efficiency: 31.72%.

For comparison, the useful power of the PWR cycle alone was about the same, and the heat input was 4,989 MW, with an efficiency of 33.28%. The impact of cogeneration on the power plant's performance is low.

7.4.3.3 Installation Variant of Nuclear Cogeneration Intended for Seawater Desalination

We now present a variant of this cogeneration cycle, where the extraction is performed on the vapor phase exiting the separator rather than on the liquid phase.

Since the specific enthalpy of steam is much higher than that of pressurized water, the steam flow rate in this case is much lower: 50 kg/s instead of 259. The desalination cycle is modified as shown in Figure 7.4.7.

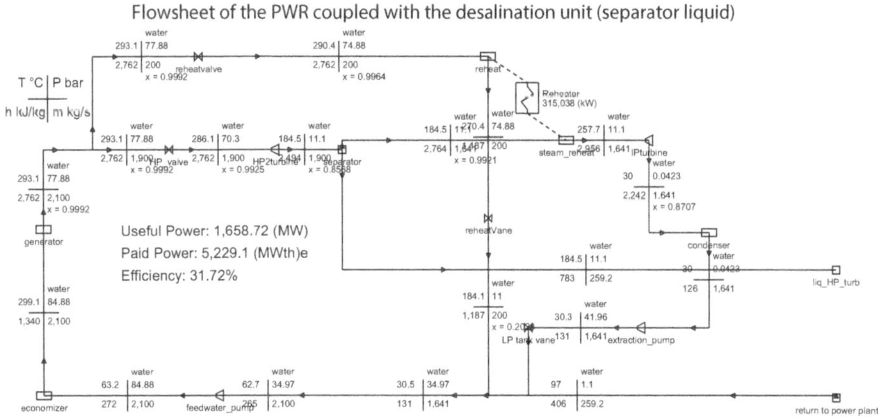

FIGURE 7.4.6 Flowsheet of the PWR reactor coupled with the desalination unit.

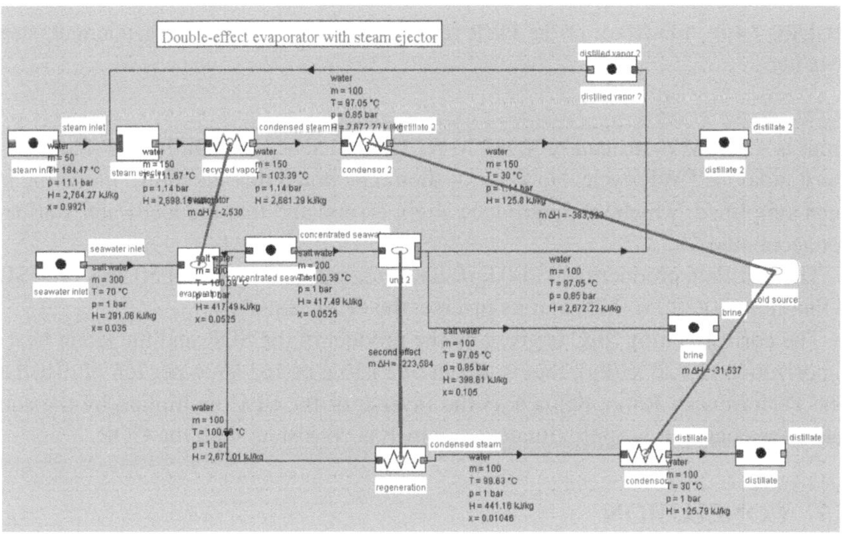

FIGURE 7.4.7 Double-effect ejector evaporator fed with steam.

As shown in Figure 7.4.8, the extraction of 50 kg/s of medium-pressure steam is made at the separator outlet, while the main flow (1.59 t/s) is directed toward the LP turbine unit.

The return of the nearly condensed steam occurs at the feedwater tank level. Note that a recovery pump to raise the pressure of this flow and a condenser for this pump to draw liquid water are also missing from the diagram.

The flow rate ratio is close to 3%, showing that the extraction is very small compared to the main flow. It would, of course, be possible to increase it, but we have determined this flow rate so that the fresh water production is the same as in the previous example.

The impact on nuclear cycle performance is slightly more significant than in the initial example. The useful power is 1,630 MW instead of 1,660 MW, and the heat

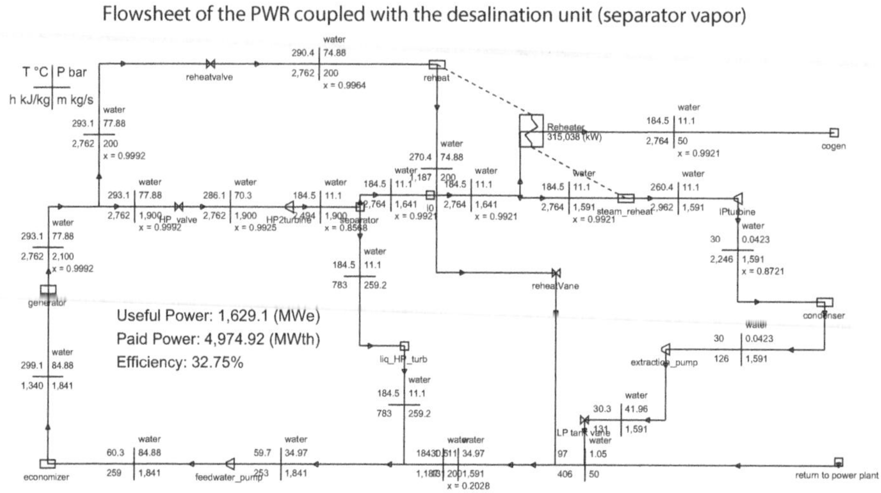

FIGURE 7.4.8 Flowsheet of the PWR reactor coupled with the desalination unit, steam coupling.

input is 4,975 MW instead of 4,989 MW. The efficiency is slightly less than that of the standalone PWR cycle. However, it should be noted that the return modeling has been simplified, which may introduce slight errors, and that the feedwater station is not accounted for.

The distillate production is 200 kg/s for a steam consumption of 50 kg/s. The SDR is therefore 50/200 = 0.25, and its inverse, the GOR, equals 4.

The corresponding SEC is given by the product of the SDR and the latent heat of vaporization (2,326 kJ/kg), that is, about 582 kJ/kg or 162 kWh per ton of distillate. The Performance Ratio, defined as the inverse of the SEC multiplied by the same reference constant, is approximately 4, which is consistent with the GOR.

7.5 CONCLUSION

This chapter examined nuclear thermodynamic cycles other than those used for conventional electricity generation. Combined cycles, integrating Brayton and Rankine cycles, achieve over 60% efficiency thanks to optimal heat recovery, with the pinch method systematically optimizing the HRSGs. Dual-pressure helium-steam configurations reach 48.73% efficiency, while supercritical CO_2 variants were also explored. Desalination technologies studied include Multi-Effect Distillation, Multistage Flash, Reverse Osmosis, MVC, and TVC. Hydrogen production via methane reforming and HTE (800°C–1,000°C) was modeled. Nuclear cogeneration applications demonstrate overall efficiencies exceeding 80%, exemplified by the Gösgen power plant (54 MW process steam) and PWR-desalination couplings that maintain efficiency while producing 200 kg/s of fresh water.

Further resources on the models presented in this chapter are available on this book's website.[6]

NOTES

1 https://direns.minesparis.psl.eu/Sites/Thopt/en/co/cycles-combines.html.
2 Some explanations about the GT-MHR are given in Section 6.3.
3 https://direns.minesparis.psl.eu/Sites/Thopt/en/co/session-s47en-district.html.
4 https://direns.minesparis.psl.eu/Sites/Thopt/en/co/cogeneration.html.
5 Opportunities for Cogeneration with Nuclear Energy, STI/PUB/1749, ISBN 978-92-0-103616-2, 91 pages, 2017.
6 Available at: https://server.s4e2.com/crc/tcnpp/chapters/chapter7/_resources/index.html.

BIBLIOGRAPHY

Danis, P. (2000). *Dessalement de l'eau de mer. Techniques de l'Ingénieur*, Article J2700, Paris.

Dostal, V. (2004). *A supercritical carbon dioxide cycle for next-generation nuclear reactors.* PhD thesis, Massachusetts Institute of Technology (MIT), January.

Dostal, V., Driscoll, M.J., Hejzlar, P. & Todreas, N.E. (2003). "A supercritical CO_2 *gas turbine power cycle for next-generation nuclear reactors." Proceedings of ICONE-10*, Arlington, Virginia, April 14–18.

Gauthier, J.C., Brinkmann, G., Copsey, B. & Lecomte, M. (2006). Antares: The HTR/VHTR project at framatome ANP. *Nuclear Engineering and Design*, 236(5-6), 526–533.

Gosset, J., Gicquel, R., Lecomte, M., & Queiros-Conde, D. (2005). Optimal design of the structure and settings of nuclear HTR thermodynamic cycles. *International Journal of Thermal Sciences*, 44(12), 1169–1179.

Leleu, R. (2000). Évaporation. Techniques de l'Ingénieur, Article J2320, Paris.

Lior, N. (2000). "Water desalination." In Kreith, F. (ed.), *The CRC Handbook of Thermal Engineering.* CRC Press, Boca Raton, FL. ISBN 0-8493-9581-X.

Maurel, A. (2006). *Dessalement de l'eau de mer et des eaux saumâtres*, 2nd ed. Éditions Tec & Doc, Paris.

NEA/ NDC (2012). *Nuclear cogeneration: Industrial and desalination applications.* OECD Nuclear Energy Agency, Report NEA/NDC(2012)22, Paris.

O'Brien, J.E. (2005). "High temperature electrolysis for hydrogen production from nuclear power plants." *Idaho National Laboratory Report,* INL/EXT-05-00793.

Pioro, I. L. (ed.) (2017). *Latest Research and Development in Thermodynamic Cycles and Heat Transfer in Nuclear and Thermal Power Industries.* InTechOpen, New York. ISBN 978-953-51-3330-7. DOI: 10.5772/intechopen.68239.

Pioro, I. L. (ed.) (2023). *Handbook of Generation IV Nuclear Reactors*, 2nd ed. Woodhead Publishing (Elsevier), New York. ISBN 978-0-12-820588-4.

Reynaud, J.F. (1984). *Concentration par évaporation et recompression mécanique de vapeur.* Éditions Eyrolles, Paris.

Span, R. & Wagner, W. (1996). "A new equation of state for carbon dioxide covering the fluid region from the triple point temperature to 1100 K at pressures up to 800 MPa." *Journal of Physical and Chemical Reference Data*, 25, 1509–1596.

Wang, D., et al. (2023). "Thermodynamics of GT-MHR-250 modular nuclear plant power conversion system using helium as the working fluid." *Thermal Science and Engineering Progress*, 39, 101686.

Werkoff, F., Maréchal, A. & Pra, F. (2003). Technico-economic study on the production of hydrogen by high-temperature steam electrolysis. In *CEA Nouvelles technologies de l'énergie, Rapport d'activités 2001–2003.* Available online: https://www.cea.fr.

Weston, K.C. (1993). "Dual gas turbine combined cycles." *Proceedings of the 28th Intersociety Energy Conversion Engineering Conference (IECEC'93),* Boston, Vol. 1, 955–958.

8 Case Studies

INTRODUCTION

This chapter presents various case studies.

Except for the Canadian supercritical water-cooled reactor (SCWR), they correspond to cycles that have been actually realized and not to theoretical cycles.

The values indicated in this chapter are in principle those of the selected Thermoptim models, so that you can find them easily in the provided flowsheets.

In all cases, we provide references to the diagrams and data from which we started, which we have attempted to respect as closely as possible. However, small discrepancies are inevitable, if only because the values of water properties are not exactly the same depending on the references chosen.

Some information used in this work originates from documents made available by the U.S. Nuclear Regulatory Commission (NRC) through its public database (ADAMS Public Documents). These nonproprietary versions are freely accessible online and may contain redactions (blank pages). The values and diagrams presented here are adapted or reconstructed from these public sources and do not reproduce the original content verbatim.

Furthermore, some of the components we used, particularly the closed feedwater heater with cascaded drains, do not necessarily reflect the reality of the technologies implemented. These discrepancies do not call into question the understanding of the operation of these cycles.

Most of the detailed flowsheets are presented divided into two parts for better readability.

The complete diagrams in SVG vector format are available at the following address.[1]

We recommend downloading them and working from these versions, which are much easier to read.

When NUSCLE models of the considered facilities were available, steam expansion trajectories corresponding to both the detailed and simplified models were plotted on the Mollier diagram.

Figure 8.1 presents the expansions of the ten thermodynamic cycles analyzed in this chapter, as depicted on the water Mollier diagram. Comprehensive information regarding the various state points is provided in the online version.[2]

8.1 DETAILED MODEL OF THE 650 MW AGR NPP

We present here the model of the Advanced Gas Reactor (AGR) cycle of the Hartlepool plant in County Durham, North East England.

See Section 2.2.9 for an overview of AGR-type reactors

The Advanced Gas-Cooled Reactor at Hartlepool exemplifies an efficient graphite-moderated CO_2-cooled nuclear power plant using high-temperature steam parameters.

DOI: 10.1201/9781003606024-9

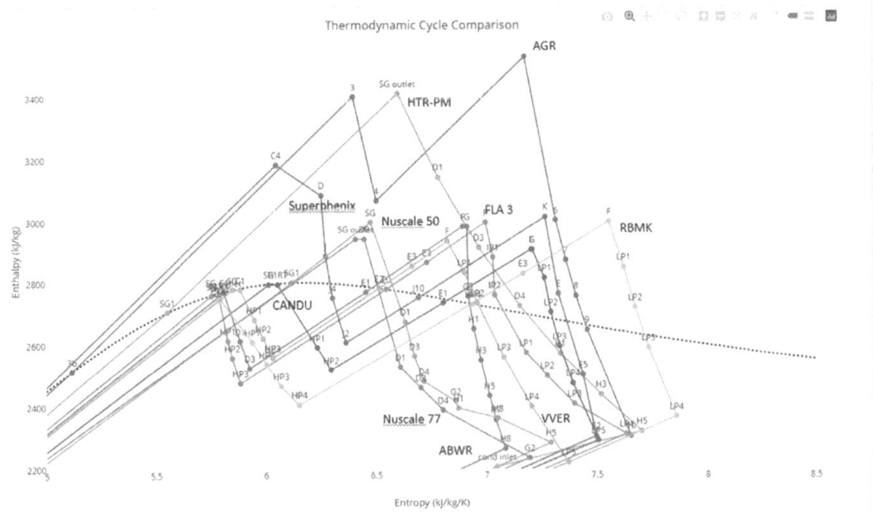

FIGURE 8.1 Case study cycles in the Mollier diagram.

8.1.1 Presentation and Sources

In Chapter 4 on high-temperature steam cycles, we used the modeling of the Hartlepool AGR plant as a guiding thread, but since our main objective was to illustrate the methodology for improving the cycles, we limited ourselves to two feedwater heaters, which were modeled in a simplified manner. In this case study, we present the complete model as we were able to establish it based on the available data.

The data used comes from two documents: Nonbel (1996) and Shengqi Zhou and Alan Turnbull (2002).

The second document was mainly used for estimating polytropic efficiencies of turbines.

AGRs are high-temperature nuclear reactors using CO_2 as a coolant. The thermodynamic cycle is a Hirn cycle with reheat, three turbine sections [high-pressure (HP), intermediate-pressure (IP), and low-pressure (LP)], and four extractions. The steam generator (SG) is heated by the CO_2 flux.

The fuel consists of uranium dioxide pellets, 2.5%–3.5% enrichment, in stainless steel clad fuel pins.

The SG is of the "once-through" type; water enters at 29 bar and ~157.6°C, is vaporized and superheated to 175 bar and 543°C.

8.1.2 Steam Circuit

The nuclear reactor is shown at the top left of Figure 8.1.1. It receives, in its lower part, 4,067 kg/s of CO_2 at 45 bar and 305°C. CO_2 exits at 43.43 bar and 639°C, and then successively preheats the reheater, exiting at 591.3°C, followed by the three heat exchangers of the SG.

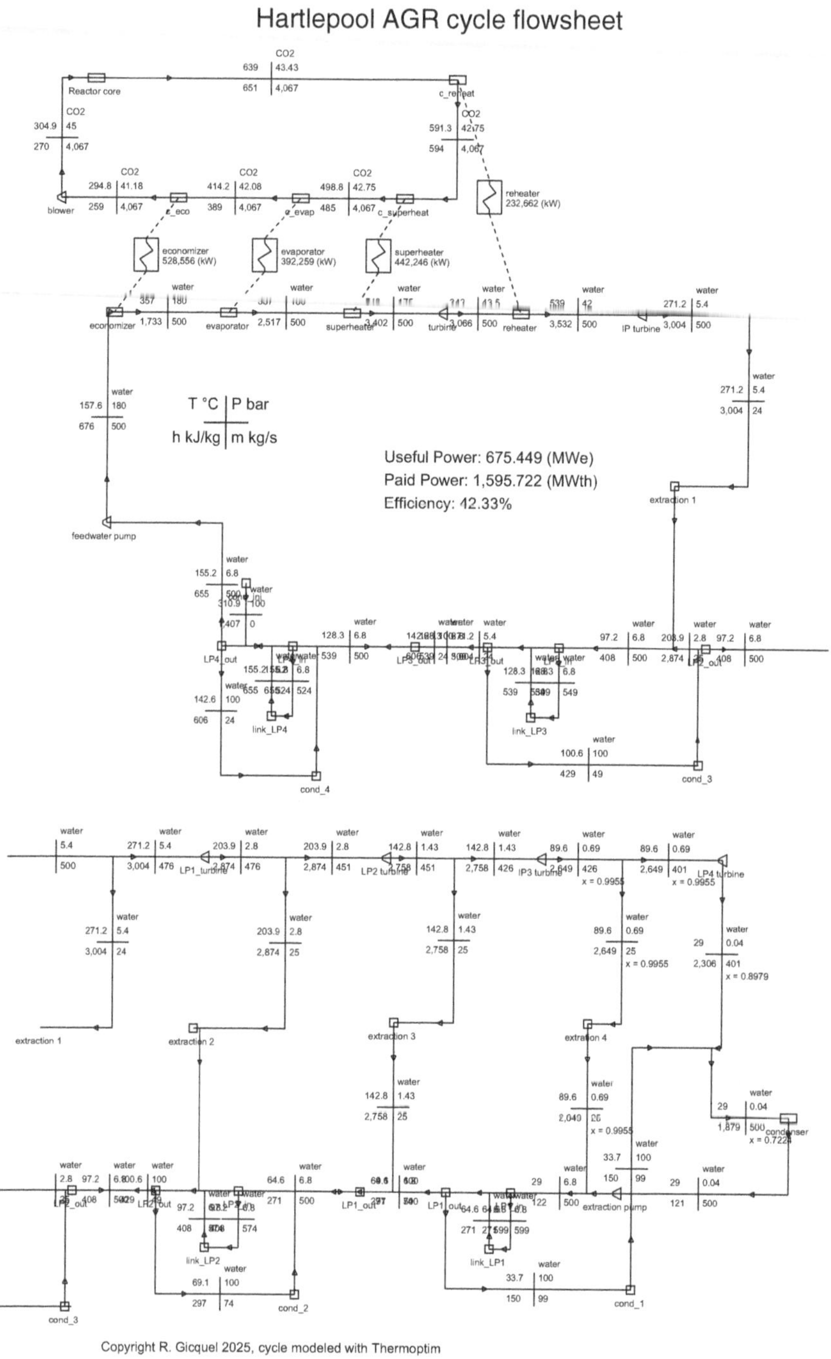

FIGURE 8.1.1 Detailed flowsheet of the Hartlepool AGR cycle.

A steam flow of 500 kg/s enters the SG at 180 bar and 157.6°C and exits at 175 bar and 543°C.

The steam is then directed to the HP turbine, exiting at 43.5 bar and 342°C. It then passes through the reheater, which raises its temperature to 539°C.

It then enters the IP turbine at 42 bar, exiting at 5.4 bar and 271.2°C. An extraction of 24 kg/s is performed to heat feedwater heater LP4.

The main flow is directed to the LP turbine, which consists of four turbine stage groups (TSGs).

It enters the first LP TSG, exiting at 2.8 bar and 203.9°C. An extraction of 25 kg/s is used to heat feedwater heater LP3.

The main flow enters the second LP TSG, exiting at 1.43 bar and 142.8°C. An extraction of 25 kg/s is used to heat feedwater heater LP2.

The main flow enters the third LP TSG, exiting at 0.69 bar and 89.6°C with a quality of 0.9955. An extraction of 25 kg/s is used to heat feedwater heater LP1.

The remaining flow passes through the fourth LP TSG and exits the turbine at a pressure of 0.04 bar, with a quality of 0.8979.

The steam expansion path on the Mollier diagram is shown in Figure 8.1.2.

8.1.3 Feedwater Circuit

All the feedwater heaters in this system are arranged with cascaded drains.

At the outlet of the condenser, an extraction pump compresses the 500 kg/s of feedwater to a pressure of 29 bar, and then it is preheated in feedwater heater LP1 to 49.2°C, then in feedwater heater LP2 to 64.6°C, then in feedwater heater LP3 to 97.2°C, and in feedwater heater LP4 to 155.2°C.

The remaining feedwater is then brought to a pressure of 180 bar by the feedwater pump before entering the SG.

The cycle is thus completed.

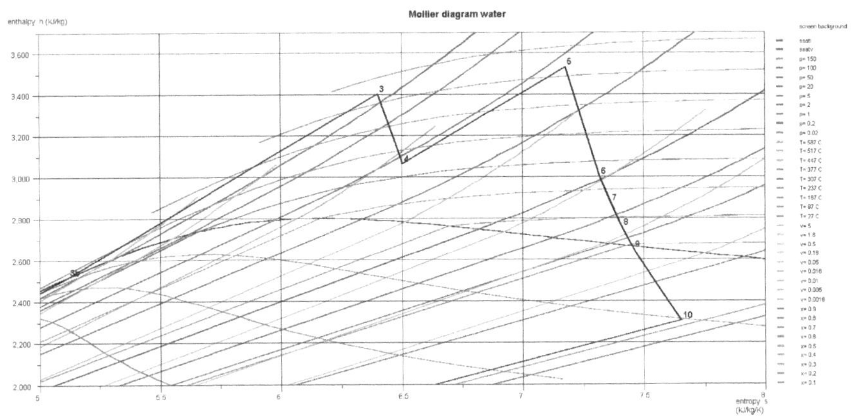

FIGURE 8.1.2 Hartlepool cycle in the Mollier chart.

8.1.4 Performance Summary Table

Parameter	Value
CO_2 flow rate	4,067 kg/s (45 bar, 305°C → 639°C)
Feedwater flow rate	500 kg/s
Steam generation	175 bar, 543°C
HP turbine inlet	175 bar, 543°C
Reheater outlet	42 bar, 539°C
Thermal power	1,596 MWth
Electrical power	675 MWe
Overall efficiency	42.33%

8.2 DETAILED MODEL OF THE 50 MW NUSCALE NPP

The NuScale US600 SMR is a pressurized water reactor from NuScale Power, LLC, whose design is based on a light water reactor developed at Oregon State University in the early 2000s. It is a natural circulation light water reactor where the core and SGs are located in a common pool within a cylindrical steel containment.

8.2.1 Presentation and Sources

The reactor containment module is almost entirely submerged in the water of the underground safety pool of the reactor building, which also serves as a heat sink for the reactor in case of malfunction. It is an Integral Pressurized Water Reactor, with natural circulation and passive safety features, which uses light water as both the coolant and moderator.

The reactor building is designed to accommodate up to 12 Small Modular Reactor (SMRs), each with a thermal power rating of 160 MWth and an electric power of 50 MWe, resulting in a total capacity of 600 MWe for 12 SMRs.

The data come from document NuScale_ML19073A048.pdf Design Certification – NuScale US600 (https://www.nrc.gov/reactors/new-reactors/smr/licensing-activities/nuscale.html downloaded on 22/08/25 at 6:40 AM).

The diagram of the model we present here is given, in British units, in p. 27 of this document. It is a cycle without a separator or reheater, but with condensate extractions: three LP feedwater heaters, one HP feedwater heater, and six extractions. The model we established under Thermoptim allowed us to generate the flowsheet in Figure 8.2.1.

The SG comprises two once-through helical coil SGs integrated inside the reactor vessel.

The fuel is made of UO_2 fuel assemblies (17 × 17 rod array), enrichment up to 4.95%.

8.2.2 Steam Circuit

The nuclear reactor is shown on the left of Figure 8.2.1. It receives, in its lower part, 67.07 kg/s of feedwater at 35.2 bar and 148.6°C. It exits at 34 bar and 307°C, slightly superheated.

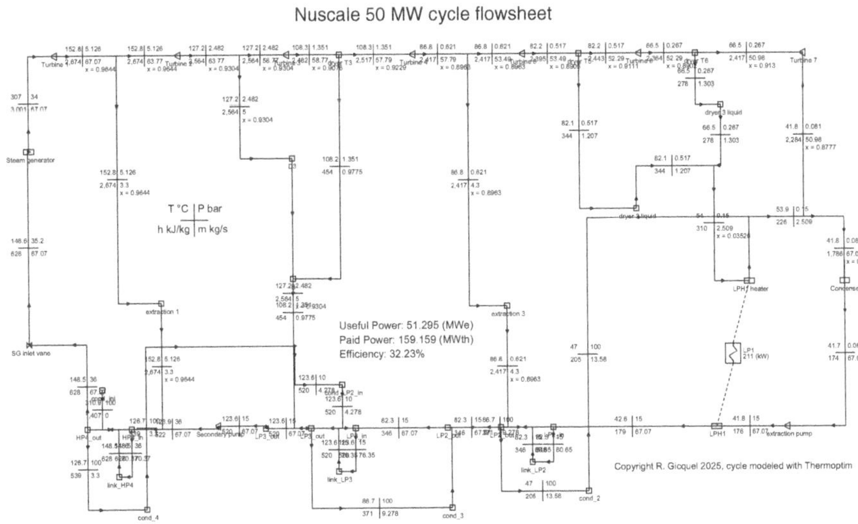

FIGURE 8.2.1 Detailed flowsheet of the 50 MW NuScale cycle.

The steam then enters the HP turbine, composed of two TSGs, and exits at 5.125 bar and 152.8°C, with a quality of 0.9644. An extraction of 3.3 kg/s is used to heat feedwater heater HP4. In our model, these two TSGs have been grouped under the name Turbine 1.

The remaining flow enters a third TSG (Turbine 2), exiting at 2.482 bar and 127.2°C, with a quality of 0.9304. An extraction of 5 kg/s is used in part to heat feedwater heater LP3.

The remaining flow enters a fourth TSG (Turbine 3), exiting at 1.351 bar and 108.3°C, with a quality of 0.9076. An extraction of condensates of 0.9775 kg/s also contributes to heating feedwater heater LP3.

The remaining flow enters a fifth TSG (Turbine 4), exiting at 0.621 bar and 86.8°C, with a quality of 0.8963. An extraction of 4.3 kg/s is used in part to heat feedwater heater LP2.

The main flow enters a sixth TSG (Turbine 5), exiting at 0.517 bar and 82.2°C, with a quality of 0.8905. An extraction of condensates of 1.207 kg/s contributes to heating feedwater heater LP1.

The remaining flow enters a seventh TSG (Turbine 6), exiting at 0.267 bar and 66.5°C, with a quality of 0.9111. An extraction of condensates of 1.303 kg/s also contributes to heating feedwater heater LP1.

The remaining flow is expanded in the final TSG (Turbine 7) before being directed to the condenser at a pressure of 0.081 bar (41.8°C).

The steam expansion path on the Mollier diagram is shown in Figure 8.2.2.

8.2.3 Feedwater Circuit

At the outlet of the condenser, an extraction pump compresses the 67.07 kg/s of feedwater to a pressure of 15 bar, and then it is preheated in feedwater heater LP1 to

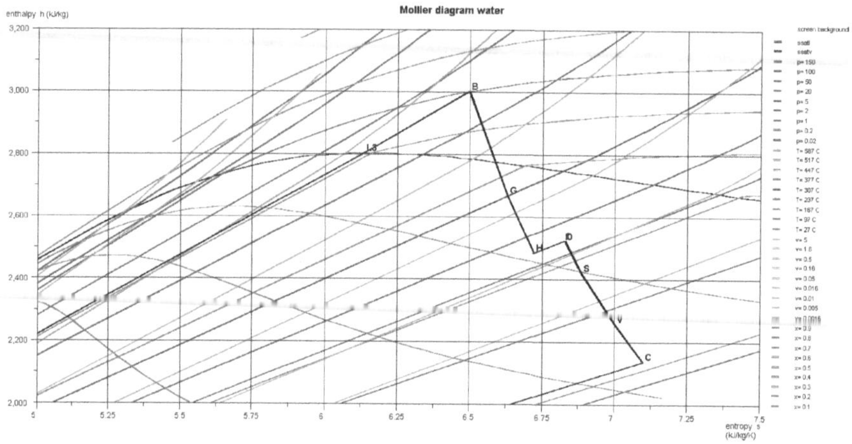

FIGURE 8.2.2 NuScale 50 MW cycle in the Mollier diagram.

42.6°C by the condensates extracted from the last TSGs, which are then directed to the condenser.

The feedwater is preheated in feedwater heater LP2 arranged with cascaded drain, to 82.3°C. Once condensed, the steam flow entering this second heat exchanger is also directed to the condenser.

The feedwater is then preheated in feedwater heater LP3 to 123.6°C before being brought to a pressure of 36 bar by a secondary pump.

It is then directed to feedwater heater HP4, exiting at 148.5°C before entering the inlet valve of the SG.

In this model, we have not accounted for pressure losses in the feedwater piping, which remain low.

8.2.4 Cycle Performance Summary

Parameter	**Value**
Thermal power (MWth)	159.16
Electric power (MWe)	51.3
Efficiency (%)	32.23
Secondary feedwater pressure	35 bar
Secondary feedwater temperature	148.6°C
Turbine inlet steam pressure	34 bar
Turbine inlet steam temperature	307°C

8.3 DETAILED MODEL OF THE 77 MW NUSCALE NPP

The NuScale US460 SMR is a pressurized water reactor from NuScale Power, LLC, whose design is a simplified and more powerful version of the 50 MW module presented in Section 8.2.

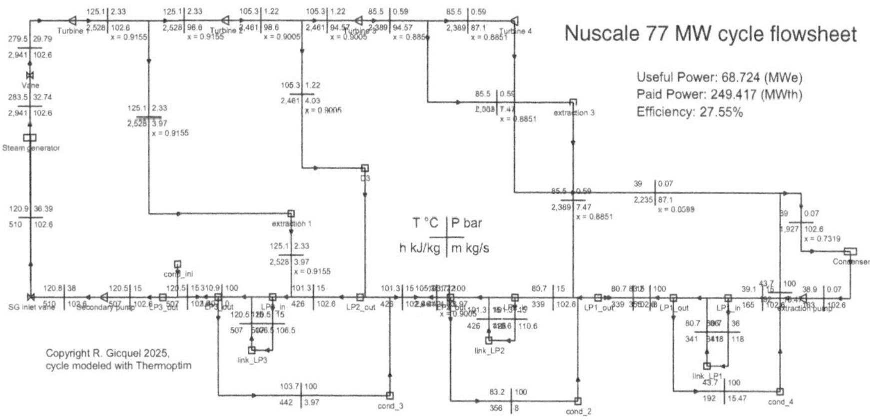

FIGURE 8.3.1 Detailed flowsheet of the 77 MW NuScale cycle.

8.3.1 Presentation and Sources

The diagram of the model presented here corresponds to a simplified schematic shown in Figures 10.1.1 and 10.1.2 of the document ML25099A261 (https://www.nrc.gov/docs/ML2509/ML25099A261.pdf, downloaded on 09/05/25 at 15:40). The numerical data are provided, in British units, in Tables 10.1.1 and 10.1.2 of the same document.

This is also a cycle without a separator or reheater, featuring three LP feedwater heaters arranged with cascaded drains and three steam extractions, but no condensate extraction.

8.3.2 Steam Circuit

The nuclear reactor is shown on the left of Figure 8.3.1. It receives, in its lower part, 102.57 kg/s of feedwater at 36.39 bar and 120.9°C. It exits at 32.74 bar and 283.5°C, superheated by approximately 45°C.

The steam then enters the HP turbine at 29.79 bar and 279.5°C, composed of four TSGs, and exits the first TSG at 2.33 bar and 125.1°C, with a quality of 0.9155. An extraction of 3.97 kg/s is used to heat feedwater heater LP3.

The remaining flow enters a second TSG, exiting at 1.22 bar and 105.3°C, with a quality of 0.9005. An extraction of 4.03 kg/s is used in part to heat feedwater heater LP2.

The remaining flow enters a third TSG, exiting at 0.59 bar and 85.5°C, with a quality of 0.8851. An extraction of 7.47 kg/s is used to heat feedwater heater LP1.

The remaining flow is expanded in the final TSG to a pressure of 0.07 bar (38.9°C) and a quality of 0.8599, before being directed to the condenser.

The steam expansion path on the Mollier diagram is shown in Figure 8.3.2.

8.3.3 Feedwater Circuit

At the outlet of the condenser, an extraction pump compresses the 102.57 kg/s of feedwater to a pressure of 15 bar, and then it is successively preheated in feedwater heater LP1 to 80.7°C, then in feedwater heater LP2 arranged with cascaded drain to 101.3°C, and finally in feedwater heater LP3 to 120.5°C, before being brought to a pressure of 38 bar by a secondary pump before entering the inlet valve of the SG.

In this model, we have not accounted for pressure losses in the feedwater piping, which remain low.

8.3.4 Model Analysis

The model we built using the manufacturer's data accurately provides all the points of the cycle, but the total power output is only 68.7 MW instead of 77 MW, without us being able to explain the origin of this difference, except if it stems from the simplification made compared to reality.

As you can see, the NuScale 77 is much simpler than the NuScale 50, and therefore likely less costly.

The available data do not provide information on condensate extractions in the various TSGs, although it is almost certain that they exist, as in the NuScale 50 circuit.

8.3.5 Performance Summary Table

Parameter	Value
Thermal power (MWth)	249.4
Electric power (MWe)	68.72
Cycle efficiency (%)	27.55
Feedwater pressure	38 bar
Feedwater temperature feed	121°C
SG outlet	32.7 bar, 283.5°C
Turbine inlet steam	29.8 bar, 279.5°C

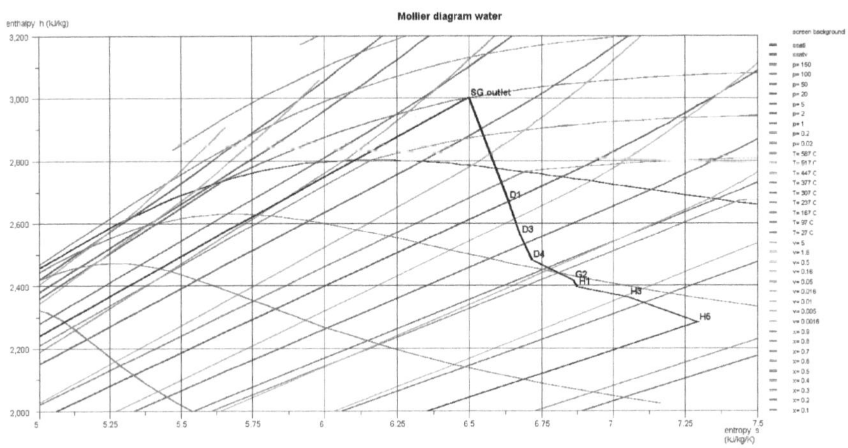

FIGURE 8.3.2 NuScale 77 MW cycle in the Mollier diagram.

8.4 DETAILED MODEL OF THE 1,350 MW ABWR NPP

Boiling Water Reactors and Advanced Boiling Water Reactors (ABWRs) use water as both a coolant and a moderator. The ABWR 1,350 MW NPP is a Generation III, forced circulation Boiling Water Reactor. In this technology, the water boils directly in the reactor core at a pressure of around 70 bar and a temperature close to 285°C, producing steam that drives a turbine without the need for a secondary circuit. A condenser and a pump complete the cycle.

See Section 2.2.6 for an overview of Boiling Water Reactor-type reactors

Safety features include multiple redundant Safety Control Rod Axe Man (SCRAM) systems and containment improvements.

The fuel comprises approximately 872 fuel assemblies, UO_2 fuel rods enriched around 3%–5% U-235.

8.4.1 Presentation and Sources

The data come from NRC Application for Certification document HitachiABWR_ML11126A128.pdf p. 71 and 72, (https://www.nrc.gov/docs/ML1112/ML11126A128.pdf, downloaded on 08/22/25 at 6:35 AM).

The U.S. ABWR design is similar to the international ABWR design, which was built at the Kashiwazaki Kariwa Nuclear Power Generation Station, Units 6 and 7 (K-6/7), by the Tokyo Electric Power Company, Inc (https://en.wikipedia.org/wiki/Kashiwazaki-Kariwa_Nuclear_Power_Plant), consulted on 08/22/25 at 6:35 AM.

This Boiling Water Reactor cycle involves 9 TSGs, 11 extractions, 2 reheaters in the moisture separator reheater (MSR), and 6 feedwater heaters: four LP simple exchangers (labeled LP1 to LP4 in Figure 8.4.1), and two HP exchangers with cascaded drains (HP1 and HP2 reheater).

The evaporator that produces sealing steam to prevent radioactive steam leakage was not modeled. An example will be provided in case study n° 5 on Reaktor Bolshoy Moshchnosti Kanalny (RBMK) reactors.

8.4.2 Steam Circuit

The nuclear reactor appears on the left side of Figure 8.4.1. It receives 2,190 kg/s of feedwater at 80.3 bar and 212.8°C. The output is at 70 bar and 285.8°C, with a steam quality of 0.9977.

Steam is then split into two flows:

- The main flow of 1,987 kg/s goes to the HP turbine.
- The secondary flow of 203.1 kg/s is used to heat the reheater.

The HP turbine consists of two TSGs.

- After the first TSG (inlet at 69.7 bar), the pressure drops to 24.4 bar with a quality of 0.9. An extraction of 120 kg/s supplies, along with the reheater condensate, feedwater heater HP6.

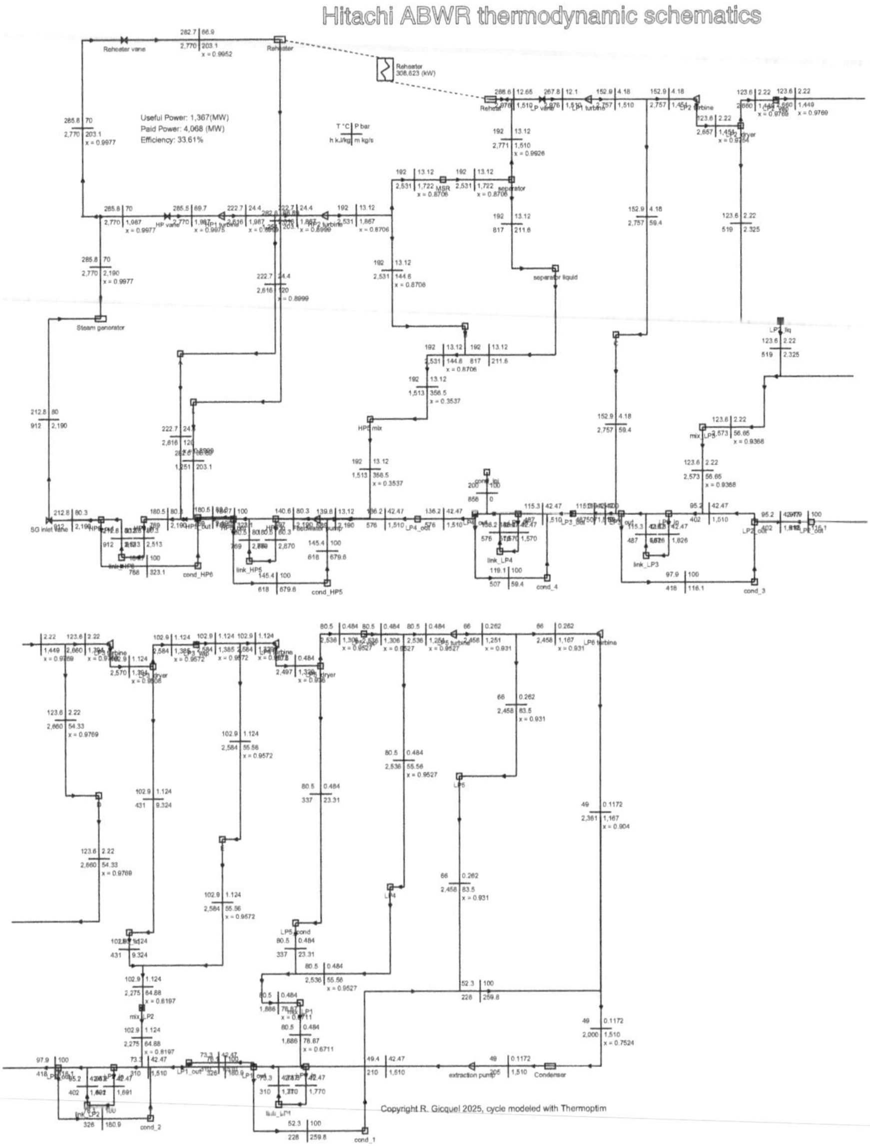

FIGURE 8.4.1 Detailed flowsheet of the 1,350 MW ABWR cycle.

- After the second HP TSG (outlet at 13.12 bar, 192°C, quality 0.8706), another extraction of 144.8 kg/s supplies steam to feedwater heater HP5, along with 211.6 kg/s of liquid phase from the separator.

The main flow then enters the MSR separator with 95% efficiency. Steam enters the reheater at 192°C and exits at 268.6°C, at a pressure of 12.65 bar.

It then enters the LP turbine composed of six TSGs:

- After the first TSG (152.9°C, 4.185 bar), a 59.4 kg/s extraction heats feedwater heater LP4.
- After the second TSG (2.22 bar, 123.6°C, quality 0.9754), there are two extractions: 2.32 kg/s of condensate and 54.3 kg/s of steam for feedwater heater LP3.
- After the third TSG (1.124 bar, 102.9°C, quality 0.9572), extractions of 9.32 kg/s of condensate and 55.56 kg/s of steam heat feedwater heater LP2.
- After the fourth TSG (0.484 bar, 80.5°C, quality 0.936), extractions of 2.5 kg/s of condensate and 83.5 kg/s of steam heat feedwater heater LP1.
- After the fifth TSG (0.262 bar, 66°C, quality 0.931 condensate and steam are directed to the condenser (not modeled in detail due to negligible impact).
- After the sixth TSG (0.1172 bar, 49°C), the flow enters the condenser.

In the Mollier diagram shown in Figure 8.4.2, the steam expansion path appears as a solid line for the detailed model and as a dashed line for NUSCLE.

8.4.3 Feedwater Circuit

At the condenser outlet, two recovery exchangers (not modeled) raise the water temperature by about 2°C. An extraction pump then compresses the 1,510 kg/s of feedwater to 42.74 bar.

The feedwater is heated sequentially:

- Feedwater heater LP1: 73.3°C
- Feedwater heater LP2: 95.2°C
- Feedwater heater LP3: 115.3°C
- Feedwater heater LP4: 136.2°C

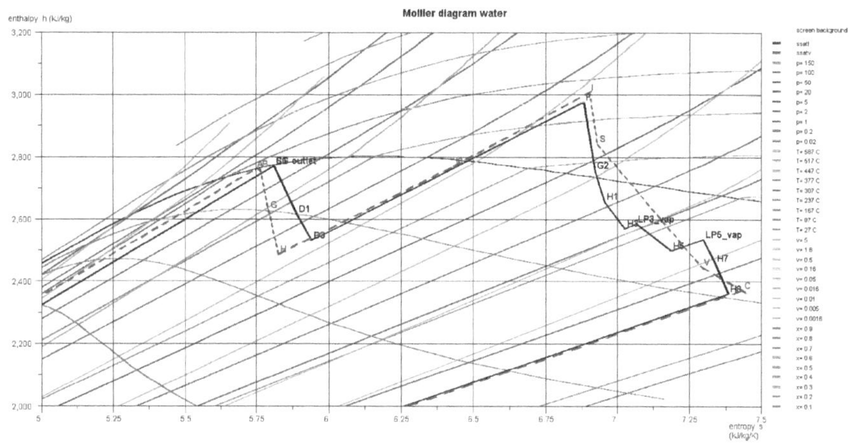

FIGURE 8.4.2 ABWR cycle in the Mollier diagram.

It is then mixed with condensate from feedwater heater HP1, reaching 139.8°C, and pressurized to 80.32 bar by the feedwater pump. It is then routed through both HP feedwater heaters, exiting at 212.8°C before entering the SG.

Compared to the official flowsheet, the initial 2°C reheating from the recovery exchangers is not included.

8.4.4 Performance Summary Table

Parameter	Value
Thermal power	3,946 MWth
Electric power	1,374 MWe
Cycle efficiency (%)	33.61
Steam pressure	~68 bar
Main steam temperature	~285°C
Steam quality at turbine inlet	~0.996

8.5 DETAILED MODEL OF THE 1,000 MW RBMK NPP

RBMK (in Russian: Reaktor Bolshoy Moshchnosti Kanalny, "high-power channel-type reactor") are boiling water-cooled, channel-type, graphite-moderated boiling water reactors primarily used in Russia. They are infamous due to the accident involving a reactor of this type at Chernobyl in 1986.

See Section 2.2.7 for an overview of RBMK-type reactors.

The fuel is naturally or slightly enriched uranium oxide (1.8%), fuel rods ~7 m long in cylindrical assemblies.

Its moderator is graphite bricks surrounding each pressure tube.

Specific features are on-load refueling, individual vertical pressure tube design, and large positive void coefficient typical of this design.

8.5.1 Presentation and Sources

The model presented here corresponds to schematic 15.3 on p. 193 of the book *Margulova and Podushko* (1982).

It features five LP feedwater heaters and one medium-pressure heater (MP feedwater heater).

One of the characteristics of RBMK reactors is the absence of HP feedwater heaters, initially decided for safety reasons related to corrosion. This choice was later reconsidered, but the Chernobyl accident led to the abandonment of this reactor type.

The model we constructed includes the five LP feedwater heaters and one MP feedwater heater as such, with the other two, corresponding to the outlets of the two reheaters, replaced by mixers, the first directly in the feedwater tank and the second at the level of feedwater heater LP5.

It also includes the evaporator designed to provide pure steam to ensure the sealing of the joints of the turbines and pumps, thus preventing the release of radioactive gases and steam exhaust. Its consumption is relatively low, but its role is significant enough to justify its inclusion.

A minor difference is that we connected the outlet of this evaporator directly to the condenser rather than to feedwater heater LP4, as the steam escaping from it is condensed according to the schematic. It would be easy to modify the cycle if desired.

The thermodynamic data come from schematic 5.2 on p. 55 of the aforementioned document, as well as from table P1 in Chapter 18, p. 239. It corresponds to the K-500-65-3000 turbine, with two such turbines providing a total nominal capacity of 1,000 MW.

The available data include the main flow rate, the extraction pressures, and the quality values at the outlet, which provide the enthalpy and thus the polytropic efficiencies of the turbines. Note, however, that we do not have the values for the extraction flow rates, which we have therefore estimated.

This cycle involves eight TSGs, eight extractions, two reheaters in the MSR, and seven feedwater heaters, five LP of the simple exchanger type (noted LP1 to LP5 in Figure 8.5.1), and one MP, of the exchanger with cascaded drains (MP1 reheater).

8.5.2 Steam Circuit

The nuclear reactor appears on the left of Figure 8.5.1. It receives, in the lower part, 1,495 kg/s of feedwater at 68 bar and 166°C. It exits at 66 bar and 281.8°C, with a quality of 0.995, and then is slightly expanded before entering the HP turbine at 64.5 bar, 280.3°C, and a quality of 0.9794.

The steam is then divided into two flows: the main flow of 1,369 kg/s, directed to the HP turbine, and a secondary flow of 125.9 kg/s, which is used to heat the second reheater.

The HP turbine is composed of four TSGs. At the outlet of the first, at 20.55 bar, 213.8°C, and a quality of 0.9, an extraction of 58.32 kg/s is performed to supply the first reheater.

At the outlet of the second HP TSG, at a pressure of 11.55 bar, 186.2°C, and a quality of 0.878, a new extraction of 15 kg/s is used to supply steam to feedwater heater MP1.

The main flow enters the third HP TSG, exiting at a pressure of 6.32 bar, 160.9°C, and a quality of 0.86. A new extraction of 86.2 kg/s is used to supply steam to feedwater heater LP5.

The main flow enters the fourth HP TSG, exiting at a pressure of 3.48 bar, 138.7°C, and a quality of 0.8486. A new extraction of 38.3 kg/s is used to supply steam to feedwater heater LP4.

The main flow then enters the separator of the MSR. The liquid phase (170.5 kg/s) also joins feedwater heater LP4, while the vapor phase enters at 138.7°C in the first reheater, exiting at 184.4°C before entering the second reheater, which raises it to 265.4°C at a pressure of 3.13 bar.

The main flow enters the first LP TSG, exiting at 1.42 bar and 189°C. An extraction of 50 kg/s is used to heat feedwater heater LP3.

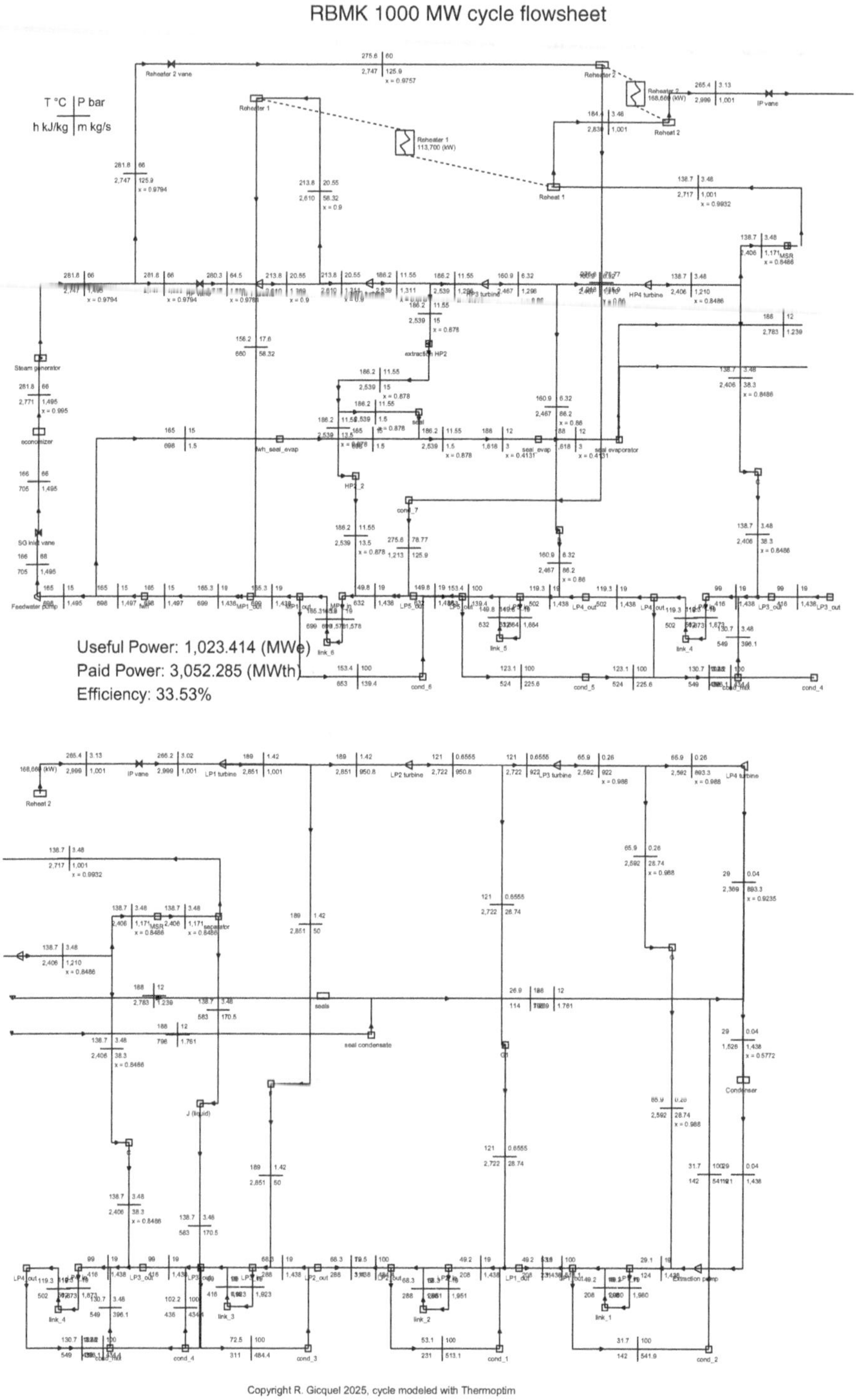

FIGURE 8.5.1 Detailed flowsheet of the 1,000 MW RBMK cycle.

The main flow enters the second LP TSG, exiting at 0.6555 bar and 121°C. An extraction of 28.74 kg/s is used to heat feedwater heater LP2.

The main flow enters the third LP TSG, exiting at 0.26 bar and 65.9°C with a quality of 0.988. An extraction of 28.74 kg/s is used to heat feedwater heater LP1.

The remainder passes through the fourth LP TSG and exits the turbine at a pressure of 0.04 bar, with a quality of 0.9235.

The condensates from the evaporator used to seal the joints and those from feedwater heater LP1 are mixed with the expanded steam before entering the condenser, from which they exit in the liquid state at 29°C.

In the Mollier diagram shown in Figure 8.5.2, the steam expansion path appears as a solid line for the detailed model and as a dashed line for NUSCLE.

8.5.3 Feedwater Circuit

All the feedwater heaters in this system are arranged with cascaded drains.

At the condenser outlet, a condensate pump compresses the 1,438 kg/s of feedwater to a pressure of 19 bar, and then it is heated in feedwater heater LP1 to 49.2°C, then in feedwater heater LP2 to 68.3°C, then in feedwater heater LP3 to 99°C, in feedwater heater LP4 to 119.3°C, in feedwater heater LP5 to 149.8°C, and finally in feedwater heater MP1 to 165.3°C.

After being mixed with the condensed steam from the first reheater, an extraction of 1.5 kg/s is directed to an evaporator to provide steam for sealing the joints to prevent radioactive water from escaping the installation. The remaining feedwater is then raised to a pressure of 68 bar by the feedwater pump before entering the SG.

The cycle is thus completed.

With these data, for a main flow rate of 1,495 kg/s, we arrive at this flowsheet.

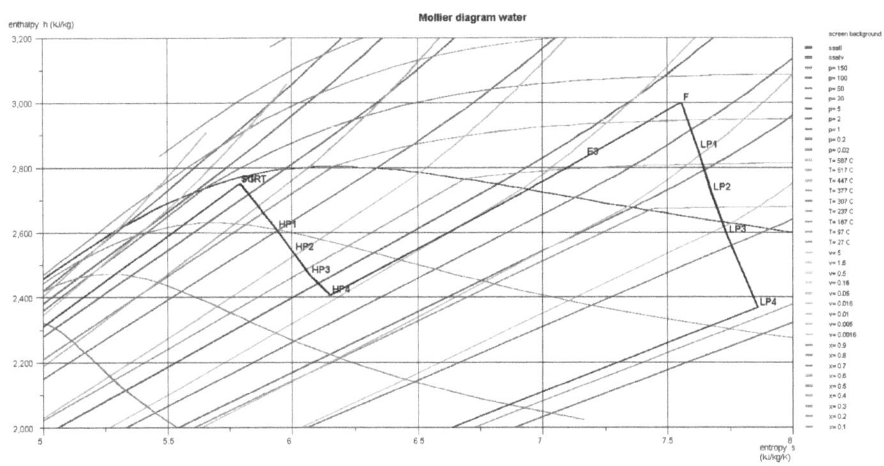

FIGURE 8.5.2 RBMK cycle in the Mollier chart.

8.5.4 Performance Summary Table

Parameter	Value
Thermal power (approx.)	3,050 MWth
Electrical power nominal	1,023 MWe
Cycle efficiency (%)	33.53
Coolant pressure at the core exit	~66 bar
Coolant temperature at the core exit	~282°C
Steam quality at the turbine inlet	~0.98

8.6 DETAILED MODEL OF THE 1,000 MW VVER NPP

VVERs (Vodo-Vodyanoi Energetichesky Reactors), also known as Russian pressurized water reactors, are used in several countries, including Russia and other former Soviet Union countries, as well as in some countries in Eastern Europe and Asia.

See Section 2.2.5 for an overview of VVER-type reactors.

Its fuel is uranium dioxide fuel rods enriched 3.3%–4.4%, arranged in 163 hexagonal assemblies.

It uses PGV-1000 horizontal once-through SGs.

8.6.1 Presentation and Sources

The schematic of the cycle is provided on p. 247 of the book (in Russian): Troïanovski et al. (1985).

The thermodynamic data for the 1982 turbine type K-1000-60/1500-2 used are available (in Russian) at this address: https://studfile.net/preview/8923253/page:2/.

A minor difference in our model is that we connected the turbine driving the feedwater pump directly to the condenser rather than to feedwater heater LP4, as the steam escaping from it is condensed according to the schematic. It would be easy to modify the cycle if desired.

This cycle involves eight TSGs, eight extractions, two reheaters in the MSR, and seven feedwater heaters, four LP of the simple exchanger type (noted LP1 to LP4 in Figure 8.6.1), and three HP, of the exchanger with cascaded drains (HP1 and HP2 reheater).

8.6.2 Steam Circuit

The nuclear reactor appears on the left of Figure 8.6.1. It receives, in the lower part, 1,750 kg/s of feedwater at 91.5 bar and 220.7°C. It exits at 65 bar and 280.8°C, with a quality of 0.9993, and then is expanded before entering the HP turbine at 58.8 bar, 274.2°C, and a moisture content of 0.5%.

The steam is then divided into two flows: the main flow of 1,675 kg/s, directed to the HP turbine, and a secondary flow of 74.81 kg/s, which is used to heat the second reheater.

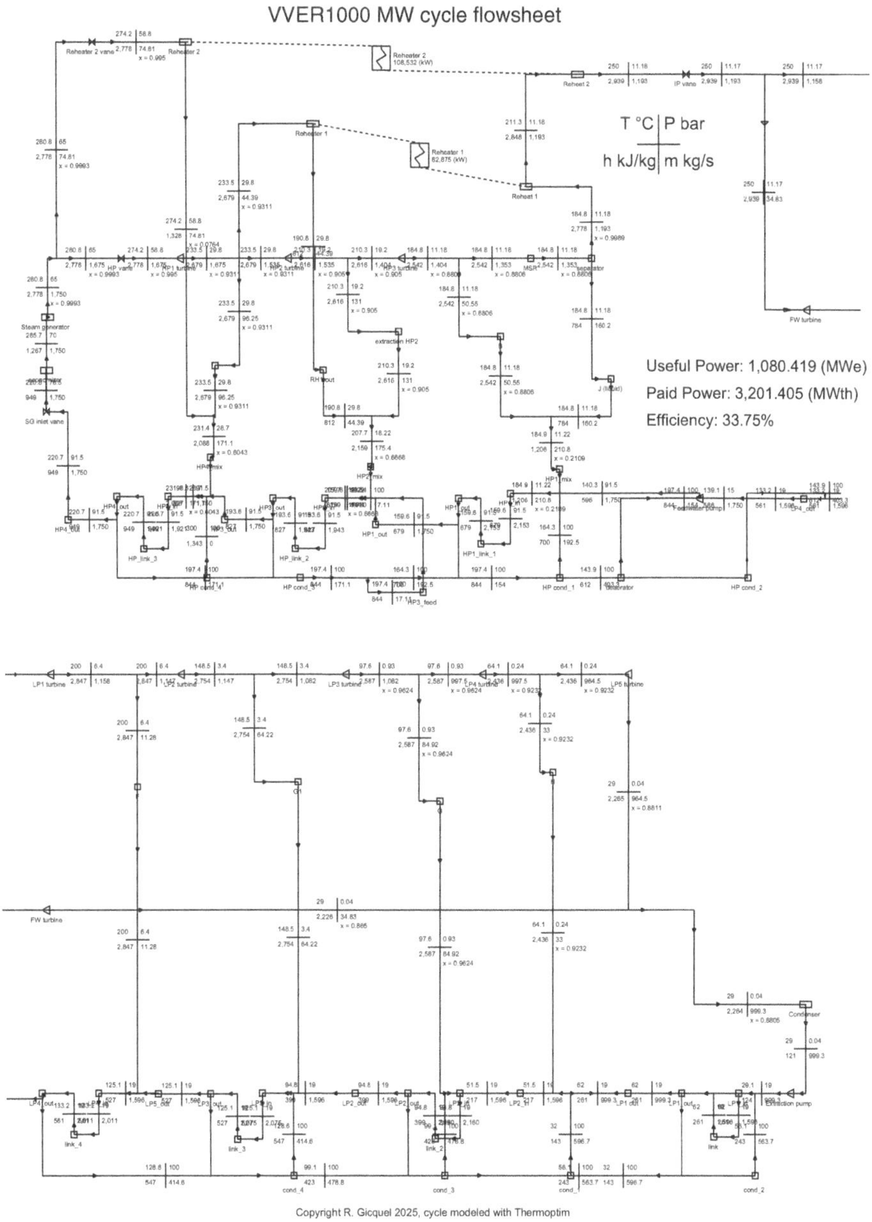

FIGURE 8.6.1 Detailed flowsheet of the 1,000 MW VVER cycle.

The HP turbine is composed of three TSGs. At the outlet of the first, at 29.8 bar, 233.5°C and a quality of 0.9311, two extractions are performed: one of 44.39 kg/s to supply the first reheater, and another of 96.25 kg/s to supply steam to feedwater heater HP3.

At the outlet of the second HP TSG, at a pressure of 19.2 bar, 210.3°C and a quality of 0.905, a new extraction of 131 kg/s is used to supply steam to feedwater heater HP2.

At the outlet of the HP turbine, at a pressure of 11.18 bar, 184.8°C and a quality of 0.8806, an extraction of 50.55 kg/s is used to supply steam to feedwater heater HP1.

The main flow then enters the separator of the MSR. The liquid phase (160.2 kg/s) also joins feedwater heater HP1, while the vapor phase enters at 184.8°C in the first reheater, exiting at 211.3°C before entering the second reheater, which raises it to 250°C at a pressure of 11.18 bar.

An extraction of 34.83 kg/s is used to supply the turbine driving the feedwater pump.

The main flow enters the first LP TSG, exiting at 6.4 bar and 200°C. An extraction of 11.28 kg/s is used to heat feedwater heater LP4.

The main flow enters the second LP TSG, exiting at 3.4 bar and 148.5°C. An extraction of 64.22 kg/s is used to heat feedwater heater LP3.

The main flow enters the third LP TSG, exiting at 0.93 bar and 97.6°C with a quality of 0.9624. An extraction of 84.92 kg/s is used to heat feedwater heater LP2.

The main flow enters the fourth LP TSG, exiting at 0.24 bar and 64.1°C with a quality of 0.9232. An extraction of 33 kg/s is used to heat feedwater heater LP1.

The remainder passes through the fifth TSG and exits the LP turbine at a pressure of 0.04 bar, with a quality of 0.8811.

The condensates from the turbine driving the feedwater pump are mixed with the expanded steam before entering the condenser, from which they exit in the liquid state at 29°C.

In the Mollier diagram shown in Figure 8.6.2, the steam expansion path appears as a solid line for the detailed model and as a dashed line for NUSCLE.

8.6.3 Feedwater Circuit

All the feedwater heaters in this system are arranged with cascaded drains.

At the condenser outlet, a condensate pump compresses the 999.3 kg/s of feedwater to a pressure of 19 bar, and then it is heated in feedwater heater LP1 to 62°C, then

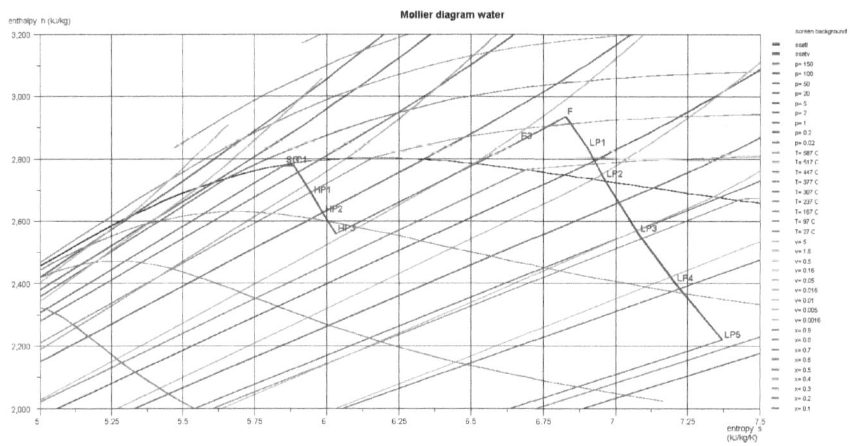

FIGURE 8.6.2 VVER 1000 cycle in the Mollier chart.

in feedwater heater LP2 to 94.8°C, then in feedwater heater LP3 to 125.1°C, and in feedwater heater LP4 to 133.2°C. This flow is mixed with the condensates from feedwater heater HP1 in the feedwater tank; then, the total flow of 1,750 kg/s at 139.1°C is raised to a pressure of 91.5 bar by the feedwater pump.

The pressurized feedwater is directed successively to the three HP feedwater heaters, exiting at 159.6°C, then 193.6°C, and finally 220.7°C, before entering the SG.

The cycle is thus completed.

In this model, we have not accounted for pressure drops in the feedwater piping, which remain low.

The overall balance is summarized in the box located in the upper left part of the flowsheet.

With these data, we obtain this schematic, for a main flow rate of 1,680 kg/s. The only changes we had to make were to slightly adjust the flow rates of the last two extractions, transferring part of the last flow to the second-to-last. Without this, feedwater heater LP1 could not function.

8.6.4 Performance Summary Table

Parameter	Value
Thermal power (MWth)	~3,200
Electric power (MWe)	~1,080
Cycle efficiency (%)	33.75
Secondary feedwater pressure	~70 bar
Secondary feedwater temperature	~220°C
Steam temperature (secondary)	~280°C

8.7 DETAILED MODEL OF THE 550 MW CANDU NPP

Pressurized Heavy Water Reactors like Canadian Deuterium Uranium (CANDU) utilize heavy water as both a moderator and a coolant. Thanks to the efficient moderation provided by heavy water, these reactors can use natural uranium as fuel.

See Section 2.2.10 for an overview of CANDU-type reactors.

8.7.1 Presentation and Sources

Moderation and cooling are made by heavy water (D_2O) used both as a moderator and a primary coolant.

The fuel is natural uranium dioxide (UO_2) in bundles, 37 elements per bundle, arranged in multiple channels.

Separate LP moderator and HP primary coolant loops contribute to unique thermal hydraulics.

Safety systems include two independent fast-acting shutdown systems and inherent safety features due to core design.

The data related to the Pickering plant come from page 130 of a Canadian report on these plants (OH-PickeringDD1969.pdf, Pickering Generating Station, Hydroelectric Commission of Ontario Atomic Energy of Canada Limited Power Projects, 1969).

https://canteach.candu.org/Content%20Library/OH-PickeringDD1969.pdf

This cycle involves 7 TSGs, 11 extractions, 2 reheaters in the MSR, and 6 feedwater heaters, all of the exchanger with condensate type (noted LP1 to LP4 and HP1 and HP2 reheater).

8.7.2 Steam Circuit

The nuclear reactor appears on the left of Figure 8.7.1. It receives, in the lower part, 813.9 kg/s of feedwater at 49 bar and 171.1°C. It exits at 45 bar and 257.4°C, with a quality of 0.9986.

The steam is then divided into two flows: the main flow of 752.2 kg/s, directed to the HP turbine, and a secondary flow of 61.7 kg/s, which is used to heat the reheater.

The HP turbine is composed of two TSGs, the first being supplied with steam at 40.33 bar, 250.8°C, and a quality of 0.9972. At the outlet of the first, at 8.649 bar, a condensate extraction of 11.5 kg/s and a steam extraction of 40.2 kg/s are performed to supply feedwater heater HP2.

At the outlet of the second HP TSG, at a pressure of 4.847 bar and a quality of 0.8929, the main flow enters the separator of the MSR. The vapor phase undergoes an extraction of 20.9 kg/s before entering at 150.7°C in the reheater, exiting at 233.9°C, at a pressure of 4.6 bar. The liquid phase (71.3 kg/s) and the extraction are mixed to supply feedwater heater HP1.

The main flow then enters the LP turbine composed of five TSGs. At the outlet of the first, at a temperature of 175°C and a pressure of 2.62 bar, an extraction of 22.67 kg/s is used to heat the feedwater tank.

The main flow enters the second TSG, exiting at 1.376 bar and 116.8°C. A steam extraction of 31 kg/s is used to heat feedwater heater LP3.

The main flow enters the third TSG, exiting at 0.655 bar and 88.2°C, with a quality of 0.9732. A steam extraction of 26 kg/s is used to heat feedwater heater LP2.

The main flow enters the fourth TSG, exiting at 0.262 bar and 86.2°C, with a quality of 0.9408. A condensate extraction of 12.2 kg/s and a steam extraction of 28.5 kg/s are performed to heat feedwater heater LP1.

The main flow enters the fifth TSG, exiting at 0.051 bar and 33.3°C with a quality of 0.921 before entering the condenser.

In the Mollier diagram shown in Figure 8.7.2, the steam expansion path appears as a solid line for the detailed model and as a dashed line for NUSCLE.

8.7.3 Feedwater Circuit

At the condenser outlet, a condensate pump compresses the 582 kg/s of feedwater to a pressure of 19 bar; then, it is heated in feedwater heater LP1 to 64.4°C, then in feedwater heater LP2 to 87°C, and in feedwater heater LP3 to 107.7°C.

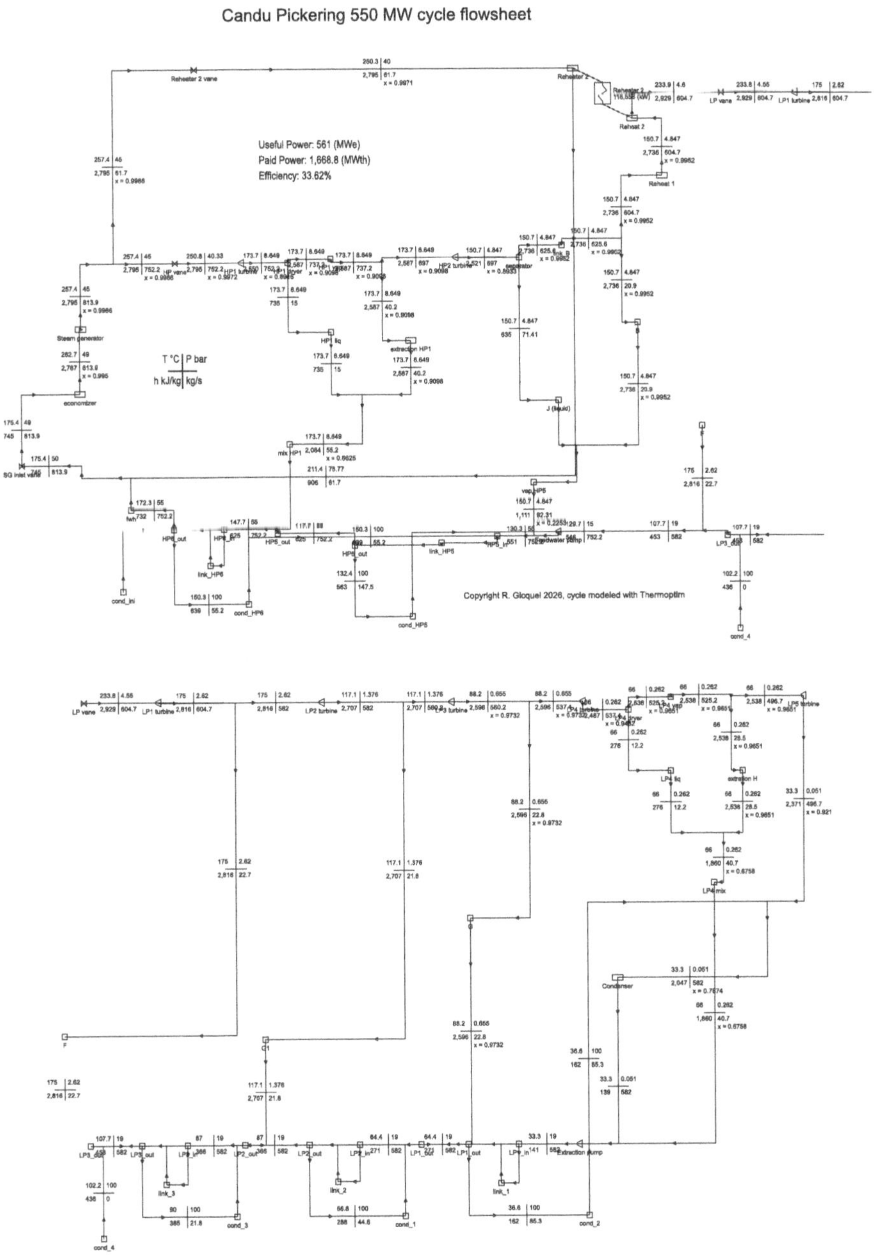

FIGURE 8.7.1 Detailed flowsheet of the 550 MW CANDU Pickering cycle.

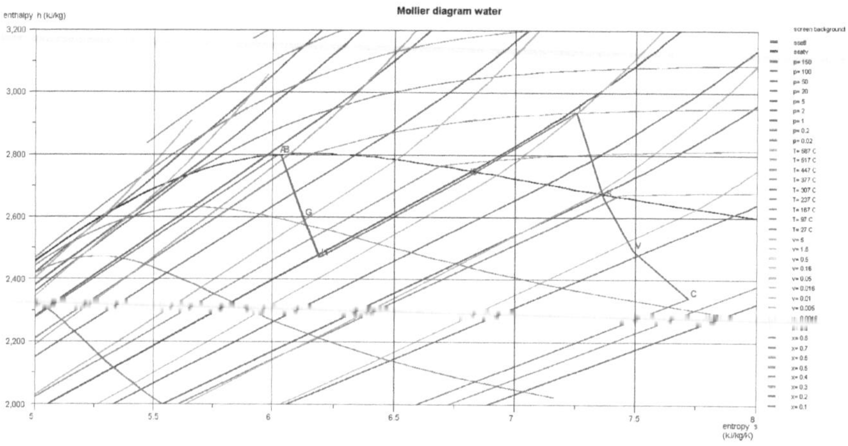

FIGURE 8.7.2 CANDU 550 cycle in the Mollier chart.

It is then mixed in the feedwater tank with the first LP extraction, raising it to 168.2°C. It is then raised to a pressure of 55 bar by the feedwater pump, and directed successively to the two HP feedwater heaters, exiting at 175.4°C before entering the SG.

It is interesting to note that, while both perform HP condensate extractions, which is rarely the case, the extractions at this level in this cycle are slightly different from those of the ABWR.

8.7.4 Performance Summary Table

Parameter	Value
Thermal power (MWth)	1,669
Electrical power (MWe)	561
Cycle efficiency (%)	33.62
Steam production pressure/temp	45 bar, 257.4°C
Steam quality at the turbine inlet	~0.9986

8.8 DETAILED MODEL OF THE 1,175 MW SUPERPHÉNIX NPP

Sodium-Cooled Fast Reactors represent a category of Fast Neutron Reactors specifically utilizing liquid sodium as a cooling fluid.

Superphénix was a French liquid-metal fast breeder reactor located at the Creys-Malville nuclear site. Decommissioned permanently for political reasons in 1997, it was originally conceived as a prototype sodium-cooled fast neutron reactor, following on from the earlier experimental facilities Phénix and Rapsodie.

See Section 2.2.11 for an overview of sodium-cooled fast reactor-type reactors.

Superphénix was the largest sodium-cooled fast breeder reactor ever operated, pioneering liquid sodium cooling with a pool-type design allowing large thermal inertia and stable operation.

8.8.1 Presentation and Sources

Technically, Superphénix had a thermal power output of 3,000 MW and an electrical output of 1,174 MW. The reactor core contained approximately 5.7 tons of plutonium, surrounded by steel reflectors for neutron economy. The fuel was mixed oxide containing approximately 80% uranium-238 fertile and 20% plutonium-239 fissile, distributed in two enrichment zones to optimize power distribution and neutron flux: about 5 tons of plutonium-239 in the core, surrounded by a uranium-238 blanket for breeding.

The reactor utilized liquid sodium as the primary coolant, circulating through an intermediate sodium loop to transfer heat safely to SGs—thereby preventing contamination in the event of leaks.

Superphénix represented the culmination of technical advancements in fast breeder technology in France, aiming at closed fuel cycle operations and advanced plutonium management; however, it faced challenges related to sodium technology and economic feasibility.

The selected model is based on the safety report of the French Nuclear Safety Authority (ASN).

The plant presented two distinctive features compared to those studied so far in this chapter (Figure 8.8.1):

1. It implemented two sodium cooling circuits for radioactive safety reasons.
2. The steam exiting the SG was highly superheated.

We only detail one of the two steam circuits here.

The SG, modeled here in three parts—economizer, evaporator, and superheater—receives, in the lower part, 680.7 kg/s of feedwater at 240 bar and 236.6°C. It exits as steam at 210 bar and 487°C.

8.8.2 Steam Circuit

The HP turbine is composed of four TSGs. At the outlet of the first, at 85.71 bar and 384°C, an extraction of 71.47 kg/s is performed to supply the reheater.

At the outlet of the second HP TSG, at a pressure of 35.95 bar and 269.3°C, an extraction of 37.09 kg/s is used to partially supply feedwater heater HP2.

At the outlet of the third HP TSG, at a pressure of 17.21 bar and 204.9°C with a quality of 0.9784, an extraction of 39.16 kg/s is used to supply feedwater heater HP1.

At the outlet of the fourth HP TSG, at a pressure of 6.82 bar and 163.9°C with a quality of 0.9262, an extraction of 48 kg/s is used to supply the feedwater tank.

A second extraction of 6.7 kg/s supplies feedwater heater LP2.

The main flow then enters the separator of the MSR. The liquid phase (33.55 kg/s) supplies feedwater heater LP2, while the vapor phase enters at 163.9°C in the reheater, exiting at 277°C, at a pressure of 6.237 bar.

An extraction of 32.88 kg/s is used to drive the feedwater pump turbine.

The main flow then enters the LP turbine composed of three TSGs. At the outlet of the first, at a temperature of 147.6°C and a pressure of 1.765 bar, an extraction of 30 kg/s is used to heat feedwater heater LP2.

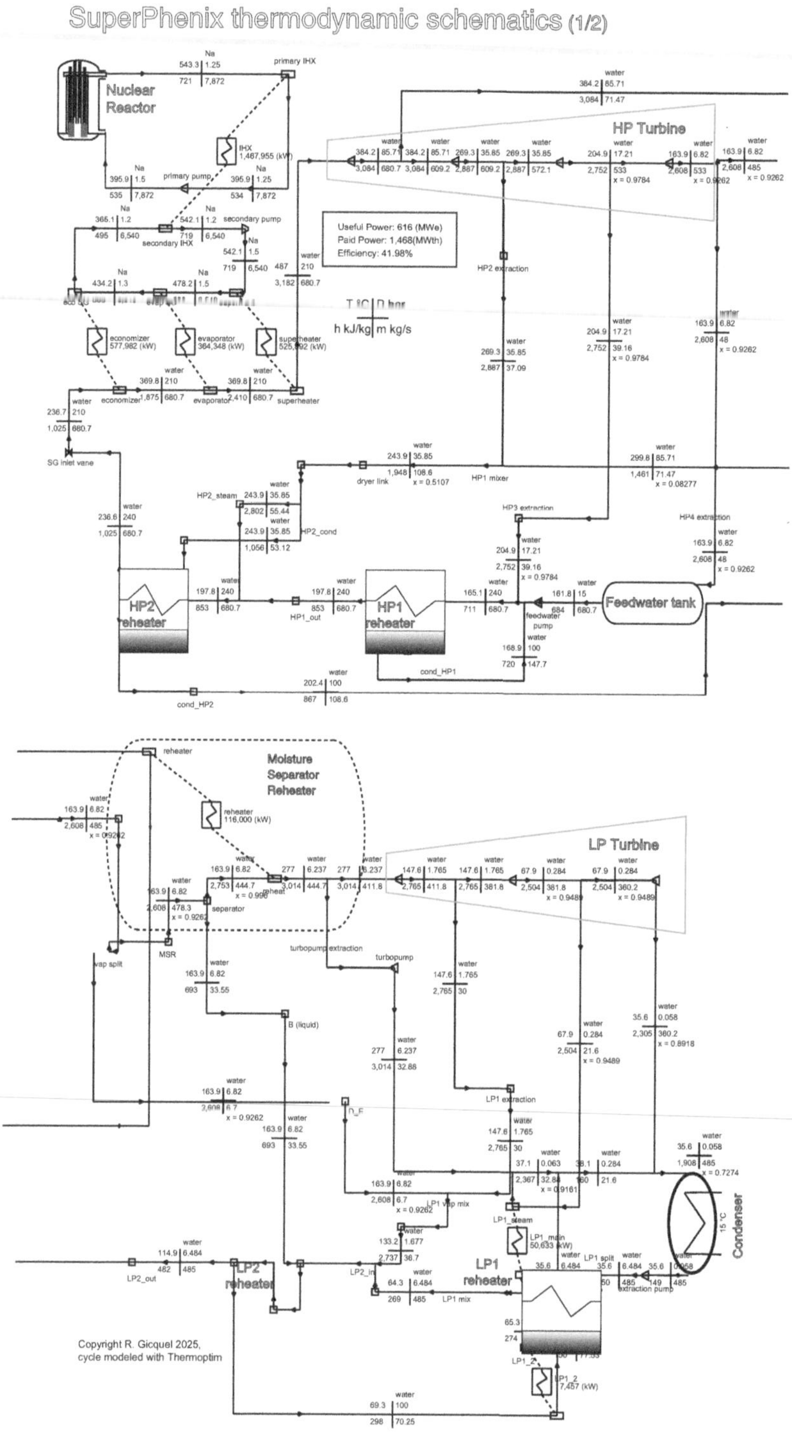

FIGURE 8.8.1 Detailed flowsheet of the 1,175 MW Superphénix cycle.

The main flow enters the second LP TSG, exiting at 0.284 bar and 67.9°C with a quality of 0.9489. A steam extraction of 21.6 kg/s is used to heat feedwater heater LP1.

The main flow enters the third TSG, exiting at 0.058 bar and 35.6°C with a quality of 0.8918, before entering the condenser.

The steam expansion path on the Mollier diagram is shown in Figure 8.8.2.

8.8.3 Feedwater Circuit

At the condenser outlet, a condensate pump raises the 485 kg/s of feedwater to a pressure of 6.484 bar, and then the majority (407.5 kg/s) is heated in feedwater heater LP1 to 64.3°C, while the remainder passes through an economizer to recover heat from the condensates. After remixing, its temperature is 69.6°C. It then enters feedwater heater LP2, exiting at 114.9°C, and is directed to the feedwater tank.

It is then mixed in the feedwater tank with the last HP extraction and the condensates from feedwater heater HP1, raising it to 161.8°C, which are compressed to a pressure of 240 bar by the feedwater pump, and then directed successively to the two HP feedwater heaters, exiting at 236.6°C before entering the SG.

8.8.4 Performance Summary Table

Parameter	Value
Thermal power	2,936 MWth
Electric power	1,230 MWe
Cycle efficiency (%)	42
Primary coolant	Liquid sodium, ~543°C
SG pressure	210 bar
Steam temperature	487°C

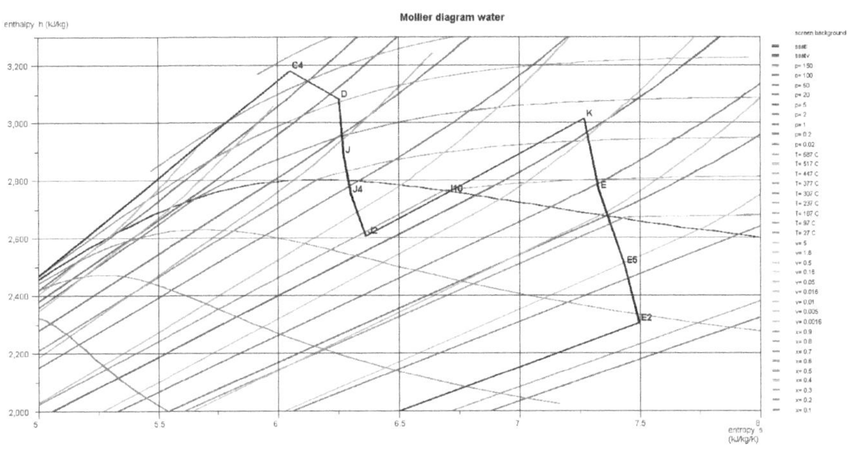

FIGURE 8.8.2 Superphénix cycles in the Mollier diagram.

8.9 DETAILED MODEL OF THE 67 MW HTR-PM NPP

The HTR-PM power plant, located in Shidao Bay, Shandong, is based on a high-temperature gas-cooled reactor coupled with a Rankine steam cycle for electricity generation.

It is the first modular high-temperature gas-cooled reactor commercial demonstration plant, combining inherent safety, modular construction, and high-temperature operation suited for cogeneration and hydrogen production.

8.9.1 Presentation and Sources

As explained in Section 4.1.6.2, it is an intrinsically safe SMR. You should refer to this section for an overall presentation.

Its reactor core is comprised of pebble bed of about 60 mm diameter fuel pebbles, each containing 7 g of uranium enriched to about 9% U-235, surrounded by the graphite moderator region. The pebble bed fuel design allows continuous refueling and high burn-up.

Safety features are based on passive decay heat removal, inherent safety due to graphite moderation and high heat capacity.

Its SG design follows proven once-through helical coil heat exchanger technology.

Very little information is available about its thermodynamic cycle. The only data available lead to much lower performance for this prototype plant: 67 MW for the first of the two modules powering a single turbine, and an efficiency of about 32.5%.

This information was deduced from the data provided in the article by Dong et al. (2023).

Specifically, we relied on Figure 5 of this paper, which presents the Human–Machine Interface of the coordinated control system of the actual plant. This operational dashboard contains a wide range of thermodynamic data, which made it possible to construct the model we present here. It should be noted, however, that this model has not been validated by the original authors. Despite several requests, it has proven impossible to obtain direct information on the secondary (steam) circuit.

The data displayed on the control screen of the plant are as follows: helium, steam and feedwater temperatures, core power, electric power, efficiency, power level, feedwater, and helium flow rates.

We were unable to find the officially announced values for the last three parameters, but all the others are available.

It is certain that the installation that was the subject of the measurements reported in this article is a first of its kind and that it is normal for its performance to be poorer than expected, but the discrepancies are still significant, as shown in Table 8.9.1.

We were able to build a model that is perfectly consistent with these results, even if, in the absence of published data, the pressures and flow rates of the extractions were assumed, as well as a polytropic efficiency of 70% for all the turbines.

This value may seem low, but it alone allows us to find the announced power of 64 MWe (in our model, we estimated this power at 67 MW to account for potential losses). Perhaps these low values come from the power level, which is only about 80% (Figure 8.9.1).

TABLE 8.9.1
Discrepancies between the Values Reported and Those Recorded on the Human–Machine Interface

	Announced	Actual	Relative Difference (%)
T helium (°C)	750	563.7	24.84
Core power (MWth)	250	197.5	21.00
Electric power (MWe)	105	64	39.05
T steam (°C)	566	519.9	
Efficiency (%)	40	34.11	14.73
Power level (%)	100	79.5	20.5
T feedwater (°C)		159.9	
m feedwater (kg/s)		72.7	
m He (kg/s)		119.4	

The cycle performance summary is as follows:

Thermal Power (MWth)	198.6
Electric power (MWe)	67
Efficiency (%)	33.76

The HTR-PM plant presents two distinctive features compared to others studied in this chapter:

1. It implements a helium cooling circuit.
2. The steam exiting the SG is highly superheated.

The SG, modeled here as a single unit, receives, in the lower part, 72.7 kg/s of feedwater at 125 bar and 159.6°C. It exits at 110.8 bar and 519.9°C.

The feedwater train comprises three LP reheaters and 1 HP.

8.9.2 Steam Circuit

The turbine is composed of six TSGs. At the outlet of the first, at 30 bar and 360.3°C, an extraction of 3.1 kg/s is performed to supply feedwater heater HP4.

At the outlet of the second HP TSG, at a pressure of 8 bar and 233.3°C, an extraction of 3.1 kg/s is used to supply the feedwater tank.

At the outlet of the third TSG, at a pressure of 2.1 bar and 129.9°C, an extraction of 3.6 kg/s is used to supply feedwater heater LP3.

At the outlet of the fourth TSG, at a pressure of 0.56 bar and 84.2°C with a quality of 0.9659, an extraction of 3 kg/s is used to supply feedwater heater LP2.

At the outlet of the fifth TSG, at a pressure of 0.15 bar and 54°C with a quality of 0.9324, an extraction of 3.5 kg/s is used to supply feedwater heater LP1.

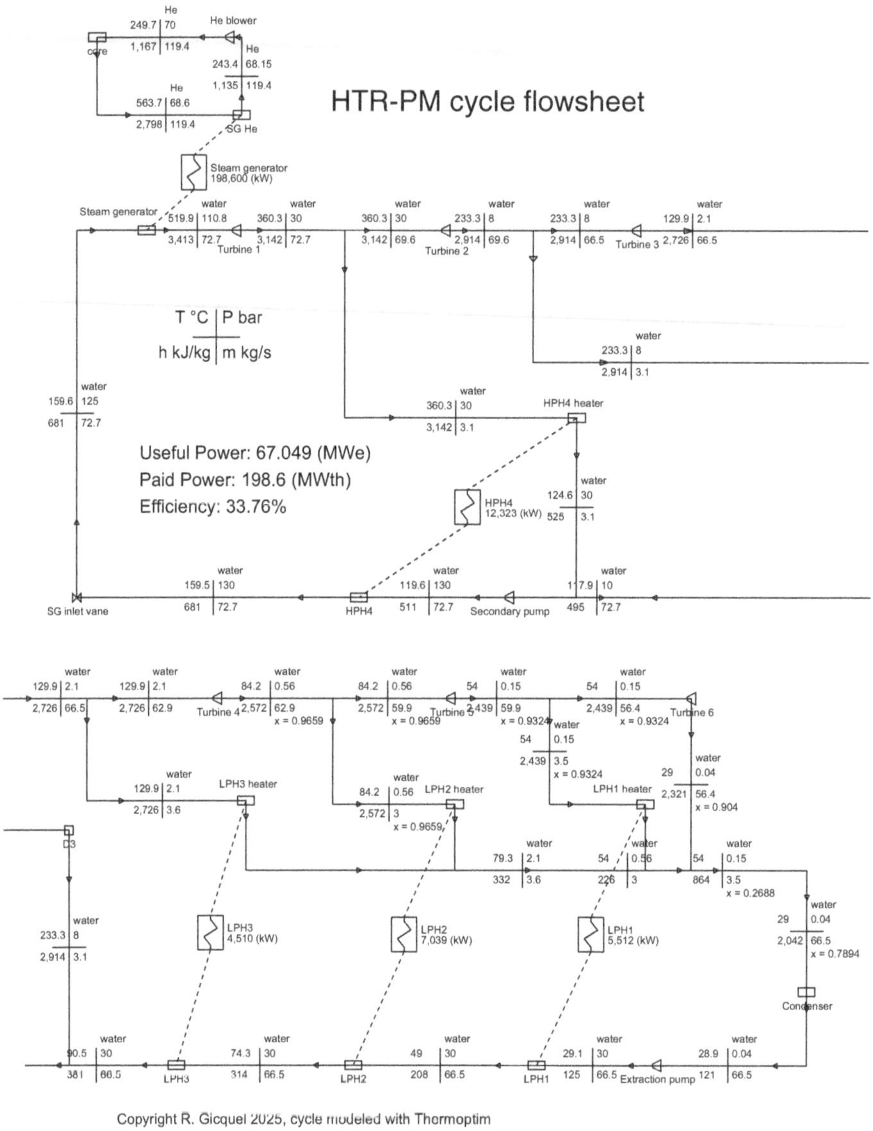

FIGURE 8.9.1 Detailed model of the 67 MW HTR-PM cycle.

The main flow enters the sixth TSG, exiting at 0.04 bar and 29°C with a quality of 0.904, before entering the condenser.

The steam expansion path on the Mollier diagram is shown in Figure 8.9.2.

8.9.3 Feedwater Circuit

At the condenser outlet, a condensate pump compresses the 66.5 kg/s of feedwater to a pressure of 30 bar. This feedwater then enters feedwater heater LP1, exiting at

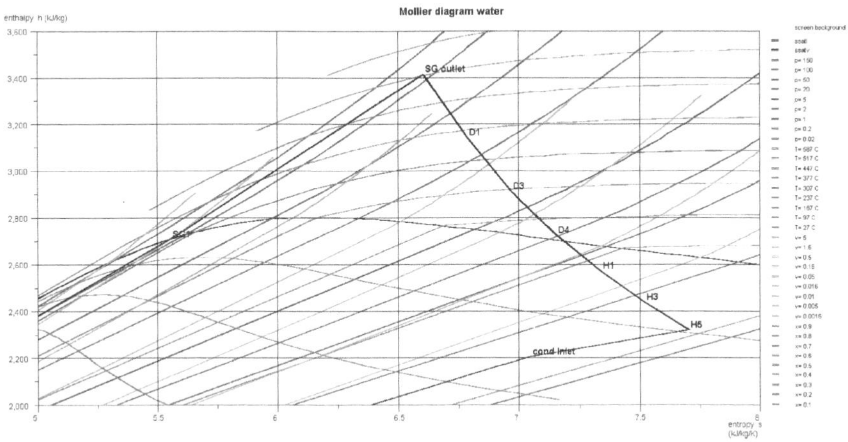

FIGURE 8.9.2 HTR-PM model cycle in the Mollier diagram.

49°C, then feedwater heater LP2, reaching 74.3°C, and then feedwater heater LP3, raising it to 90.5°C, before being directed to the feedwater tank.

It is then mixed in the feedwater tank with the extraction from the second TSG and the condensates from feedwater heater HP4, raising it to 117.9°C. It is then compressed to a pressure of 130 bar by the feedwater pump, and directed to the HP feedwater heater, exiting at 159.5°C before entering the SG.

8.9.4 Performance Summary Table

Parameter	Value
Total thermal power	500 MWth (2 × 250 MWth modules)
Electrical power output	210 MWe
Cycle efficiency (%)	33.76
Helium coolant pressure	7 MPa
Helium inlet/outlet temperature	250°C/750°C
Steam pressure	13.5 MPa
Steam temperature	~538°C superheated
Core dimensions	~3 m diameter × 11 m height
Fuel enrichment	~9% U-235 by weight
Reactor pressure vessel height	~24 m

8.10 DETAILED MODEL OF THE CANADIAN 1,200 MW SCWR NPP

The Canadian SCWR is a project of 2,540 MWth pressure tube type reactor with a steam flow of 1,176 kg/s entering the reactor at 300 bar and 315°C, and exiting at 250 bar and 625°C. It is reported to have an output power of 1,196 MW, corresponding to an efficiency of 47.1%. It integrates pressure tube reactor technology with supercritical water coolant, combining benefits of heavy water moderation and high cycle efficiency.

See Section 2.2.8 for an overview of SCWR-type reactors.

As we indicated in the introduction, this cycle has never been built before. It is simply a project envisioned for the future.

Reactor type is pressure tube SCWR, heavy water moderated, thermal spectrum.

Its core design includes 145 fuel assemblies with flask-like fuel bundle structures, featuring fuel rods arranged in rings with wire-wrapped spacers and central flow tubes to improve moderation and optimize radial power distribution.

Its fuel is low-enriched uranium fuel bundles of 5 m active length.

8.10.1 Presentation and Sources

A schematic of the cycle is provided on p. 264 of the G4NR Handbook: Igor L. Pioro (éd.). (2023).

It has the rare characteristic for this type of cycle of using an MSR with two steam reheaters between the IP and LP turbines.

The cycle involves eight feedwater heaters, four at LP and four at HP.

The model we have established based on the data provided in this schematic gives very similar results. Some differences arise from the values of the water properties, which are slightly different. Our model leads to a delivered power and efficiency that are slightly higher, equal to 1244 MWe and 48.2 respectively.

This SCWR cycle involves 10 TSGs, 12 extractions, 2 reheaters in the MSR, and 8 feedwater heaters with condensate drain: four LP and four HP.

8.10.2 Steam Circuit

The nuclear reactor appears on the left side of Figure 8.10.1. It receives 1,176 kg/s of feedwater at 376.6 bar and 314.6°C. The outgoing steam enters the first HP TSG at 230.4 bar and 620.5°C. After the first TSG (outlet at 114.5 bar and 497.7°C), an extraction of 85.4 kg/s supplies feedwater heater HP4.

After the second HP TSG (outlet at 70 bar, 420.8°C), another extraction of 72.3 kg/s supplies steam to feedwater heater HP3.

After the first IP TSG (outlet at 45 bar, 359.4°C), two extractions take place. One of 22.96 kg/s is directed to the second steam reheater, while the second of 72.79 kg/s supplies steam to feedwater heater HP2.

After the second IP TSG (outlet at 25.74 bar, 287.8°C), two extractions take place. One of 54 kg/s is directed to the first steam reheater, while the second of 42.7 kg/s supplies steam to feedwater heater HP1.

After the third IP TSG (outlet at 13.71 bar, 217°C), another extraction of 47.6 kg/s supplies steam to feedwater heater LP4.

After the fourth IP TSG (outlet at 6.2 bar, 160.1°C, quality of 0.9789), another extraction of 42.76 kg/s supplies steam to feedwater heater LP3.

The main flow of steam enters the MSR separator with 86% efficiency.

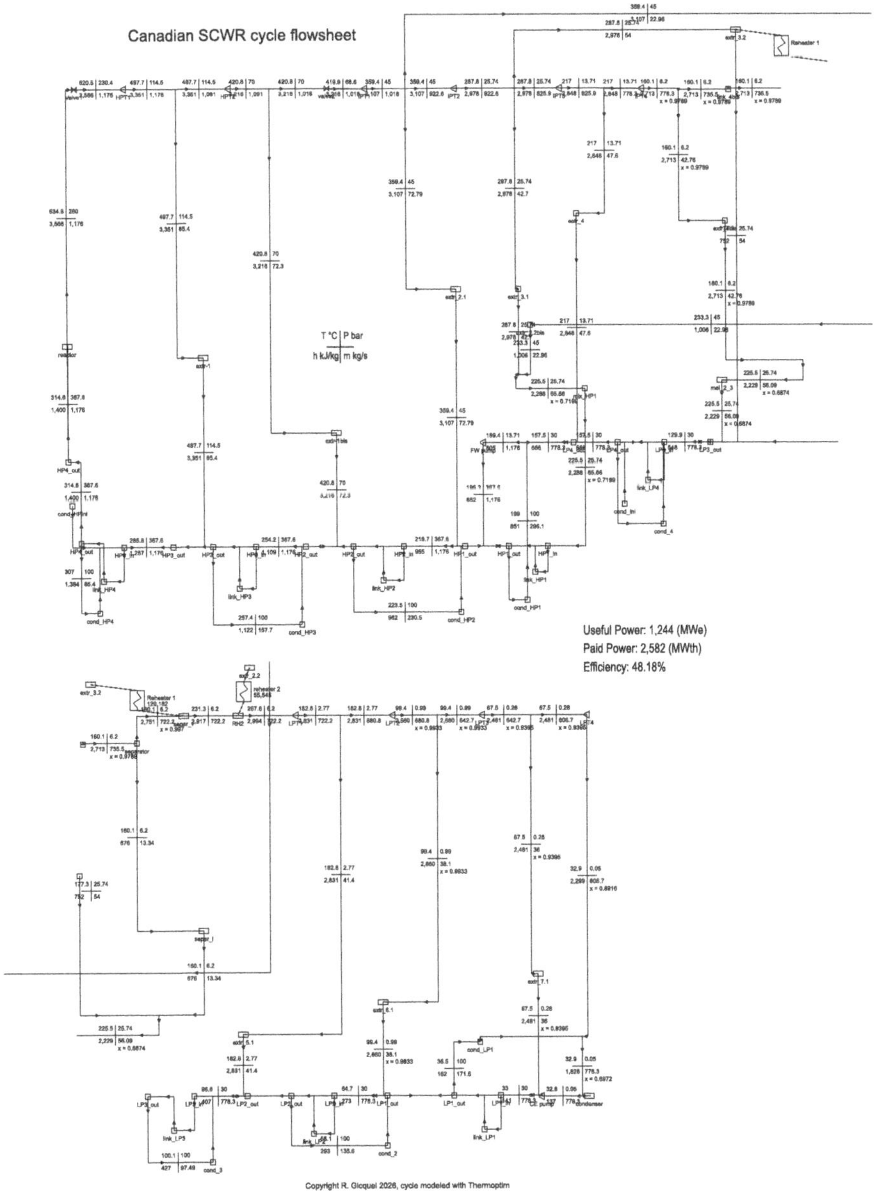

FIGURE 8.10.1 Detailed model of the Canadian 1,200 MW cycle.

722.2 kg/s of steam enters the first reheater at 160.1°C and exits at 231.3°C, while 13.34 kg/s of liquid phase from the separator is mixed with the previous extraction.

The steam then enters the second reheater and exits at 267.6°C and then enters the LP turbine composed of four TSGs:

After the first TSG (182.8°C, 2.77 bar), a 41.4 kg/s extraction heats feedwater heater LP3.

After the second TSG (0.99 bar, 99.4°C, quality 0.9933), an extraction of 38.1 kg/s of steam is used to heat feedwater heater LP2.

After the third TSG (0.28 bar, 67.5°C, quality 0.9395), an extraction of 36 kg/s of steam is used to heat feedwater heater LP1.

After the fourth TSG the flow enters the condenser.

The steam expansion path on the Mollier diagram is shown in Figure 8.10.2.

8.10.3 Feedwater Circuit

At the condenser outlet, an extraction pump then compresses the 778.3 kg/s of feedwater to 30 bar.

The feedwater is heated sequentially:

- Feedwater heater LP1: 64.7°C
- Feedwater heater LP2: 96.6°C
- Feedwater heater LP3: 129.9°C
- Feedwater heater LP4: 157.5°C

It is then mixed with flows entering the feedwater tank, and pressurized to 367.6 bar by the feedwater pump. It is then routed through the four HP feedwater heaters, exiting at 314.6°C before entering the SG.

The cycle performance summary is as follows:

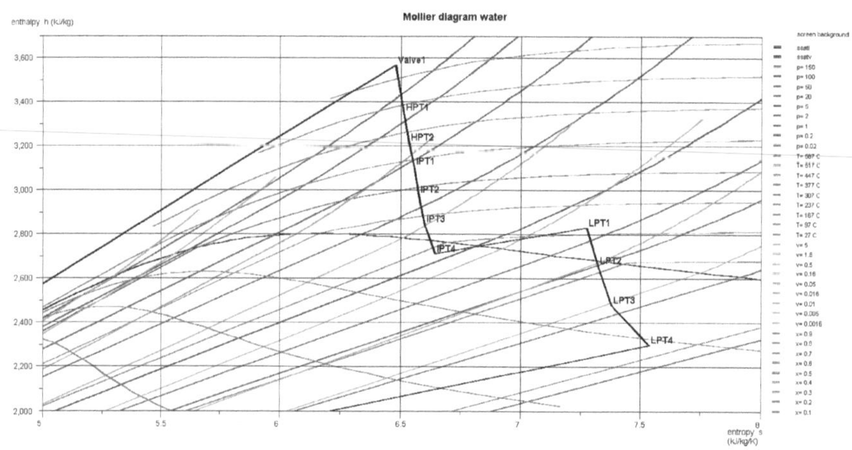

FIGURE 8.10.2 Canadian SCWR cycle in the Mollier diagram.

8.10.4 Performance Summary Table

Parameter	Value
Thermal power (MWth)	2,582
Electric power (MWe)	1,244
Cycle efficiency (%)	48.18
Operating pressure	280 bar

8.11 FLAMANVILLE 3 EPR SG MODEL

Achieving high performance in power plant design requires advanced optimization studies. To this end, a set of specifications is established, which considers a wide range of criteria, including, most notably, the definition of the design operating conditions—typically those under which the plant will operate for the longest duration.

When the operating conditions deviate from the design specifications, the system enters what is referred to as an "off-design regime." To clarify, the off-design regime corresponds to the stabilized operation of the plant under conditions different from those for which it was designed. It does not refer to the rapid transient regime, which may result from the activation of control system actuators. For this reason, the term "sliding steady-state regime" is also sometimes used.

In the case of a PWR nuclear power plant, the two main parameters subject to variation are the temperature of the cold source and, more importantly, the electrical power produced by the plant.

8.11.1 Variation of the Cold Source Temperature

The temperature of the cold source directly impacts the condensation pressure of the steam exiting the last stage of the LP turbine, and this pressure, in turn, affects the entire cycle.

However, plant operators cannot adjust this pressure arbitrarily. Its value is dictated by the thermal balance within the condenser, a heat exchanger of fixed surface area. As the thermal power to be dissipated in the condenser increases, the temperature difference between the cold source and the condensing steam must also increase. Because the steam pressure is directly related to this temperature, it varies with both the temperature of the cold source and the thermal power to be dissipated, which is approximately proportional to the plant's electrical output.

Figure 8.11.1 illustrates the evolution of the condensation pressure in Flamanville 3 as a function of the percentage of electrical power produced, for a cold source temperature of 17°C.

Just as with the condenser, the pressure in the SG is governed by the thermal balance within the component. It can be modified to some extent by adjusting the conditions in the primary circuit, but the degree of flexibility is limited due to specific constraints of this circuit.

The main variables used to control power output in the SG are the flow rates of the primary and secondary circuits, as well as the pressure in the secondary circuit.

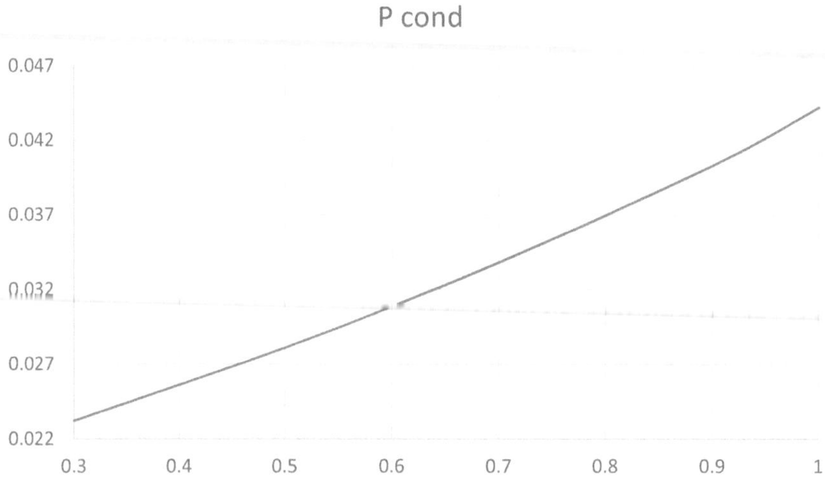

FIGURE 8.11.1 Condensation pressure function of the load rate.

As a result, plant operation is primarily adjusted by regulating these flow rates and the secondary pressure.

As the heat exchange in a large part of the SG occurs in a two-phase regime, varying the pressure of the secondary circuit affects the average temperature difference between the two fluids and thus the thermal power exchanged.

In order to study the interdependence between these different parameters, we created a model of the off-design behavior of a SG, and we used it to analyze flowsheets[3] related to the Flamanville 3 EPR, which is a third-generation nuclear plant with a capacity of 1,650 MW.

The results presented in this case study are based on the detailed data provided by EDF to the French Nuclear Safety Authority, in order to obtain authorization to commission the Flamanville EPR power plant.[4]

This page of this book's website gives you access to the flowsheets of the five cycles, created in SVG format: https://server.s4e2.com/crc/tcnpp/resources/epr/index.html

The purpose of this case study is to allow the reader to become familiar with using an external controller for studying the off-design behavior of a pressurized water reactor SG modeled with Thermoptim. We assume you already are knowledgeable about Thermoptim and its external class mechanism. Otherwise, you will find explanations in the software reference manual, especially in volume 3. We recommend that you also refer to Chapters 6 and 7 of the first volume of this set of books for further developments on the issue of component sizing and off-design operation, or on the related pages of the Thermoptim-UNIT portal.[5]

8.11.2 Principle of Calculation of Heat Exchangers in Off-Design Mode

The sizing of components requires to refine the phenomenological models used in Thermoptim's core, supplementing them to reflect the off-design operation mechanisms if any. The software has been equipped with screens for this, called component sizing, that define the geometric characteristics representative of the different technologies used and the parameters needed to calculate their performance.

This environment, developed as external classes must be able to work both complementary to the core components, and at the same time fully consistent with them. At times, the calculations are actually performed by the software package kernel, or made by the sizing classes, the controller ensuring synchronization between the two modes.

The heat exchanger calculations in the simulator are done using the NTU method (presented in Annex A1), with UA as an unknown intermediate, while the calculations in the sizing screens are carried out in detail, leading, for a set of inlet and outlet values of a given exchanger and taking into account the corresponding U value, to an estimate of the required area. Consistency between the two calculations is provided when the value of UA is such that the total exchange area is equal to that of sizing.

We have written a Thermoptim external controller (PWR_SG_controller.java) to solve the coupled equations that govern the operation of the SG. On the secondary side, this controller uses the TechnoSteamGenerator external class dedicated to calculating the heat transfer during the vaporization of water, taking into account nucleate boiling.

The complete documentation of the controller is provided in the Thermoptim modeling library. We will limit ourselves here to showing how it can be used for our purpose.

8.11.3 External Controller Screen

The controller screen is given in Figure 8.11.2.

It makes it possible to modify the surface of the exchanger, and the temperature, as well as the pressure and flow of the secondary circuit. The inlet temperature of

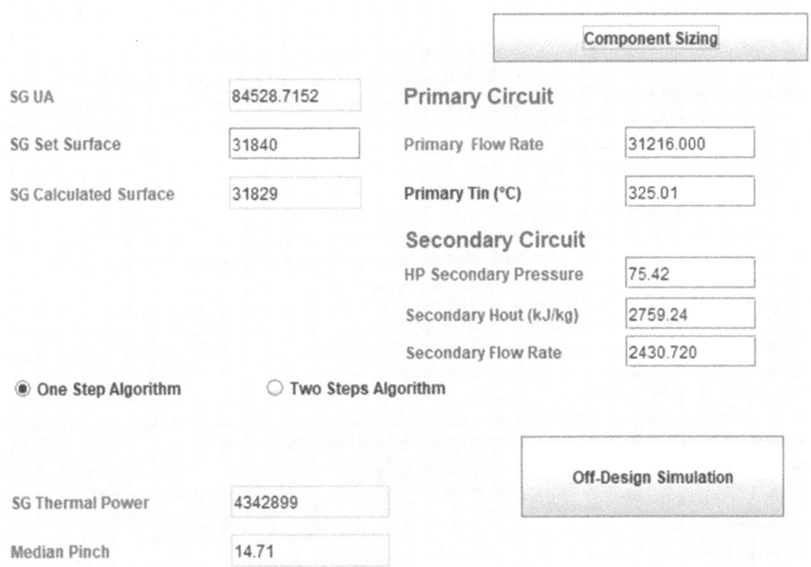

FIGURE 8.11.2 Controller screen.

the primary circuit is recalculated, but you may provide an initial value in this field in order to help the algorithm convergence. The outlet steam enthalpy is given for comparison with that calculated.

Start by clicking "Component sizing" to instantiate the TechnoSteamGenerator and make an initial geometrical setting, in this case calculate the surface of the SG corresponding to the project file setup, on the basis of default values for TechnoDesign parameters.

Among the data available for Flamanville 3 are the SG surface (31,840 m²) and the primary flow rate (31,216 kg/s). Thanks to the ASN document, we also know the temperatures of the feedwater entering the SG and the states of the steam at the outlet, namely T = 227.6°C, h = 2,759.24 kJ/kg, P = 75.44 bar at rated power. However, we do not know the internal configuration of the SG. After several trials, we end up with this setting which seems to be consistent with the data available (Figure 8.11.3).

For the secondary circuit, we chose the correlations of Kim & Mudawar for the NTU, and that of Sun & Mishima for the pressure drop. For the primary circuit we selected the Colburn correlation. The free flow area, hydraulic diameter, and length of the heat exchanger were set as indicated.

8.11.4 Using the Controller

Knowing at partial load the temperatures of the feedwater entering the SG and the states of the steam at the outlet, it is then possible to simulate the behavior of the SG in order to find the primary inlet temperatures at 30%, 50%, 70%, and 90% of the full capacity.

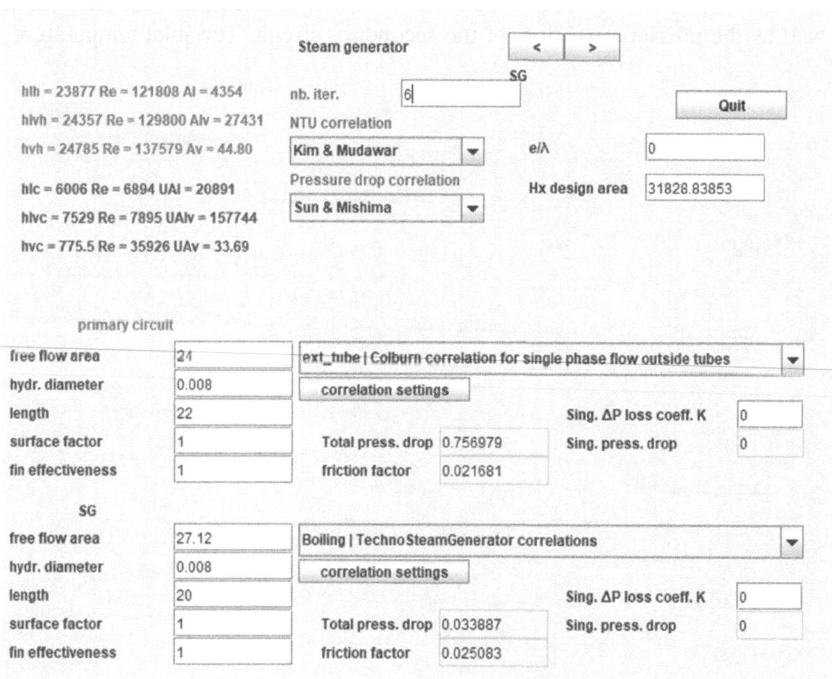

FIGURE 8.11.3 SG technological screen.

All calculations done, we obtain the curves of Figure 8.11.4, which provides the evolution of the temperatures in the SG. It clearly shows that the average temperature of the primary circuit varies much less than the other ones.

The temperature difference between the primary outlet and the saturation temperature varies little, while the difference between the inlet and outlet of the primary circuit varies greatly, the primary flow being constant.

Figure 8.11.5 provides, related to the rated power value, the evolution of the flows in the SG, its thermal power, the temperature of the feedwater, the steam outlet

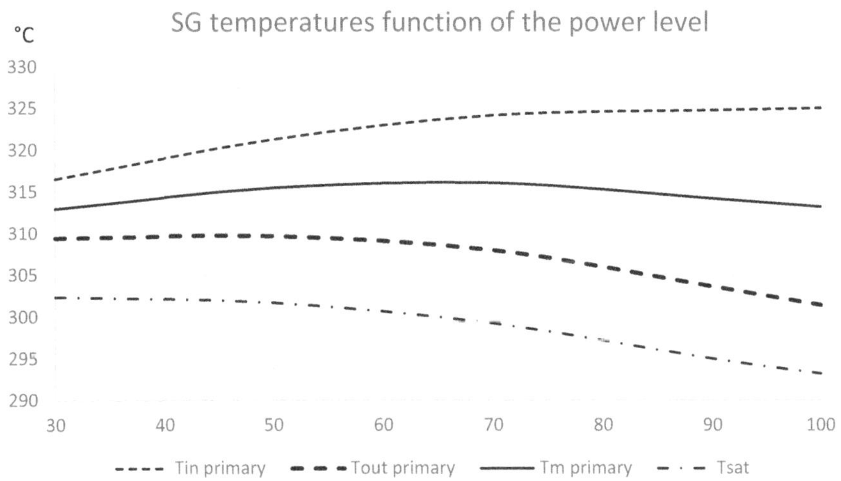

FIGURE 8.11.4 SG temperatures.

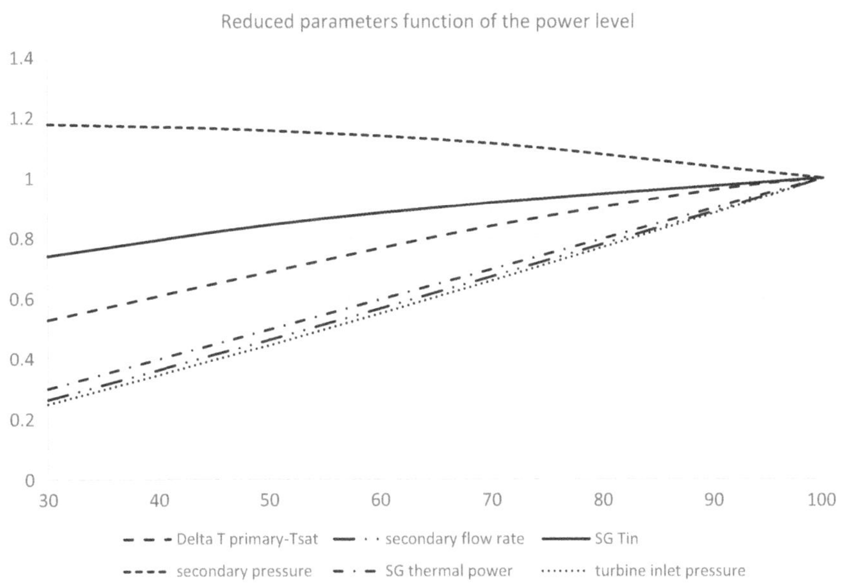

FIGURE 8.11.5 SG reduced parameters.

pressure, the inlet pressure into the HP turbine, and the difference ΔTsat between the primary circuit temperature and the saturation temperature.

The secondary circuit flow rate, the turbine inlet pressure, and the SG thermal power are functions of the load rate that are very close to linear, while the SG inlet temperature, the difference ΔTsat, and the outlet pressure follow more complicated laws.

The constraints governing the regulation of the primary circuit are numerous, particularly those related to the need to control the reactivity and the criticality of the nuclear reactor, so it is not surprising that the primary inlet temperature variation according to the load is not linear. This model shows how the steam outlet pressure provides a second control parameter.

Note that the inlet steam pressure in the turbine varies in the opposite direction of that of the SG outlet: it follows a quasi-linear increasing law.

8.12 CONCLUSION

This chapter presented 11 case studies of detailed nuclear thermodynamic cycles, all modeled using Thermoptim. The studies covered various technologies: Gas-Cooled Reactors (AGR 650 MW achieving 41% efficiency), Pressurized Water Reactors (SMR NuScale, VVER 1000, EPR Flamanville 3), Boiling Water Reactors (ABWR 1,350 MW), Pressure Tube Reactors (RBMK 1000, CANDU 550), Fast Reactors (Superphénix 1,174 MWe sodium-cooled), High-Temperature Reactors (HTR-PM), and supercritical water reactors (Canadian SCWR reaching 48.73%). Each case study detailed the steam-feedwater circuits, turbine configurations, extractions, and heaters. The models accurately reproduce operational parameters validated by regulatory authorities. A comparative analysis using the Mollier diagram reveals the specific performance characteristics and variations for each reactor type.

Further resources on the models presented in this chapter are available on this book's website.[6]

NOTES

1 https://server.s4e2.com/crc/tcnpp/resources/svg/index.html
2 https://s4e2.com/diag/fla3/HS-cycles-plotter.html.
3 We would like to thank Mr. J. M. Monteil, senior expert at EDF now retired, for his valuable help in interpreting certain subtleties of the flowsheets.
4 https://www.asn.fr/content/download/194137/file/10.%20Circuit%20vapeur%20et%20transformation%20de%20l%E2%80%99%C3%A9nergie%20-%20Rapport%20de%20s%C3%BBret%C3%A9.pdf.
5 https://direns.minesparis.psl.eu/Sites/Thopt/en/co/Dimensionnement_regime_non-nominal.html.
6 Available at: https://server.s4e2.com/crc/tcnpp/chapters/chapter8/_resources/index.html.

BIBLIOGRAPHY

Dong, Z., Zhang, Z., Dong, Y. et al. (2025). "Testing the feasibility of multi-modular design in an HTR-PM nuclear plant." *Nature Communications*, 16, 2778. DOI: 10.1038/s41467-025-58194-7.

Margulova, T. Kh. & Podushko, L.A. (1982). *Nuclear Power Plants M 25: A Textbook for Technical Schools.* Energoizdat, Moscow. 264 p., ill. (in Russian).

Nonbel, E. (1996). *Description of the Advanced Gas Cooled Type of Reactor (AGR).* Risø National Laboratory, Roskilde, Denmark.

Pioro, I.L. (ed.) (2023). *Handbook of Generation IV Nuclear Reactors*, 2nd ed. Woodhead Publishing (Elsevier), New York. ISBN 978-0-12-820588-4.

Troïanovski, B.M., Filippov, G.A. & Boulkine, A.E. (1985). *Steam and Gas Turbines of Nuclear Power Plants.* Energoatomizdat Publishing, Moscow. 256 p.

Zhou, S. & Turnbull, A. (2002). *Steam turbine operating conditions, chemistry of condensates, and environment assisted cracking – a critical review.* NPL Report MATC (A) 95, May.

9 Systemic Modeling
A Strategic Imperative for Energy Engineering Education

INTRODUCTION

Having reached the end of this work, it seems appropriate to reflect on the value of the systemic thermodynamic analysis methods presented and applied throughout this volume.

In the introduction, we noted that this book was designed to fill a gap: to our knowledge, there is no existing reference devoted to the detailed modeling of thermodynamic cycles of nuclear power plants (TCNPPs). This observation naturally raises two questions:

- Does this absence suggest that the topic lacks sufficient interest to justify its inclusion in engineering education?
- Alternatively, might it instead reflect the inadequacy of traditional teaching methods in this area?

As we have pointed out in the first chapter, the equation-based pedagogy commonly used in the Anglo-Saxon world quickly reaches its limits as system complexity increases, whereas the systemic approach we advocate enables learners, as demonstrated in this work, to explore realistic cycle models that closely mirror those used by professionals in industry.

9.1 STRATEGIC VALUE OF THERMODYNAMIC EXPERTISE

The methodologies developed throughout this work extend far beyond purely academic interest, providing essential capabilities across multiple dimensions of professional practice in energy systems engineering.

Obviously, advanced thermodynamic modeling first serves as a catalyst for innovation while simultaneously reducing development risks. Virtual simulations of innovative architectures enable concept validation before constructing expensive prototypes, significantly reducing development iterations and costs while identifying feasibility constraints early in the design process. This capability becomes increasingly valuable as organizations explore novel reactor concepts and advanced energy conversion systems.

Those who master these dual analytical/systemic methods are able to carry out rigorous and critical evaluations of both existing and emerging technologies.

DOI: 10.1201/9781003606024-10

Detailed thermodynamic modeling allows for an objective assessment of technological claims, distinguishing realistic projections from overly optimistic promises and quantifying the gap between theoretical potential and operational reality.

Our analysis of the high-temperature reactor pebble bed module (HTR-PM) reactor illustrates this point clearly: while the reported efficiency is about 40%, detailed modeling revealed that actual commercial operation achieved closer to 34%. This difference highlights how real-world constraints—such as material limitations, safety margins, and operational stability—inevitably temper theoretical performance expectations.

This capability becomes increasingly essential as the nuclear industry invests in advanced reactor concepts while simultaneously working to optimize existing plant performance. Organizations possessing sophisticated expertise in thermodynamic modeling enjoy significant advantages in technology selection, system optimization, cost reduction, and strategic decision-making.

In large-scale energy systems, even modest efficiency improvements can generate substantial annual savings, while optimized designs reduce capital costs and enhance operational reliability. The ability to accurately model complex thermal systems enables identification of performance bottlenecks and opportunities for enhancement that might otherwise remain hidden.

The methods developed in this work facilitate objective comparison of competing technologies, supporting informed decision-making regarding technology selection and system design. This expertise strengthens credibility when engaging with suppliers, partners, regulatory authorities, and investors, while enabling meaningful participation in technically demanding projects. In an industry where equipment lifespans extend over decades, the economic implications of sound thermodynamic design decisions provide benefits significantly over time.

Although we have focused here solely on secondary cycles of nuclear reactors, it should obviously be noted that expertise in thermal sciences, fluid mechanics, and thermodynamics enables simulation of critical scenarios and validation of system robustness under extreme conditions, which is essential for safety applications, particularly in nuclear energy, where reducing failure risks requires thorough analysis of thermal behavior during accident scenarios.

In highly regulated sectors such as nuclear energy, advanced modeling capabilities ensure that installations meet strict safety requirements while supporting validation of emergency systems and accident mitigation strategies. The ability to model complex thermal phenomena supports regulatory compliance and enables demonstration of system safety margins.

For advanced reactor concepts, sophisticated thermodynamic analysis becomes essential for licensing and regulatory approval, providing the technical foundation necessary to demonstrate safety and performance under various operating conditions.

9.2 AN ENDURING PROFESSIONAL ASSET

The development of high-temperature nuclear reactors offers valuable lessons about the gap between theoretical calculations and achievable real-world performance.

While theoretical analyses may indicate significant performance advantages, practical implementation must balance multiple competing constraints:

- High-temperature operation demands materials capable of maintaining structural integrity and resisting corrosion over extended periods.
- Novel designs must demonstrate adequate safety margins and gain approval from conservative regulatory frameworks.
- Theoretical performance gains must justify the additional capital and operational expenses.
- Sophisticated systems require specialized expertise and may face extended outages.

This evolution from concept to reality does not diminish the value of advanced thermodynamic analysis: rather, it emphasizes its critical importance for realistic technology assessment. The most successful nuclear technologies emerge from careful balancing of thermodynamic performance with materials limitations, safety requirements, regulatory acceptance, and economic competitiveness.

Understanding these trade-offs requires precisely the types of expert capabilities developed here. They provide strategic advantages extending well beyond nuclear applications. The skills acquired through mastering complex thermodynamic modeling represent enduring professional capabilities applicable across diverse energy system applications.

Industrial energy optimization, renewable energy integration, combined heat and power installations, emerging energy storage technologies, etc., all benefit from the same rigorous methodological approaches developed here for nuclear applications. The ability to construct detailed models, validate predictions against operational data, and optimize performance within real-world constraints constitutes a transferable skill set valuable across the entire energy sector.

In an era where energy transition demands both bold innovation and rigorous pragmatism, the mastery of advanced thermodynamic modeling represents more than technical competence: it constitutes a strategic differentiator. Such an expertise provides lasting professional value in a world where energy system performance increasingly determines economic competitiveness and environmental sustainability.

This analytical capability constitutes a strategic asset for individuals and organizations committed to developing energy solutions aligned with the complex realities of modern technological progress. In a highly constrained environment where multiple objectives must be met simultaneously, the ability to perform advanced thermodynamic analyses stands out as both a major professional differentiator and an essential tool for effectively addressing the energy challenges of the 21st century.

The journey through this volume, from fundamental principles to the detailed modeling of TCNPPs, has equipped readers with skills that extend well beyond specific applications. The methods and approaches presented here, firmly grounded in rigorous thermodynamic principles yet pragmatically oriented toward real-world implementation, offer a foundation for lifelong professional development and meaningful contributions to the evolving energy landscape.

Taken together, these arguments confirm the relevance of the pedagogical approach adopted in this book. Through the use of a simulation environment such as *Thermoptim*, emphasis is placed on fostering learners' systemic understanding rather than on their ability to manually derive every equation governing the behavior of individual components.

A useful parallel can be drawn with the evolution of structural mechanics education. Where teaching once focused primarily on analytical solutions for beams and surfaces, modern curricula—after covering fundamental principles—now emphasize mastery of finite element codes that enable analysis of complex structures.

This pedagogical shift recognizes that professional competence requires both theoretical understanding and practical capability with contemporary tools. The systemic approach to thermodynamic modeling presented in this volume reflects a similar evolution, equipping learners with methods appropriate for addressing the complexity of real-world energy systems.

The question posed at the outset—whether the absence of TCNPP modeling in curricula reflects lack of relevance or pedagogical inadequacy—can now be answered definitively: it is the latter.

Appendix 1
Reminders

INTRODUCTION

This appendix offers a review of key concepts concerning thermodynamic diagrams and the calculation of heat exchangers.

A1.1 THERMODYNAMIC PROPERTY DIAGRAMS OF PURE SUBSTANCES

A pure substance of given mass, being a bivariant system, can have its state represented on a two-axis diagram on which a number of iso-values or "contours" are plotted.

Among the many possible coordinate systems, we will consider three:

- the (h, ln(P)) diagram, with enthalpy as the abscissa and pressure as the ordinate (usually with a logarithmic scale to cover a wide range of values);
- the entropy diagram (T, s), with entropy as the abscissa and temperature as the ordinate;
- the Mollier diagram (h, s), which is a variant of the former, with entropy as the abscissa and enthalpy as the ordinate.

Thermoptim provides electronic versions of these diagrams, called interactive charts, which can be configured as desired and on which cycles can be represented.

The visualization of cycles in thermodynamic diagrams provides a synthetic representation that proves to be highly informative once a certain familiarity with them has been acquired.

The (h, ln(P)) diagram is traditionally used in the study of reversed cycles (heat pumps or refrigeration), but it can also be used to represent power cycles. Its advantage is to combine the essential mechanical constraint (P) of compressors with the energy variable (h).

One of the advantages of the (T, s) entropy diagram is that any perfect cyclic process results in a contour (Γ) in the (T, s) plane, whose area A measures both the amount of heat Q involved and the work provided or received τ. Indeed, $Q = \int T\,ds$ by definition. Since the cycle is closed, $Q + \tau = 0$, and $|\tau| = A$. The entropy diagram thus makes it possible to directly compare cycles with each other, as is done with the Carnot cycle in Section 3.3.

The (h, s) Mollier diagram, which is a variant of the former, directly shows the energy transfers in an open system, with the advantage that the enthalpy involved can be read without difficulty. Moreover, it provides a global view of the thermodynamic domain, including the saturation dome, and clearly distinguishes between saturated and superheated regions.

Like the entropy diagram, it is imprecise in the liquid zone, which is almost never represented. On the other hand, it allows excellent visualization of the entire expansion zone of turbines, which makes it valuable for the studies carried out in this book.

In this sense, it is highly complementary to the other two diagrams, and in particular to the enthalpy diagram, which allows clear visualization of the part of the cycle where water is in the liquid state.

It should also be noted that one of the limitations of visualizations on diagrams is that flow rates do not appear, whereas they vary, particularly according to extractions when present.

A1.1.1 (h, ln(P)) Diagram

In the (h, ln(P)) diagram (Figure A1.1.1), the abscissa is enthalpy and the ordinate is pressure, usually on a logarithmic scale. This diagram includes isentropic, isothermal, and isovolume curves, and in the two-phase zone, constant quality curves.

The critical point C is located at the maximum of the vaporization curve.

A1.1.2 (T, s) Entropy Diagram

In the entropy diagram (Figure A1.1.2), entropy is the abscissa and temperature is the ordinate. The vaporization curve again separates the plane into two parts, defining

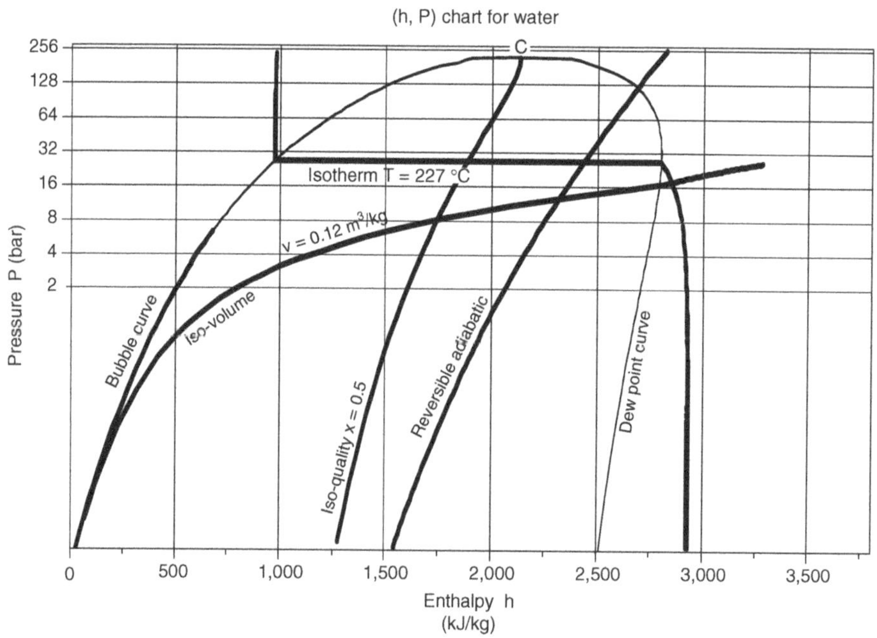

FIGURE A1.1.1 (h, ln(P)) diagram.

the two-phase zone and the single-phase fluid zone. The critical point is still at its maximum (C in the figure).

For some substances, the right branch of the vaporization curve (saturated vapor) presents a point of maximum entropy. This is particularly the case for certain hydrocarbons and certain chlorofluorocarbons (Figure A1.1.3).

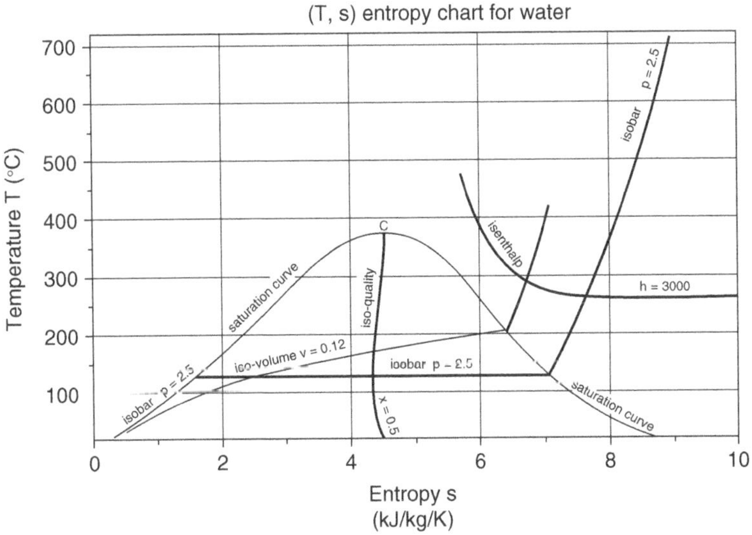

FIGURE A1.1.2 Entropy diagram.

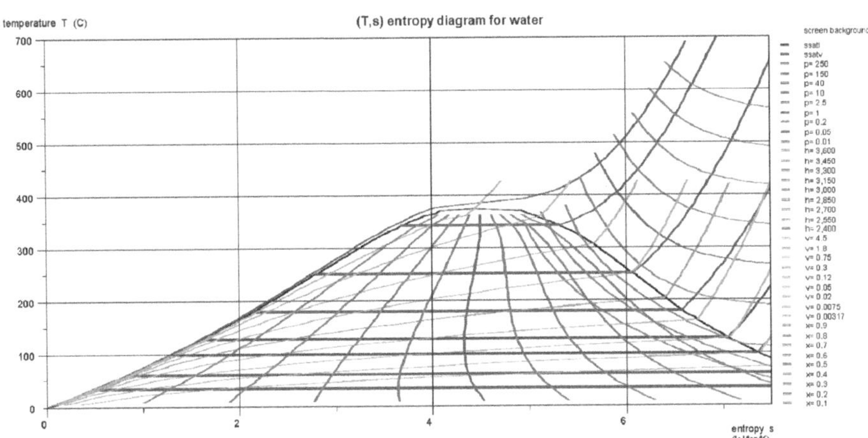

FIGURE A1.1.3 Water entropy diagram.

A1.1.3 (h, s) Mollier Diagram

In the Mollier diagram (Figure 4.1.14), entropy is the abscissa and enthalpy is the ordinate. The advantage is that enthalpies can be directly measured on the axis. The critical point C is located to the left of the maximum of the vaporization curve (Figure A1.1.4).

The synoptic view of a reheat and extraction cycle modeled in Thermoptim is given in Figure 4.1.14. Its representation in the water (h, s) Mollier and enthalpy diagrams is given in Figure A1.1.6 (isovolumes have not been plotted to increase diagram readability).

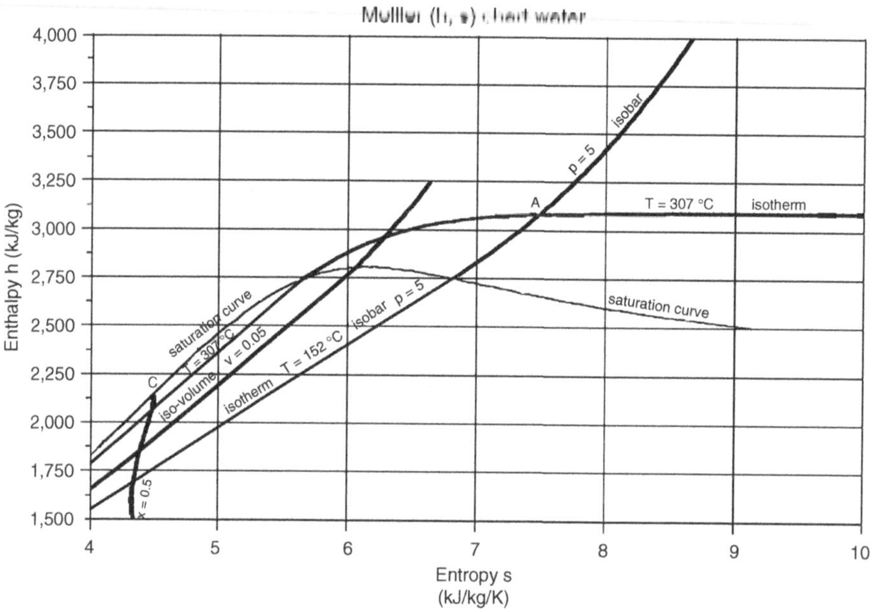

FIGURE A1.1.4 Mollier diagram.

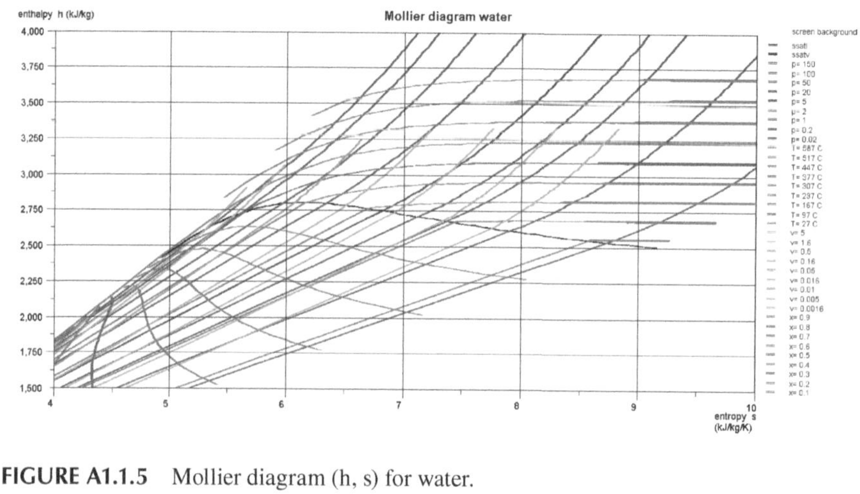

FIGURE A1.1.5 Mollier diagram (h, s) for water.

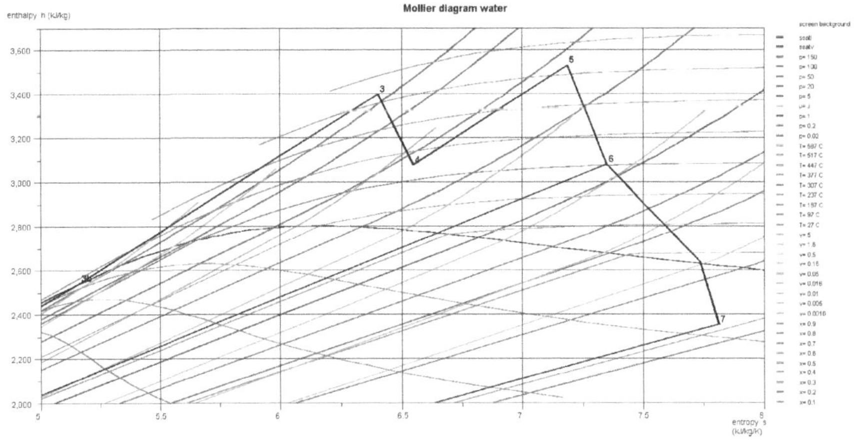

FIGURE A1.1.6 Cycle with reheat and extraction in the Mollier diagram.

Note that in the Mollier diagram, all turbine stage groups have been represented, so that the expansion curve is modeled much more precisely than in the synoptic view. This type of highly detailed expansion curve is characteristic of the turbine used in the cycle and is therefore a kind of signature of this turbomachine.

A1.2 CALCULATION OF HEAT EXCHANGERS BY THE NTU METHOD

Heat exchangers are devices for transferring heat between two fluids at different temperatures. In most cases, the two fluids are not in direct contact, and the transfer occurs through a separating wall.

We will limit ourselves here to a few reminders on how to calculate heat exchangers using the NTU (Number of Transfer Units) method. In-depth explanations on this topic are available in volume 1 of this set of two books or on the Thermoptim-Unit portal.

We will denote the hot fluid by index h, and the cold fluid by index c. Besides the geometric configuration, exchanger sizing or performance depends on many parameters and variables:

- mass flows $\dot{m}_h$ and $\dot{m}_c$ passing through them;
- inlet temperatures T_{hi} and T_{ci} and outlet temperatures T_{ho} and T_{co} of both fluids;
- heat exchange coefficients U_h and U_c of each fluid;
- thermal resistance of the wall;
- surface A of the exchanger;
- pressures of both fluids, almost constant;
- thermophysical properties of fluids, which are used to determine coefficients U_h and U_c. It is essentially the heat capacity c_p, density ρ, thermal conductivity λ, and viscosity μ.

In what follows, we will assume that the heat transfer coefficients U_h and U_c and thermophysical properties of fluids maintain constant values throughout the entire heat exchanger.

If this assumption is not verified, then in order to study the performance of the exchanger, it is necessary to divide it into small volume elements in which these properties can be considered constant. The calculations then become much more cumbersome (Figure A1.2.1).

Finally, we always assume that the heat exchanger is globally adiabatic, meaning there is no heat exchange with the surroundings.

In the following, the notations and equation numbers are those of Chapter 4 of Volume 1 of this set.

A1.2.1 Heat Flux Exchanged, UA-LMTD Method

It can be shown, where fluid flow is pure (parallel or counter-flow), that the heat flux exchanged between the two fluids is given by the formula (Figure A1.2.2):

$$\phi = UA\Delta T_{ml} \tag{4.1.1}$$

$$\text{with } T_{ml} = \frac{\Delta T_0 - \Delta T_L}{\ln \dfrac{\Delta T_0}{\Delta T_L}} \tag{4.1.2}$$

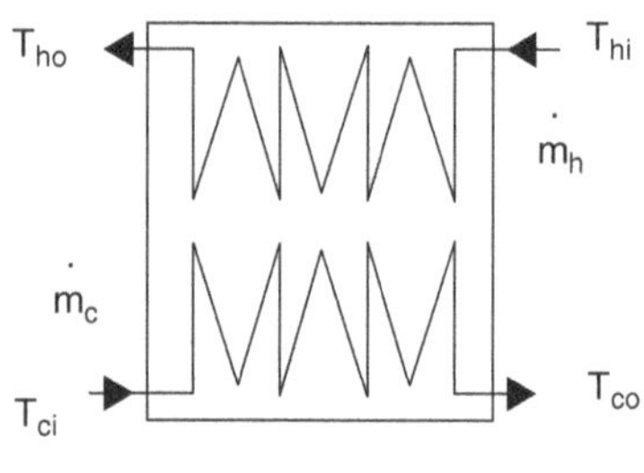

FIGURE A1.2.1 Heat exchanger sketch.

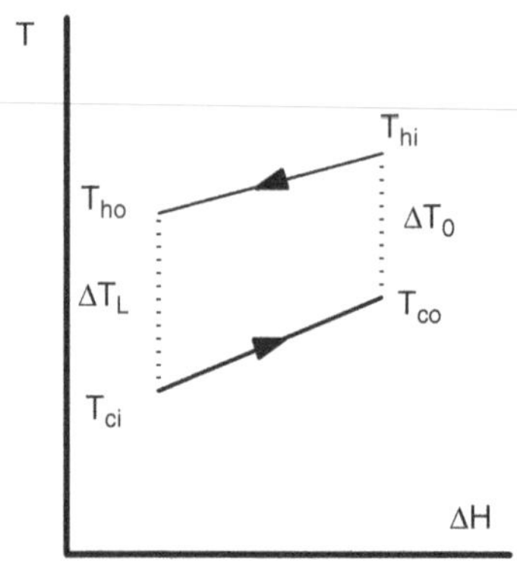

FIGURE A1.2.2 LMTD notations.

where A is the exchange surface, U is the overall heat transfer coefficient (see the next section), δT_{ml} the logarithmic mean temperature difference (LMTD), and δT_0 and δT_L are the temperature differences between fluids at one end and the other end of the exchanger, respectively, with the convention: $\delta T_0 > \delta T_L$.

Care must be taken that we refer to temperature differences between the two fluids at the entrance and exit of the exchanger (positions 0 and L), not the differences between inlet temperatures $(T_{hi} - T_{ci})$ and outlet temperatures $(T_{ho} - T_{co})$ of the fluids.

In the example in Figure A1.2.3, we have:

$$\Delta T_0 = T_{hi} - T_{co}$$

$$\Delta T_L = T_{ci} - T_{ho}$$

When the flow configuration is more complex, we introduce a correction factor F (less than 1), which is given by charts that are calculated or determined experimentally, most often provided by manufacturers.

This formula shows that to transfer a given flux φ, if we wish to reduce the irreversibilities and therefore ΔT, the product UA must be as large as possible, which can be achieved either by increasing the surface area, though this directly affects cost, or by increasing the U value, which is an objective sought by all heat exchanger designers.

In this section, the design of the heat exchanger will be limited to determining the product UA of the exchanger surface A by the overall heat transfer coefficient U, as the accurate estimation of the latter depends on the internal constructive details of the exchanger.

In general, the design of heat exchangers is a compromise between conflicting objectives, the two main ones being:

- A large exchange area is desirable to increase the effectiveness of heat exchangers, but it results in high costs.
- Small fluid flow sections increase the values of the heat transfer coefficients U_h and U_c, and thus reduce the required area, but they also increase pressure drops.

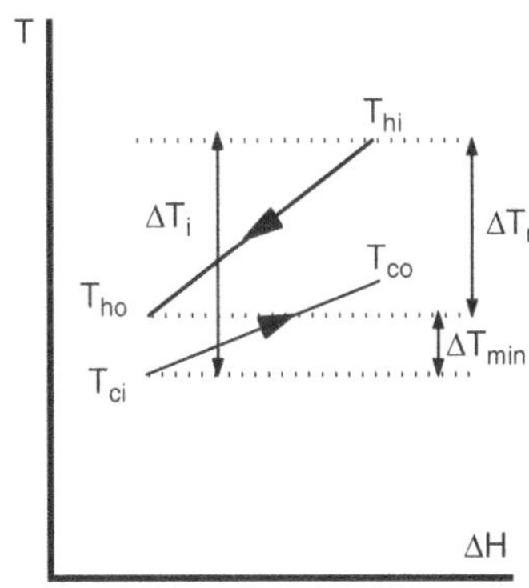

FIGURE A1.2.3 Effectiveness notations.

To size a heat exchanger, it is possible to use relations (4.1.1) or (4.1.3), but they assume the LMTD is known, which is rarely the case.

A1.2.2 NTU Method

A simpler and more general method is the NTU method developed by Kays and London (1984). This method has the advantage of requiring knowledge of fluid inlet temperatures, not those at the outlet.

By definition, NTU is defined as the ratio of the product UA of the heat exchanger to the minimum heat capacity rate.

$$NTU = \frac{UA}{(\dot{m} \cdot c_p)_{min}} \tag{4.2.1}$$

We call R the ratio (less than or equal to 1) of the heat capacity rates:

$$R = \frac{(\dot{m}c_p)_{min}}{(\dot{m}c_p)_{max}} \leq 1 \tag{4.2.2}$$

and ε the effectiveness of the exchanger, defined as the ratio between the heat flux actually transferred and the maximum possible flux:

$$\varepsilon = \frac{\phi}{\phi_{max}} \tag{4.2.3}$$

We can show that ϕ_{max}, which would be obtained for a counter-flow heat exchanger of infinite length, is:

$$\phi_{max}\big| = (\dot{m}c_p)_{min} \Delta T_i$$

where δT_i is the difference between inlet temperatures of both fluids. Figure A1.2.3 shows the temperature profile in a heat exchanger. Note that the abscissa used here is the enthalpy exchanged, not the length of the exchanger, which explains, with the assumption that the heat capacities are constant, why this profile is linear rather than exponential.

Note also that δT_{max} and δT_{min} are the temperature differences within each of the two fluids, not between them as in Figure A1.2.2.

$$\text{As } \phi = (\dot{m}c_p)_{min} \Delta T_{max} = (\dot{m}c_p)_{max} \Delta T_{min} \tag{4.2.4}$$

$$\text{We have}: \varepsilon = \frac{\Delta T_{max}}{\Delta T_i} \tag{4.2.5}$$

$$\text{Since } \phi = UA\Delta T_{ml} = (\dot{m}c_p)_{min} \Delta T_{max} = (\dot{m}c_p)_{max} \Delta T_{min}$$

$$\text{NTU} = \frac{UA}{\dot{m}c_{p\,min}} = \frac{\Delta T_{max}}{\Delta T_{ml}}$$

NTU measures the ability of the exchanger to change the temperature of the fluid whose heat capacity rate is lower.

This is indeed an indicator of the equipment quality: the more efficient the heat exchanger, the more it allows heating or cooling a fluid with a small temperature difference relative to the other fluid.

When U varies along the surface of the exchanger, $\text{NTU} = \frac{1}{\dot{m}c_{p\,min}} \int_0^A U\,dA$

For example, in the example of Figure A1.2.1, where $(\dot{m}c_p)_h < (\dot{m}c_p)_c$ we have:

$$(\dot{m}c_p)_{min} = (\dot{m}c_p)_h$$

$$\Delta T_i = T_{hi} - T_{ci}$$

$$\Delta T_{max} = T_{hi} - T_{ho} \quad \text{and}$$

$$\varepsilon = \frac{T_{hi} - T_{ho}}{T_{hi} - T_{ci}} \tag{4.2.6}$$

With these definitions, it is possible to show that there is a general relation of type:

$$\varepsilon = f(\text{NTU}, R, \text{flow pattern}) \tag{4.2.7}$$

In practice, it suffices to have a set of relationships corresponding to flow patterns representative of the exchangers studied, and the design of a heat exchanger is based, on the one hand, on balance equations (4.2.4) and, on the other hand, on the internal equation (4.2.7).

If we know the flow rates of both fluids, their inlet temperatures, and the heat flux transferred, the procedure is as follows:

- We start by determining the outlet temperatures of fluids from equation (4.2.4).
- We deduce the fluid heat capacity rates $\dot{m}cp$ and their ratio R.
- The effectiveness ε is calculated from equation (4.2.5).
- The value of NTU is determined from the appropriate (NTU, ε) relationship.
- UA is calculated from equation (4.2.1).

If UA is known, equation (4.2.1) gives the value of NTU, and ε can be determined from the appropriate (NTU, ε) relationship. Equation (4.2.5) allows calculation of the heat transferred.

In both cases, the enthalpy balance provides the outlet temperatures. This method therefore connects the UA product to the exchanger performance, and the surface area is deduced from an estimate of the value of the exchange coefficient U.

As shown in equation (4.2.6), ε measures the "temperature" effectiveness of the exchanger, i.e., the fraction of the maximum temperature difference between the fluids that is achieved by the fluid with the lower heat capacity rate.

The value of the method proposed by Kays and London lies in its generality due to the formal appearance of $(\dot{m}c_p)_{min}$ and $(\dot{m}c_p)_{max}$.

Phenomenological (ε, NTU) models provide only the exchanger UA product. The calculation of U depends on many factors not taken into account by these models.

Given its advantages, this method has been very successful and is now widely used.

Experience shows that, at a thermodynamic level, (ε, NTU) models generally represent well the behavior of many types of exchangers.

A1.2.3 Relationship between NTU and ε

Here, we give relations between ε and NTU function of R for counter-flow heat exchangers, and their inverses which can be expressed simply, as well as the corresponding graph in which curves represent ε function of NTU, for R varying from 0 to 1, top to bottom, by steps of 0.25.

For a counter-flow heat exchanger, analytical expressions are, for R ≠ 1:

$$NTU = \frac{1}{1-R} \ln \frac{1-\varepsilon R}{1-\varepsilon} \qquad (4.2.8)$$

$$\varepsilon = \frac{1-\exp(-NTU(1-R))}{1-R\exp(-NTU(1-R))}$$

$$\text{For } R = 1 \quad \varepsilon = \frac{NTU}{1+NTU}$$

The corresponding chart is given in Figure A1.2.4.

A1.2.4 Relationship with the LMTD Method

The LMTD method (LMTD) starts from equation (4.1.3):

$$\phi = UAF\Delta T_{ml}$$

It is possible to switch from one calculation method to another using the relationship:

$$\Delta T_{ml} = \frac{\varepsilon \Delta T_i}{NUT}$$

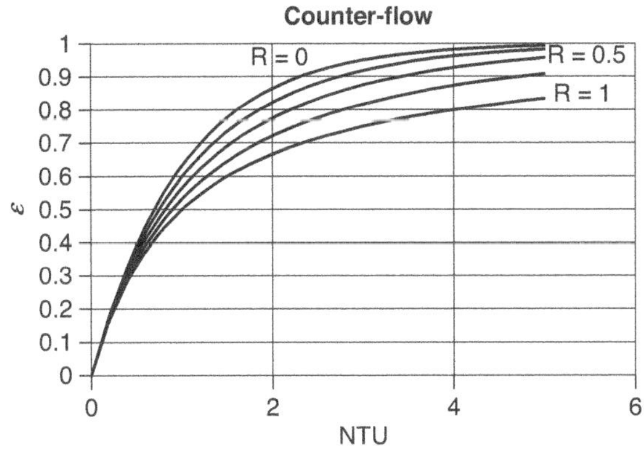

FIGURE A1.2.4 Counter-flow heat exchanger chart.

A1.2.5 Calculation in Thermoptim

In Thermoptim, a heat exchanger connects two « exchange » processes. One of them is a hot fluid which corresponds to a fluid that gets colder, while the other is a cold fluid that gets hotter. Once the coupling of the processes is made, the design problem is the following: it is necessary on the one hand to conserve enthalpy in the heat exchanger, and on the other hand to respect some temperature constraints.

Given that there are four temperatures (two for each fluid) and two flow rates, the problem has five degrees of freedom once enthalpy is conserved. It can furthermore be shown that at least one of the two flow rates must be set; otherwise, the problem is indeterminate.

For temperatures, one can either set explicit constraints (for example the fluid inlet temperatures), or implicit constraints (the value of the heat exchanger effectiveness, or the pinch minimal value).

Most generally, the inlets of the exchanger are known, that is to say the temperatures and the flow rates of the two fluids, T_{ci}, mc, T_{hi}, and mh, that is to say four constraints. It therefore remains to set the fifth one.

It can be depending on the case:

- an outlet temperature, for example, in the study of the steam generator of a steam power plant cycle, the state of each point of the steam cycle is known, including the outlet temperatures of the three elements of the exchanger, namely the economizer, the evaporator, and the superheater, and the design consists in determining the value of the temperature of the hot fluid at the outlet of each of these elements
- the fifth constraint can also be an effectiveness, as for example in the study of a gas turbine regenerator
- or the pinch value, i.e., the minimum temperature difference between the two fluids

A1.2.6 Calculations in Off-Design Mode

In design mode, if we know the flow rates of both fluids, their inlet temperatures, and the heat flux transferred, the procedure is as follows:

- We first determine the outlet temperatures of the fluids from equation (4.2.4).
- We deduce the fluid heat capacity rates $\dot{m}cp$ and their ratio R.
- The effectiveness ε is calculated from equation (4.2.5).
- The value of NTU is determined from the appropriate (NTU, ε) relationship.
- UA is calculated from equation (4.2.1).

In off-design mode, knowing inlet temperatures and flow rates of both fluids, exchanger area A and its geometry (flow patterns and technological parameters), calculation proceeds in three steps:

- Determining U by correlations depending on exchanger flow pattern and geometry. Refer to Chapter 7 of this set's first volume for explanations on the calculation of U.
- Calculating UA, the product of U and A, and then NTU.
- Determining exchanger effectiveness ε by the NTU method, and calculating hot and cold fluid outlet temperatures, with equations (4.2.20) and (4.2.22) determining how the heat exchanger adapts to operating conditions.

A1.2.7 Thermocouplers

Thermocouplers complement heat exchangers by allowing components other than exchange processes to be connected to one or more exchange processes in order to represent a thermal coupling. This mechanism does not replace the exchanger mechanism: two exchange processes cannot be directly connected by a thermocoupler.

This mechanism has a number of benefits, because it can be used to represent many thermocouplers that do not constitute a heat exchange in the traditional sense, such as, for example, providing heat to a high-temperature electrolyzer (Figure A1.2.5) as explained in Section 7.3.3.

A1.2.8 Regenerator Equations

As we have pointed out several times, Thermoptim allows exporting the equations it implements. As an example, here are those corresponding to the exchanger used in the modeling of the regenerative gas turbine, Section 6.1.2.1.

```
//Heat exchanger: regenerator
//Equation: 28
mCp_HPregenerator = (h_5 - h_4)/(T_5 - T_4)*m_dot_
HPregenerator // mCpf = deltaH/deltaT - HP regenerator
//Equation: 29
mCp_LPregenerator = (h_2 - h_1)*m_dot_LPregenerator/(T_2
- T_1) // mCpc = -deltaH/deltaT - LP regenerator
// Comment = mCpc<mCpf
```

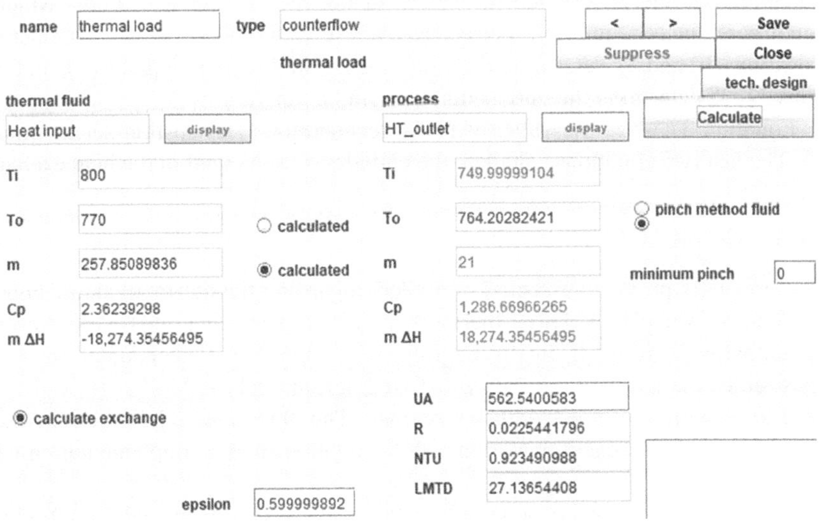

FIGURE A1.2.5 High-Temperature Electrolyzer thermocoupler screen.

```
//Equation: 30
UA_regenerator = NTU_regenerator *mCp_LPregenerator // Hot
fluid - LP regenerator
//Equation: 31
R_regenerator = mCp_LPregenerator /mCp_HPregenerator // Hot
fluid - LP regenerator - Cold fluid - HP regenerator
//Equation: 32
T_2 = T_1*(1 - epsilon_regenerator) + epsilon_regenerator*T_4
// Hot fluid outlet temperature
//Equation: 33
h_2 = calcH_TP("He";T = T_2;P = p_2)// Enthalpy
//Equation: 34
h_5 = m_dot_LPregenerator /m_dot_HPregenerator*(h_1 - h_2) +
h_4 // Hot fluid - LP regenerator - Cold fluid - HP
regenerator
//Equation: 35
T_5 = calcT_PH("He";P = p_5;H = h_5)// Cold fluid outlet
temperature
// Comment = epsilon given value
//Equation: 36
epsilon_regenerator = 0.9090906773363381 // Given value
//Equation: 37
NTU_regenerator = epsilon_regenerator/(1 - epsilon_regenerator)
// Counterflow heat exchanger mCpc = mCpf
//Equation: 38
Q_dot_LPregenerator = m_dot_LPregenerator*(h_2 - h_1) //
DeltaH hot fluid
//Equation: 39
Q_dot_HPregenerator = m_dot_HPregenerator*(h_5 - h_4) //
DeltaH cold fluid
```

Equations 28 and 29 provide the values of the mC_p of the two fluids, which are equal here since helium is a perfect gas whose C_p is independent of temperature. Equations 30 and 31 determine the values of UA and R (here equal to 1), and equations 32–35 determine the outlet enthalpies and temperatures.

Equation 37 gives the value of NTU_regenerator, with epsilon set by equation 36. The last two equations calculate the enthalpies exchanged in the heat exchanger.

A1.3 CONCLUSION

This appendix provided essential reminders regarding the thermodynamic concepts of steam power plants. Three thermodynamic property diagrams were presented: the enthalpy-pressure (h, ln(P)) diagram, which combines mechanical and energy constraints, the temperature-entropy (T, s) diagram, and the Mollier (h, s) diagram, used to visualize turbine expansion regions. The NTU method for heat exchanger calculation was presented, including ε-NTU relations and its implementation within Thermoptim.

Further resources on the models presented in this annex are available on this book's website (https://server.s4e2.com/crc/tcnpp/chapters/appendix1/_resources/index.html).

BIBLIOGRAPHY

Incropera, F.P. & DeWitt, D.P. (1996). *Fundamentals of Heat and Mass Transfer*, 4th ed. John Wiley & Sons, New York.

Kakac, S. & Liu, H. (2002). *Heat Exchangers: Selection, Rating, and Thermal Design*, 2nd ed. CRC Press, Boca Raton, FL. ISBN 0-8493-0902-6.

Kays, W.M. & London, A.L. (1984). *Compact Heat Exchangers.* McGraw-Hill, New York.

Shah, R.K. & Bell, K.J. (2000). Heat exchangers. In Kreith, F. (ed.), *The CRC Handbook of Thermal Engineering.* CRC Press, Boca Raton, FL. ISBN 0-8493-9581-X.

Smith, E.M. (1997). *Thermal Design of Heat Exchangers: A Numerical Approach.* Wiley, Chichester.

Appendix 2
Turbines

A2.1 STEAM TURBINES

The general layout of a turbine is shown in Figure A2.1.1. Steam at high pressure and temperature passes through a stator followed by a rotor in series. The stator consists of guide nozzles that accelerate the fluid, while the wheel (rotor) converts at least part of the available enthalpy into mechanical energy. U represents the tangential velocity of the rotor.

In a turbine, the fluid undergoes expansion. The compressible flow equations, which we will not develop here, indicate that for a subsonic regime, the section of the flow path should decrease while velocity increases. This evolution occurs in two steps (Figure A2.1.1): in the stator, the absolute velocity C increases, while in the impeller, the relative velocity W increases and the absolute velocity C decreases.

For a given rotation speed N, the performance map of a turbine has the appearance shown in Figure A2.1.2, where a group of variables called corrected mass flow, proportional to the Mach number in the flow, appears on the ordinate, and the inverse of the expansion ratio on the abscissa. The index 0 refers to the turbine inlet conditions.

As can be seen, there is a corrected mass flow rate limit corresponding to the establishment of sonic flow at the nozzle throat: the flow is said to be choked.

This point is very important because it means that the flow rate passing through a turbine cannot exceed a certain limit when the expansion ratio exceeds a value of approximately 2.

We will not develop the study of compressible flow equations and similarity relations in this book, as they are not necessary for our purpose. Interested readers may refer to specialized literature or to Sections 3.3.3 and 9.1.5 in the first volume of this set.

A2.1.1 Degree of Reaction of a Stage

Most steam and gas turbines are multistage axial turbines. The study of the energy balance of a stage shows that the enthalpy change occurs partly in the nozzle and partly in the rotor. The degree of reaction ε is defined as the fraction of the enthalpy change that takes place in the rotor.

By definition, ε is between 0 and 1.

Steam turbines can be grouped into two broad classes:

- **impulse turbines**, in which the degree of reaction ε is 0: all expansion of the fluid occurs in fixed blades or nozzles upstream of the wheel, and pressures upstream and downstream of the rotor are equal;
- **reaction turbines**, where $\varepsilon = 0.5$: expansion is evenly distributed between the nozzle and the wheel.

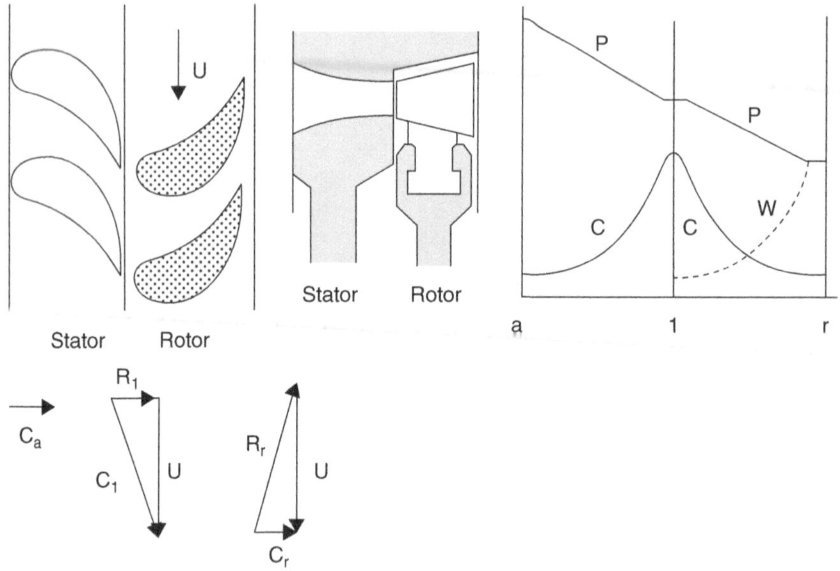

FIGURE A2.1.1 Velocity and pressure profile in an axial turbine stage.

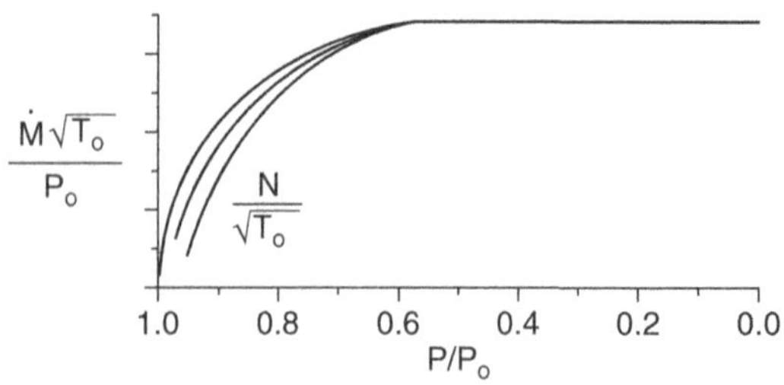

FIGURE A2.1.2 Characteristics of a turbine.

Each of these two turbine types has its own advantages and disadvantages: impulse turbines are generally used for multistage turbine head stages or for small capacity units, while reaction turbines are well adapted for low-pressure turbines.

A2.1.2 Different Types of Steam Turbines

Depending on their use, there are four broad categories of turbines (Figure A2.1.3):

- **condensing turbines**, where steam is completely expanded to a pressure of about 0.02–0.04 bar and then liquefied in a condenser cooled by ambient air or water. This type of turbine is mainly used in power production facilities;
- **back-pressure turbines**, in which vapor is expanded from High Pressure (HP) pressure (> 40 bar) to low pressure (about 4 bar). This type of turbine allows production of mechanical power or electricity by exploiting the high

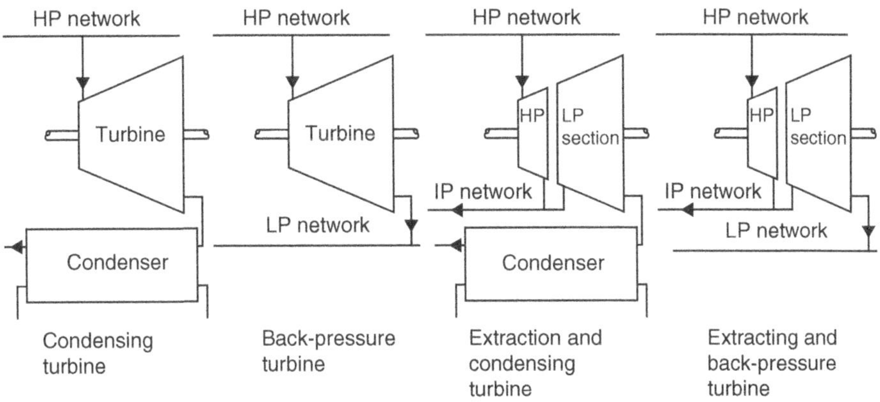

FIGURE A2.1.3 Different types of steam turbines.

temperature and pressure that can be obtained in a boiler, while using the residual enthalpy for various processes;

- **extracting and condensing turbines**, in which vapor undergoes partial expansion to an intermediate pressure (about 20 bar) in a high-pressure section. Part of the steam is directed to a user network, while the remainder is expanded in a low-pressure section, as in a condensing turbine. This type of turbine finds important application in cogeneration plants whose heat requirements are likely to vary considerably over time;
- **extracting and back-pressure turbines**, in which steam is discharged at low pressure into an Low Pressure (LP) network instead of being condensed.

A2.1.3 Isentropic and Polytropic Efficiencies

It is possible to characterize a real process by introducing an imperfection factor, often called **effectiveness** or **efficiency**, which expresses its performance compared to that of the corresponding **reference process**.

To determine the performance of compressions and expansions, we first assume that they are perfect, that is to say, follow reversible adiabatic processes as a reference, and then we account for the irreversibilities in the machines through their isentropic or polytropic efficiency.

A2.1.3.1 Concept of Isentropic Efficiency

The concept of isentropic efficiency makes it possible to characterize the performance of compressors and turbines, which in reality are not perfect, so that compression and expansion follow irreversible adiabatic processes.

In the case of a compressor, the isentropic efficiency η is defined as the ratio of the work of the reversible compression to the actual work, and in the case of a turbine as the ratio of the actual work to the work of the reversible expansion. It should be noted that η is less than or equal to 1 for both compression and expansion (Figure A2.1.4).

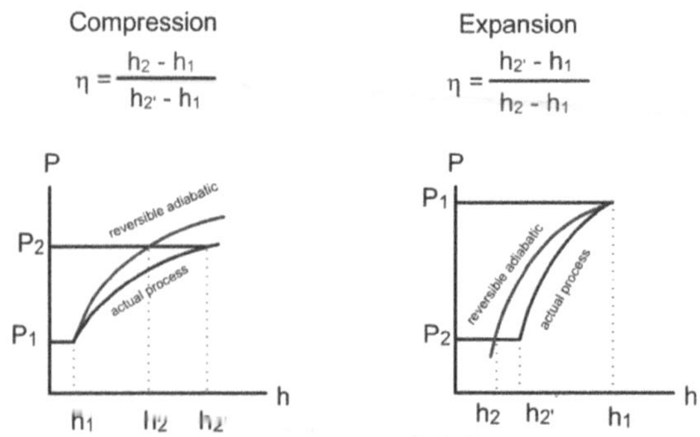

FIGURE A2.1.4 Isentropic efficiency definition.

To remember its definition, it suffices to remember that η is always equal to the smallest work divided by the largest work, and that irreversibilities have the effect of increasing the compression work and reducing the expansion work.

The values of these efficiencies, which depend greatly on the technologies used, are typically between 0.75 and 0.9 in current applications.

A2.1.3.2 Polytropic Concept

Note that knowing the isentropic efficiency η_s provides no indication of the law followed by the fluid during the irreversible expansion. To determine this law, additional hypotheses must be made.

One of the most common hypotheses leads to the widely used notion of polytropic efficiency, which can have slightly different definitions depending on the authors.

The hypothesis here is to consider that irreversibilities are uniformly distributed throughout the expansion, which amounts to assuming that during any infinitesimal stage of the process, the isentropic efficiency maintains a constant value, by definition equal to the polytropic efficiency η_p which thus appears to be an infinitesimal isentropic efficiency (Figure A2.1.5 where τ_{ra} is the work value for the reversible adiabatic).

We can then show that, for a perfect gas, the law of the process followed by the fluid is of the type $Pv^k = \text{Const}$, but it is not as simple for a vapor.

For multistage machines with large numbers of similarly constructed stages, such as a steam turbine, this efficiency has a clear physical meaning: it is, in a sense, the isentropic efficiency of a single stage.

Modeling a multistage turbine according to a polytropic approach is therefore more realistic than assuming that its isentropic efficiency is constant regardless of the expansion ratio.

Knowing the polytropic efficiency of the machine and the expansion ratio, it is possible to determine the isentropic efficiency of the machine.

Thermoptim allows compression or expansion to be configured by choosing either approach, as shown in guided exploration C-M1-V3.

Let us return to the Advanced Gas-cooled Reactor (AGR) power plant simple model example. Figure A2.1.6 shows the turbine screen. Its configuration for the expansion from 163 bar to 0.04 bar is as follows: polytropic reference and "Set the efficiency and calculate the process"; the polytropic efficiency is equal to 0.805.

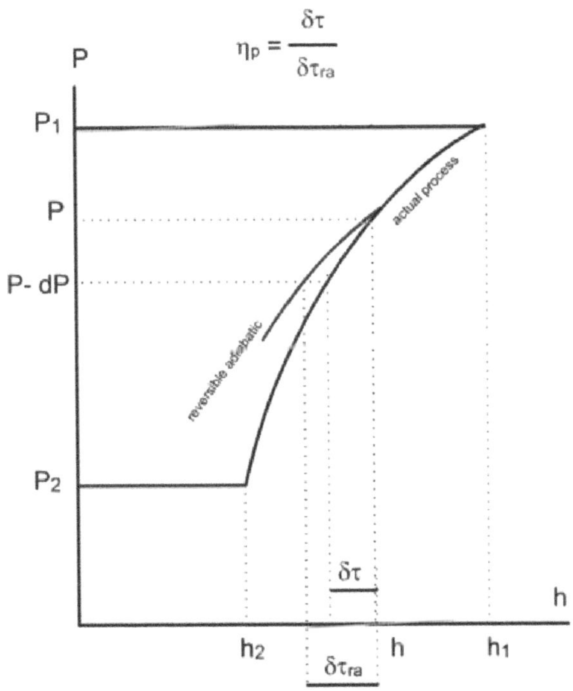

FIGURE A2.1.5 Polytropic efficiency definition.

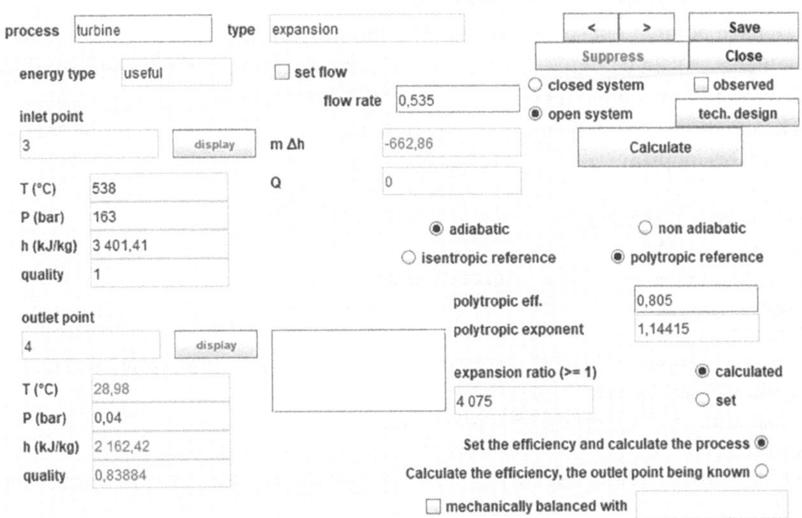

FIGURE A2.1.6 Turbine set with a polytropic reference.

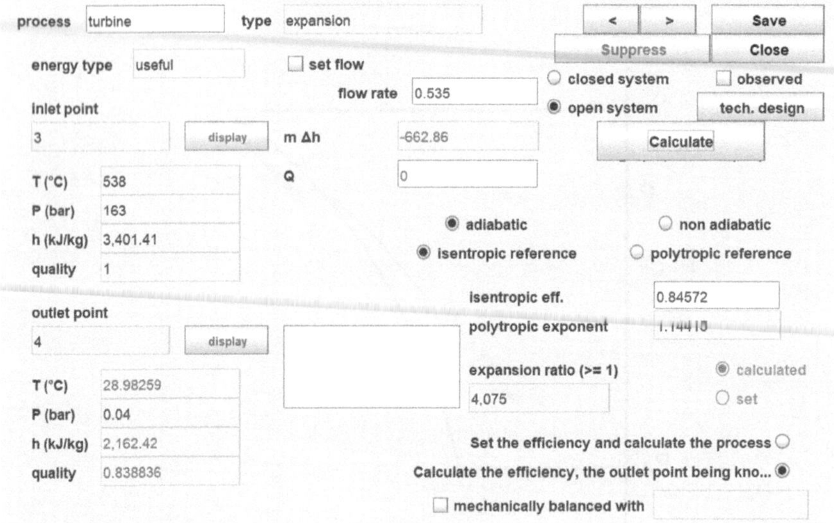

FIGURE A2.1.7 Turbine set with an isentropic reference.

To calculate the isentropic efficiency corresponding to this value, choose isentropic reference and "Calculate the efficiency, the outlet point being known," and then click on "Calculate."

The value of the isentropic efficiency of the machine is determined as 0.84572 (Figure A2.1.7).

You can thus switch from one calculation mode to another without difficulty.

A2.1.3.3 Polytropic Model Equations

As we have pointed out several times, Thermoptim allows exporting the equations it implements. As an example, here are those corresponding to the turbine used in the AGR power plant simple model.

The equations corresponding to this polytropic model are given in the file generated by Thermoptim:

```
//Process: turbine
//Equation: 10
m_dot_turbine = m_dot_superheater // Upstream process
- superheater
//Equation: 11
s_3 = calcS_PH("water";P = p_3;H = h_3) // Upstream point - 3
- Downstream point - 4
// Comment = Polytropic reference
//Equation: 12
ds_4 = -(1 - etaT_turbine)*8.314/M_3*ln(p_4/p_3) // Upstream
point - 3 - Downstream point - 4
```

```
//Equation: 13
s_4 = s_3 + ds_4 // Entropy - 4
//Equation: 14
M_3 = 18.01528 // Molar mass - 4
//Equation: 15
etaT_turbine = 0.805// Polytropic efficiency
//Equation: 16
xl_4 = 0.// Saturated liquid quality
//Equation: 17
Tl_4 = T_4- 0.01// Saturated liquid temperature
//Equation: 18
xv_4 = 1.// Saturated vapor quality
//Equation: 19
Tv_4 = T_4+ 0.01// Saturated vapor temperature
//Equation: 20
sl_4 = calcS_TP("water";T = Tl_4;P = p_4)// Saturated liquid
entropy
//Equation: 21
sv_4 = calcS_TP("water";T = Tv_4;P = p_4)// Saturated vapor
entropy
//Equation: 22
x_4 = (s_4 - sl_4)/(sv_4 - sl_4)// Quality
//Equation: 23
T_4 = calcTsat("water";P = p_4 ;X = x_4) // Downstream point
- 4
//Equation: 24
h_4 = calcH_TPx("water";T = T_4;P = p_4;X = x_4) // Enthalpy
// Comment = Given outlet pressure
//Equation: 25
p_4 = 0.04// Outlet pressure
//Equation: 26
W_dot_turbine = m_dot_turbine*(h_4 - h_3) // DeltaH
```

Equation 10 represents the propagation of the flow rate from the upstream process.

Equation 11 calculates the entropy of the inlet point, and equations 12–14 enable calculation of the entropy change at the outlet point corresponding to the polytropic evolution.

Equation 15 provides the value of the polytropic efficiency, and equations 16–24 determine the actual enthalpy of the outlet point.

Equation 22 provides the turbine outlet quality.

Equation 23 provides the temperature of the downstream point.

Equation 25 gives the downstream pressure, and equation 26 gives the useful work.

A2.1.4 Stodola Law

The Stodola law is an equation that relates the flow passing through a turbine stages group to its inlet and outlet conditions. Several versions exist, varying in complexity depending on the assumptions made about the properties of the fluid flowing through the turbine. The simplest form is (9.4.12) of this set's first volume, but it is not precise enough for flows in the wet zone that characterize water-cooled reactors.

$$\frac{\dot{m}\sqrt{T_{in}}}{P_{in}} = \text{Const} \tag{9.4.12}$$

One of the most complex forms is given by (9.4.14). It involves not only the volume and upstream pressure, but also the exponent of the $Pv^k = \text{Constant}$ law characterizing the flow.

$$\dot{m}\frac{\sqrt{(k+1)v_{in}}}{k\,P_{in}} = K_0\sqrt{1-\left[\frac{P_{out}}{P_{in}}\right]^{(k+1)/k}} \tag{9.4.14}$$

Detailed explanations can be found in Section 9.4.3 of volume 1 of this set of books, or at this address.[1]

A2.1.5 Degradation of Expansion Isentropic Efficiency as a Function of Steam Quality

When the expansion of a fluid (usually steam) extends beyond the dew point curve, vapor begins to condense and droplets appear. The phenomenon is not immediate because equilibrium conditions are not met, but it occurs gradually and increasingly as steam quality decreases. The presence of these droplets leads to:

- mechanical damage to the leading edges of turbine blades;
- reduction of the expansion isentropic efficiency.

For these two reasons, the permissible quality limit at the end of expansion should not fall below approximately 0.8.

To account for expansion isentropic efficiency degradation, relation (A2.1.1) can be used.

It provides the isentropic efficiency of a humid stage η_{hum}, with α being the Baumann coefficient, close to 1, and η_{dry} the isentropic efficiency for dry steam.

$$\frac{\eta_{hum}}{\eta_{dry}} = 1-\alpha(1-x) \tag{A2.1.1}$$

This relationship shows that an increase of 1 point in quality reduces the isentropic efficiency by approximately 1 point. This is not currently accounted for in wet expansion calculations by Thermoptim. Readers may refer to the book by de Vlaminck and Wauters (1988) for detailed explanations on this subject.

A2.2 GAS TURBINES

Closed-cycle gas turbines are much less developed than those used in open cycles, most often downstream of a combustion chamber. Over the years, the latter have seen their performance increase, mainly driven by the needs of aeronautical propulsion, particularly military applications. Today their main constraints are at the level of the first expansion stages, which are subjected to very high temperature gas flows.

We will limit ourselves here to specifying the main differences between the latter and those used in closed cycles.

These technological differences arise from the properties of the working fluids, the operating conditions, and the specific applications.

Closed-cycle gas turbines use an inert gas as a working fluid, often helium or CO_2.

Combustion gas turbines operate at relatively low pressures (about 10–30 bar), whereas closed cycles with helium operate at higher pressures (up to 60–100 bar). Helium, being a monoatomic gas with low viscosity and density, can be compressed to higher pressures without major problems of efficiency loss or heat transfer.

These high pressures increase the overall efficiency of the thermodynamic cycle, even when temperatures are lower than those of combustion turbines.

Due to the high temperatures generated by combustion, combustion gas turbines require the use of heat-resistant materials such as nickel-based superalloys (Inconel, Hastelloy) for the blades and other turbine components. Blade cooling systems (often using compressed air) are essential to prevent rapid thermal degradation of materials.

The more moderate temperatures and properties of helium allow for the use of more standard materials, although heat-resistant materials are still required. The low density and high fluidity of helium minimize mechanical wear.

Less active cooling is required compared to combustion turbines, simplifying mechanical design.

Closed-cycle gas turbines require less maintenance of components exposed to high temperatures, as helium does not chemically react with materials. However, controlling helium leaks is crucial due to its small molecular size.

The risks are reduced since the working fluid is not flammable.

Combustion turbines must handle flue gases, which often flow unevenly due to the turbulence created by internal combustion. The blades are contoured to handle this unstable flow and must be optimized to withstand thermal shocks, variations in gas flow, and corrosion due to combustion residues.

In closed-cycle turbines, the flow of helium is generally more stable and uniform, as there is no combustion. This allows for smoother blade profiles optimized for aerodynamic efficiency rather than handling unstable flue gases.

The blades are often larger to compensate for the low helium density, and the geometry is adjusted to minimize aerodynamic losses while maintaining high pressure. In CO_2 turbines, however, the blades are much smaller.

In these turbines, component life is generally longer, as there is no internal combustion to cause significant thermal wear. Helium turbines require less intensive maintenance than combustion turbines.

The main maintenance concern is the detection of helium leaks due to the small size of the molecules, but the mechanical components are less subject to extreme conditions.

A2.3 CONCLUSION

This appendix addressed the operation of turbines in steam and gas power plants, covering design principles, performance characteristics, and operational considerations. Velocity and pressure profiles in axial turbine stages were analyzed: expansion occurs in two steps, namely acceleration in the stator nozzles and conversion of enthalpy into mechanical energy in the rotors. Performance maps illustrate the phenomenon of choked flow when the expansion ratio exceeds 2. The degree of reaction ε distinguishes between impulse turbines and reaction turbines. The concepts of isentropic and polytropic efficiency have been presented. Stodola's law and Baumann's rule quantify flow rates and efficiency degradation as a function of steam quality, respectively. The gas turbine section compared closed-cycle systems using inert fluids and combustion turbines, emphasizing their differences in materials, cooling, and maintenance requirements.

Further resources on the models presented in this annex are available on this book's website.[2]

NOTES

1 https://server.s4e2.com/crc/tcnpp/resources/stodola/.
2 Available at: https://server.s4e2.com/crc/tcnpp/chapters/appendix2/_resources/index.html.

BIBLIOGRAPHY

de Vlaminck, M. & Wauters, P. (1988). *Thermodynamique et Turbines.* Ciaco, Louvain-la-Neuve.

El-Sayed, A.F. (2008). *Aircraft Propulsion and Gas Turbine Engines.* CRC Press, Boca Raton, FL. ISBN 978-0-8493-9196-5.

Gorla, R.S.R. & Khan, A.A. (2003). *Turbomachinery Design and Theory.* Marcel Dekker, New York. ISBN 978-0824740651.

Stodola, A.B. (1927). *Steam and Gas Turbines.* English translation by L.C. Loewenstein. McGraw-Hill Book Company, New York.

Vivier, L. (1965). *Turbines à vapeur et à gaz.* Albin Michel, Paris.

Index

D

E

F

G

H

I